25.

RESTORING FRASER RIVER SALMON

RESTORING FRASER RIVER Salmon

by John F. Roos

A HISTORY OF THE INTERNATIONAL PACIFIC SALMON FISHERIES COMMISSION

1937 - 1985

Published by The Pacific Salmon Commission Vancouver, Canada

Published by
The Pacific Salmon Commission
Vancouver, Canada

Available through
The Pacific Salmon Commission
600 - 1155 Robson Street
Vancouver, B.C.
V6E 1B9

Price in Canadian currency
Hardcover $25.00
plus postage and applicable taxes

Prices subject to change without notice

Canadian Cataloguing in Publication Data

Roos, John F. (John Francis), 1932 -
Restoring Fraser River Salmon

Includes bibliographical references and index.
ISBN 0-9694947 - 0 - X (bound)
ISBN 0-9694947-1-8 (pbk)

1. International Pacific Salmon Fisheries Commission — History.
2. Pacific slamon fisheries — British Columbia —
Fraser River Watershed — History.
I. Pacific Salmon Commission. II. Title
SH349.R65 1991 354,7110082'363 C91-091526-1

Cover Photography Credits:
Top photograph: Point Roberts Historical Society, Corbett.
Bottom left photograph: British Columbia Archives and Records Service HP12278
Bottom right photograph: Provincial Archives of Alberta Ernest Brown Collection B9852

The Peace Arch on the international boundary between Canada and the United States, is symbolic of the good will that resulted in the formation of the International Pacific Salmon Fisheries Commission in 1937. American seiners and gillnetters pass this area headed for Point Roberts fishing, near the mouth of the Fraser River.

In August 1943, after the Commission had completed six years of investigations, a major commercial fisheries magazine editorialized:

> The Fraser River sockeye salmon enterprise stands for the world to see as a glowing example of how international intelligence and cooperation between Canada and the United States can advance Pacific North America.

> The manner in which these neighbor nations deal in concord with a joint resource must be regarded as foretelling their approach to other problems. Their success in this enterprise, in which they unite their interests and energies for the future common good, whilst retaining in full their national independence, proves their ability to act in amity for the solution of undertakings affecting the people of both countries *(Pacific Fisherman 1943)*.

PACIFIC SALMON

River-born fugitives, red muscled under sheathing silver,

Alive with lights of ocean's changing colors,

The range of deeps and distances through wild salt years

Has gathered the sea's plenty into your perfection.

Fullness is the long return from dark depths

Rendering toll of itself to the searching nets

Surging on to strife on brilliant gravel shallows

That opened long ago behind the failing ice.

In violence over the gravel, under the burn of fall,

Fullness spends itself, thrusting forth new life

To nurse in the stream's flow. The old life,

Used utterly, yields itself among the river rocks of home.

- Roderick Haig-Brown
Copyright Valerie Haig-Brown

T A B L E   O F   C O N T E N T S

L I S T   O F   T A B L E S

L I S T   O F   F I G U R E S

Except where noted, all figures are derived from IPSFC data and reports, updated from Pacific Salmon Commission (PSC) for 1986-1989.

This work is dedicated to members of the International Pacific Salmon Fisheries Commission, its Advisory Committee, and Commission staff who were challenged by the enormity, complexity and urgency of the Fraser River sockeye salmon depletion and to those who faithfully performed scientific and engineering investigations as well as the fishery administration required by the Convention. Special appreciation is given to the governments of the United States and Canada and the fishing industries of both countries for their many years of support in ensuring that this international body preserved, protected and extended the Fraser River sockeye and pink salmon resource.

An era of research and management of Fraser River sockeye and pink salmon stocks spanning 48 years came to an end on December 31, 1985 with the dissolution of the International Pacific Salmon Fisheries Commission following the signing of the new United States/Canada Pacific Salmon Treaty. While the Commission was created by treaty between Canada and the United States in 1937, its roots went back as early as 1892 and in 1905 when the Prince Commission examined international problems of stock protection and management of the stocks. There followed many years of frustrated efforts to reach agreement on a means of international management. Changing circumstances in the commercial fisheries eventually led to an agreement on the Sockeye Salmon Convention and the creation of the Commission. With such a protracted and contentious genesis, the Commission inherited differing political and ideological viewpoints, along with the responsibilities prescribed by the Convention. Despite the fact that the Commission carried out its responsibilities successfully and had the goal of full restoration of the stocks in reach, and in addition had attained worldwide recognition as an excellent model of a fishery management system, the political and ideological inheritance lingered on and eventually resulted in termination of the Convention. This is not interpreted as a consequence of failure of the Commission, for both signatories repeatedly acknowledged the success of the Commission's work. Rather, the termination was the end result of forces within each country seeking greater autonomy in managing and controlling their own resources.

Much of the success of the Commission was attributable to the relatively uncomplicated terms of the Convention, for which the founders should receive credit. These terms specified three Commissioners from each country, who bore the responsibility to their respective governments for carrying out the mandate of the Convention. They performed this responsibility, without pay, as an adjunct to their vocation, and were called upon once a week or more during fishing seasons, often at irregular and inconvenient times, to deal with regulatory matters. They deserve much credit for the successful functioning of the Commission. An Advisory Committee of five to seven members from each country was also established to represent specific facets of the fishing industry. The Committee, whose members also served without pay, provided much valuable assistance to the Commission over the years. This compact Commission could reach decisions quickly when necessary for regulation of the fishery. It could also consider staff recommendations, approve programs, and set policies much more readily than would be possible with a large complicated organization. Not unimportantly, this compact structure minimized operational costs of the Commission.

The Commission brought together a combination of biologists and engineers to support its functions. At the time, this combination of disciplines was unique in Canadian Fisheries management and the practice was soon followed in other agencies. Many employees of the Commission took advantage of the opportunities so created with the result that there is a fairly large alumni of former Commission employees in various aspects of fisheries management in Canada, the United States and abroad. The Commission continued to apply the multi-discipline approach throughout its existence and was in the forefront of research and technology to protect and enhance the fishery resource.

As a result of its studies of the timing of runs of the various races of Fraser River sockeye in conjunction with data on size and age composition of the various races, the Commission was able to promulgate fishery regulations designed as much as practicable to suit the management objectives for each race. Many years of such data coupled with environmental data led to refinement of forecasts of the size of returning runs which fishermen could use to plan their operations. All of these data were essential to the Commission for management of the resource and for equal division of the allowable catch each year between the fishermen of the two countries, a task that was accomplished with remarkable accuracy.

Because of the challenge in the Commission's work and the successes obtained, the Commission elicited enthusiasm and dedication amongst its employees. Theirs was no Monday-through-Friday, nine-to-five job, but often involved long hours and physical discomfort in the field. The esprit de corps which developed was a trademark of the Commission and former employees continued to follow the Commission's work with interest.

The Commission's objectives in its later years were challenged by court and government decisions which caused confusion and serious discourse amongst the Commission, governments and fishing industries.

Perhaps the ultimate tribute to the Commission was the respect it earned from fishermen. All was not peace and tranquility when it came to fishing regulations, but nevertheless, the fishermen knew their concerns would be heard and considered. But more than that, they knew the Commission was working to protect and expand the very fishery resource from which they obtained their livelihood. This regard was shown repeatedly at annual open meetings with fishermen and Advisory Committee representatives. May they be served as well in the future.

A.C. Cooper, P.Eng.

Commercial harvesting and canning of Fraser River sockeye salmon began in British Columbia in the 1870s. Prior to this, Indians relied on the salmon. They fished by a variety of methods taking young fish migrating seaward and adult fish returning from the Pacific Ocean. Large annual returns in some years led people to believe the supply was inexhaustible. Early evidence, however, indicated that unknown factors affected the size of returns. Hudson Bay Company journals in the early 1800s clearly described good and bad years. Records of commercial fishing in the late 1800s also showed that on some years exceptional numbers returned, and on others, returns were much less abundant.

The rapid development of the commercial industry in British Columbia was followed by increased interest and participation by Washington State. Both countries moved quickly to develop methods of harvesting and processing. Initially, little concern was directed at understanding factors affecting the abundance of the species harvested. Nevertheless, in 1892, with a view to establishing a joint system of international regulation, officials from the two countries met and formally discussed problems developing in the sockeye fisheries. Although several proposals for an international agreement between Canada and the United States were put forward, an acceptable convention covering Fraser River sockeye salmon was not adopted until 1930, and then, not ratified until 1937.

While negotiations were being conducted prior to any agreement, the annual harvest of Fraser River sockeye was increasing in both countries. At the same time, mining, logging and railroad construction projects were taking place in the Fraser River watershed. These played a major role in future production of Fraser River sockeye and pink salmon. The most noted decline of runs was due to railroad construction in the Fraser Canyon from 1911-1912. Fraser sockeye production in 1913 had reached its pinnacle, but in that year accumulated rock dumped at Hells Gate during railway construction prevented millions of fish from reaching the spawning grounds. The blocked conditions in 1913 were further exacerbated by a large slide in 1914. As a result, the drastic decline of all runs, on all cycles, along with increased fishing pressure in both countries, created an urgent need for international cooperation. Canadian government engineers and scientists had observed and studied the problem for many years prior to 1937, but the run sizes continued to be only a small fraction of those that occurred up through 1913. This occurred despite the fact that large amounts of slide material had been removed. Pink salmon returns also declined.

The task facing the Commission was enormous. Hundreds of millions of dollars had been lost by the fishing industries of both countries and many stocks faced extinction. Through its well-planned scientific programs, engineers and biologists laid the foundation for one of the most intriguing and productive restoration programs in the history of any fishery. From 1937 to 1957, the Commission's investigations were restricted to sockeye salmon. In 1957, both governments agreed that Fraser River pink salmon should also be included in the Commission's responsibilities.

This report is the record of the Commission's achievements and how they were developed. It is about dedicated international administrators, engineers, scientists, and technical staff committed to obtaining the facts and acting without political or government interference - but with government approval. It is a success story, however, that was not without its problems.

In large measure, the restoration of Fraser River sockeye and pink salmon stocks can be traced directly to those who founded the Sockeye Salmon Convention and to the simplicity of the Convention's mandate.

The termination of the IPSFC on December 31, 1985 ended what many consider the finest fishery research and management organization ever established. The record of Commission investigations has been accurately documented in its annual reports to the governments of Canada and the United States and in many scientific publications.

All of those who participated in the formation of the IPSFC are now deceased. These include original Commission members Dr. William A. Found, A.L. Hager, Senator Thomas Reid, and (one year later) A.J. Whitmore for Canada; and Edward W. Allen, B.M. Brennan and Charles E. Jackson for the United States. Many others from both countries participated in the negotiations that preceded the 1937 Convention. Most notable of these was Miller Freeman for the United States who spent more than 30 years promoting a treaty. Governor Martin (State of Washington), Dr. David Starr Jordan, E.A. Sims, and Henry O'Malley also played significant roles in drafting the treaty. For Canada, John Pease Babcock (British Columbia) was a strong treaty advocate for over 30 years and one of the first in Canada to push for international control. He was assisted by Major J.A. Motherwell, Richard J. Gosse, J.S. Eckman and R. Payne. Dr. Found was probably the principal Canadian government individual responsible for the treaty.

V

I am indebted to Milo Bell, A.C. Cooper, Doug Hager, Roy Jackson and Loyd A. Royal for time spent in consultation and to Dave Blackbourn, Jim Cave, Jim Gable, John Gilbert, Ian Guthrie, Steve Hoag, Robin Kent, Bert Larkins, Ken Medlock, Donovan Miller, John Plancich, Wayne Saito, Per Saxvik, Paul Steere, Doug Stelter, Marjorie Stevens, Ross Stewart, Teri Tarita, Ian Todd, Glenna Westwood and Jim Woodey for technical assistance. Initial editorial reviews were made by former Commissioners W.R. Hourston and D.R. Johnson, particularly for Chapter 15. The editorial comments and funding support through former Commissioners R.A. Schmitten and C.W. Shinners, acting as trustees for the governments for this project are appreciated. I am also extremely grateful for the extensive editorial assistance and advice rendered by Dr. Richard B. Thompson. He gave considerable encouragement for which I extend special gratitude and appreciation. Susan Stitt was contracted by Pacific Salmon Commission for editorial review. Fred Andrew was contracted by Pacific Salmon Commission to assist in final preparation of figures, research for photos, other technical matters as well as editorial assistance. The typing of the manuscript by Greta Grant involved many hours and her special efforts (through the Pacific Salmon Commission) were greatly appreciated. Thanks also to Annettee Althoff (National Marine Fisheries Service) for typing assistance with Chapter 15. Teri Tarita developed the index. I utilized previous Commission publications (data, figures and unpublished photographs taken by Commission employees) extensively and hereby acknowledge the collective contributions of a large number of former Commission employees. Interpretations and additional analyses, however, remain my responsibility.

The Early Fishery on the Fraser River

DESCRIPTION OF SOCKEYE AND PINK SALMON LIFE HISTORIES

The annual Fraser River sockeye (*Oncorhynchus nerka* Walbaum) run comprises more than thirty separate races (populations, runs) that return to their stream or river of origin from late June through October of each year.

Terms such as stock, population, run, and race have been used extensively. Stock identifies those sockeye returning to a general geographic area such as Quesnel. Within the Quesnel stock there are specific discrete races or populations such as the Horsefly River and Mitchell River runs spawning in separate areas of the Quesnel Lake system. The early and late Nadina River sockeye races are distinct runs spawning in the Nechako system (Nechako stock) in separate locations of the Nadina River with different spawning times. Total Fraser River sockeye return is often referred to as the Fraser River run. Royal (1953) defined race as follows: "The term 'race' as used here assumes that homogeneity exists in each population spawning in a particular area which is subject to the same general reproductive environment". The seasonal run is made up of early, summer, and late fall races. Usually, more than 90 percent of each year's run are fish in their fourth year (4_2's: a fish which spent one full year in a lake as fry-fingerling, migrated to sea as a smolt in its second year, and returned to the Fraser River as a mature adult after two winters at sea).

Mortality rates are high. Each female sockeye lays about 3,500 eggs in the gravel. Following fertilization and hatching after a six to seven month incubation period, alevins emerge from the gravel. About 10 percent survive to the fry stage. Most fry stay in a lake to feed and grow for one year with about 25 percent surviving to become seaward migrating smolts. On average, about 10 percent of the smolts survive to be adults. Of the ten mature sockeye returning to spawn, about eight are taken by the commercial and Indian food fisheries, leaving two adults with about 50:50 sex ratio for the spawning grounds.

Some races produce precocious three-year-old fish which are commonly referred to as "jacks" (3_2's - one year in the lake and one year (winter) at sea). Other races produce a higher percentage of five-year-olds due to the genetic characteristics of the particular race.

Each sockeye run has its own distinctive migration timing and behavior. Most of the early and summer sockeye runs migrate without delay up the Fraser River toward their spawning grounds. Some races such as the early Stuart travel about 670 miles. They move upstream at about 30 miles a day and live about two weeks on the spawning grounds. Some runs have developed unique behavior characteristics. For example, the peak of the upper Pitt River run (Figure 1) arrives at the mouth of the Fraser River about July 25, which is similar in timing to the Gates Creek run. But where the upper Pitt River population spawns in the lower Fraser River area, a short migration of 55 miles, the Gates Creek population spawns above Hells Gate, a migration of about 175 miles.

Peak spawning for the Gates Creek run is about September 1 while the upper Pitt River run has a peak spawning date of about September 15. Thus, the Pitt River run has genetically adapted itself to provide for a delay period of about one month in Pitt Lake before moving into the upper Pitt River spawning grounds. It is believed that this adjustment to the later spawning period is correlated to the spawning ground water temperature which provides for optimal production of fry.

Other runs such as the late-run Adams River population delay for up to a month off the mouth of the Fraser River in the Strait of Georgia and only a few days off the mouth of Adams River in Shuswap Lake. These variations in migration behavior maintain the integrity of the individual runs.

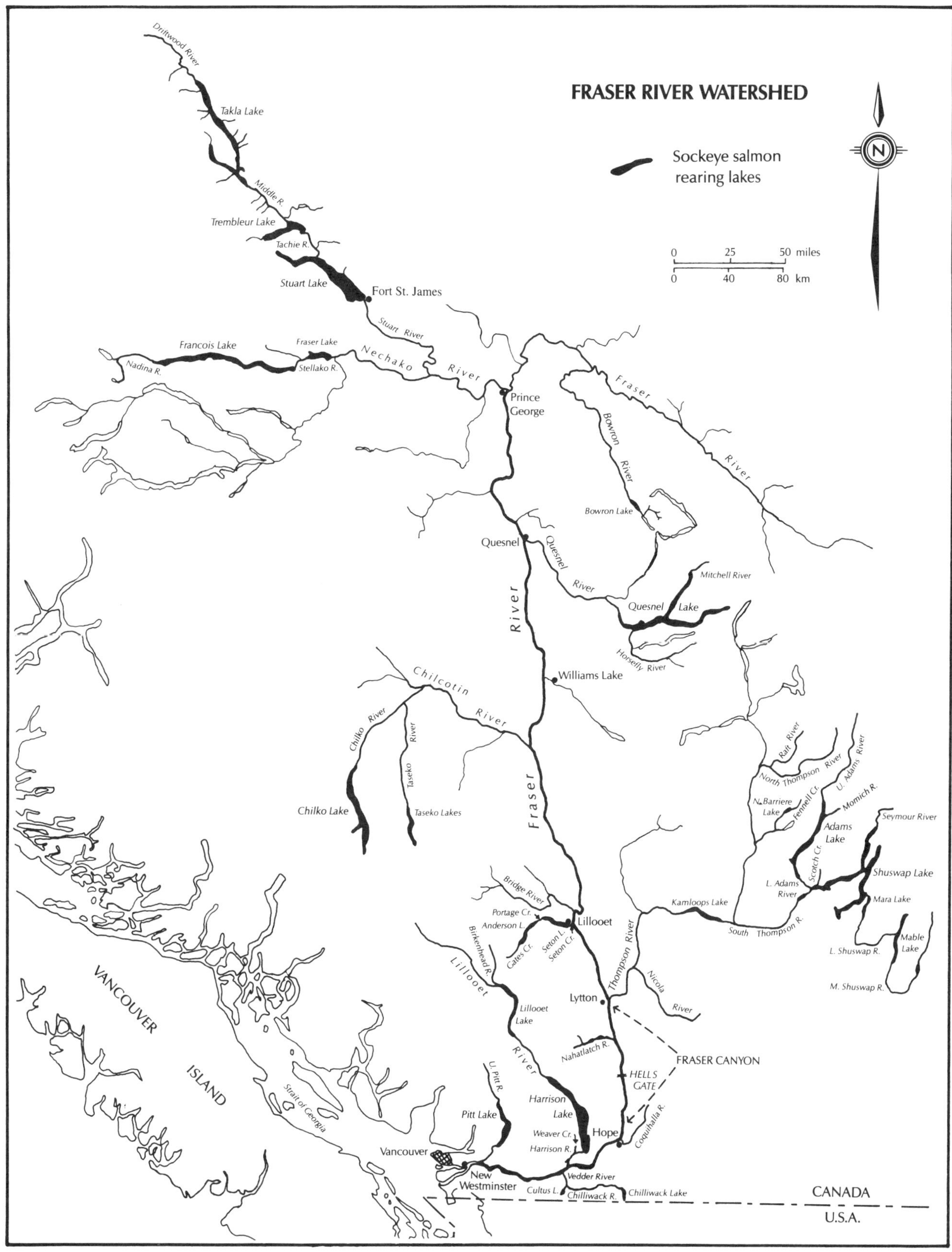

Figure 1. *Fraser River watershed, showing lakes utilized by major sockeye populations.*

Sockeye eggs and newly hatched alevins.

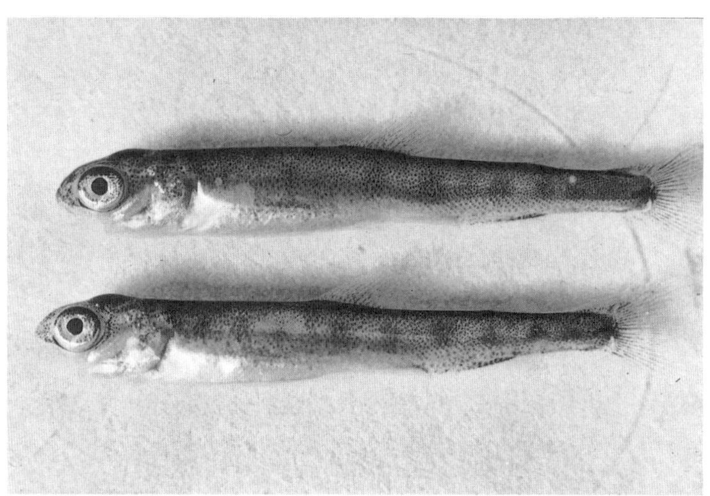

Sockeye fry.

Adult male and female sockeye on the spawning ground.

Fry migrations of the different races range from very simple to complex, as do smolt migrations.

Sockeye return to all streams each year but in varying abundance, a phenomenon called cyclic dominance. For example, the Adams River run on the dominant cycle will produce about 10,000,000 sockeye whereas the smallest off-year return (the year preceding the dominant run) may consist of only 10,000 adult fish. The average weight of a four-year-old sockeye is usually about five and a half to six pounds but can vary between races from five to seven pounds.

Dramatic changes in color take place during the sockeye's journey back to the spawning grounds. Their ocean silver turns to a deep, brilliant red the closer they come to spawning.

For pink salmon, *(Oncorhynchus gorbuscha)* there are two distinct groups which make up the Fraser River pink salmon run. These are the early and late runs. Within the early run there are three stocks, namely the Thompson River, Seton Creek, and main Fraser River. The late run component is comprised of the Harrison River and Chilliwack-Vedder River stocks. As with sockeye, there are also many races of pink salmon spawning in specific locations in the Fraser River watershed.

Pink salmon spawn in fewer locations in the Fraser River than sockeye. They return to the Fraser only on odd-numbered years, in August and September, and are two years old at maturity, weighing from five to six pounds. Upon emergence in the following spring, the young fry go directly to sea and thus have a much simpler life history.

All sockeye and pink salmon die after spawning. Although all of the Fraser River sockeye and pink salmon originate from within Canada (British Columbia), the international flavor of the fishery stems from the fact that a high percentage of the returning runs of both species migrate through United States waters before they reach the Fraser River. Young pink salmon also rear in U.S. waters before entering Juan de Fuca Strait on their way to sea.

Small numbers of non-Fraser sockeye and significant numbers of non-Fraser pink salmon stocks are also taken in the Convention area fisheries.

THE HISTORIC FRASER RIVER INDIAN SUBSISTENCE FISHERY

One of the earliest references to salmon in the upper Fraser River was made by Alexander Mackenzie on August 6, 1793. He observed "The salmon were now driving up the current in such large schools, that the water seemed, as it were, to be covered with the fins of them" (Mackenzie 1801).

Salmon was the main food of the Indians in British Columbia. The sockeye were taken by spears, hooks, gaffs, baskets, weirs, nets, traps and any other method available. Both adult and young salmon (in nursery areas) were taken, the latter being considered a delicacy. However, Babcock voiced strong opposition to the capture of young sockeye (Babcock 1902-1932). Adult sockeye salmon were taken in various locations along the entire length of the Fraser and in its many tributary streams (Figure 1). Salmon was an important item for barter with other tribes and became much more valuable with the arrival of the new settlers.

Historically, the impact of the Fraser River Indian fishery on sockeye stocks appears to have been minimal. Both Indians and fish coexisted without serious problems. The Indian population was large but then became greatly reduced. Examination of Hudson Bay records indicates that sockeye returned on a consistent cyclic abundance basis dating back to at least 1793, very similar to the abundant pre-1913 cyclic run characteristic. No factual basis (catch records) is available to document the magnitude of the total Indian catch in these years. However, early records indicate

August 22, 1891. A trap used by Indians for catching sockeye on Stuart River at Fort St. James.

August 22, 1891. Early records indicate that a fence built each year across Stuart River at the outlet of Stuart Lake was the main method used by Indians for harvesting sockeye. The trap shown in the upper photo was probably used for capturing sockeye that had been blocked by the fence. Both photos were provided by the Hudson Bay Company from the records of their trading post at Fort St. James.

An Indian fisherman at Bridge River Rapids in 1943 using a dip net of an early design. When the fisherman felt a sockeye hitting the net, he released a cord and the net slipped down to the bottom of the hoop, trapping the fish.

September 7, 1944. Many thousands of sockeye are taken by Indians for personal use each year at Bridge River Rapids and at many other popular fishing sites along the sockeye migration routes.

August 20, 1947. Anatole Duncan and his wife and children in front of their smokehouse at Trembleur Village in the Stuart River watershed.

August 29, 1945. An Indian couple drying their sockeye on Raft River, tributary to the North Thompson River.

July 25, 1945. Mending sockeye gillnets at the Tachie Reserve on Stuart Lake.

After the Commission had been formed and its efforts to protect, preserve and enhance the resource became known, some of the Indian people in British Columbia showed their support and appreciation by honoring the chairman, Senator Tom Reid.

that large numbers of salmon were taken by the Indian fisheries in certain years. For example, an extract of the Hudson Bay Company Fraser Lake (Figure 1) Journal for January 19, 1824 stated:

> …the weather was so excessively cold that the Indians could not leave their houses to go to where their salmon was, but the weather becoming milder we went on briskly afterwards and this evening the men brought the last loads here the whole amounting to 22,000 Salmon…. (Hudson Bay Company Archives 1822-1824)

On September 16, 1899 the Journal record stated: "Indians catching salmon by the hundred-from 4 up to 600" (op. cit. 1898-1902). At another location (Stuart Lake-Fort St. James, Figure 1) the entry for November 11, 1852 stated: "Mr. Wm. Manson with his brother John and J.M. Bouche arrived from Fraser Lake when I was much gratified to learn that all were well there and that notwithstanding the very gloomy aspect of affairs some time ago, in regard to provisions, they had succeeded in trading and storing 44,000 Salmon a supply sufficiently great to meet all our wants until next Autumn" (op. cit. 1851-1856).

Journal extracts indicate that sockeye salmon were the focal point in lifestyle and economy of the early settlers. Cyclic abundance trends in the runs were evident long before the advent of the commercial fisheries, and there were good years and poor years when starvation occurred. Even though weirs were used in some locations to attempt to block the upstream migrants (the weirs were seldom, if ever, fish-tight) and similar attempts were made to take downstream migrants (smolts) in the spring, there is no evidence that catches by Indians had any measurable effect on subsequent runs.

Prior to the Fraser River commercial fisheries, salmon was a primary food source of the Indians and early settlers for the entire year. It was salted, sun-dried, and smoked. There were even homemade attempts at canning. Early reports stated appetites were dulled by the constant consumption of salmon. Occasionally sturgeon, trout, and whitefish were available. The Thompson River District Report to the Governor and Council Northern Department Ruperts Land, 1827 (page 227 of "Simpson's 1828 Journey to the Columbia") stated: "Dried salmon is the staff of life, and fortunately seldom fails" (Lyons 1969).

The first exporting of salmon began in 1830 with the packing and shipping of about 200 barrels of cured salmon (*ibid.*). By 1835, over 3,000 barrels of salted fish were shipped from Fort Langley in the lower Fraser River. In 1836, Hudson Bay Company trading posts in the upper Fraser River took in 67,510 salmon, 11,941 smaller fish, 781 sturgeon and 346 trout (Morice 1904). Some of the fish went to England but most went to Hawaii. Although fur trading was British Columbia's first export item and salmon the second, the value of the fur trade soon became negligible compared with the value of the salmon industry.

When the great "Gold Rush" swept through British Columbia in the 1850s and 1860s, much of the fishing industry activity focused on curing salmon. Others, however, concentrated on the development and operation of another major industry—the canning of Pacific Salmon.

....
CANADA

Although the canning process was developed in the late 1700s, it was not until almost 100 years later that commercial salmon canning began in British Columbia with the first cans of salmon produced in 1863 on the Fraser River (Doyle 1920). The first cannery on the Fraser was built in 1866 at New Westminster, B.C. This profitable venture was then entered into in a substantial manner in 1870 near New Westminster (Lyons 1969).

From 1870-1890, the industry developed rapidly and was very competitive. Sockeye runs to the Fraser in 1877, 1881 and the subsequent four year cycles into the early 1900s were immense. Other cycle years were consistently far less abundant. It soon became evident that financing the industry, marketing the pack, regulating the fishery, and protecting the environment through government controls would play major roles in determining the health and stability of the industry.

The sockeye fishery took place primarily in the Fraser River area and the catch was taken almost exclusively by gillnets. The earliest record of the salmon pack on the Fraser River was in 1873 when 8,125 cases were put up (Rathbun 1899).

During the development of the Canadian Fraser sockeye fishery, there was both excitement and apprehension concerning the long-term stability and viability of the industry. The problems later created by the Americans were non-existent. Americans did not become involved in the commercial catch and harvest of Fraser River sockeye until the 1890s; however, participation of the United States industry by 1900 created continuous dissatisfaction to Canada even though the two would become long-term partners in the restoration of the runs.

....
UNITED STATES

The first U.S. cannery was built on Puget Sound at Mukilteo in 1877 (Rounsefell and Kelez 1938). The northern waters of Puget Sound, including the San Juan Islands and Point Roberts, were lightly fished until the first cannery in the area was constructed at Blaine in 1891. Canneries were soon established at Point Roberts and at Friday Harbor on San Juan

Gillnetters off the mouth of the Fraser River in 1904. The boats relied on sails and oars and many were towed to and from the fishing grounds by steam-powered vessels. From Pacific Fisherman, 1904.

Part of the early fleet of sail-powered gillnetters at the mouth of the Fraser. From Pacific Fisherman, 1905.

One of the original purse seine outfits as used on Puget Sound. A large skiff was used to carry the net and a small scow with a hand winch was used to purse and haul in the net and to transport the catch. From Pacific Fisherman, 1905.

American salmon traps at "Cannery Point", the southeast tip of Point Roberts. Similar traps in many locations among the San Juan Islands and at Point Roberts in American waters and in Canadian waters at Sooke on Vancouver Island proved to be an efficient and economical method of harvesting salmon. From Pacific Fisherman, 1905.

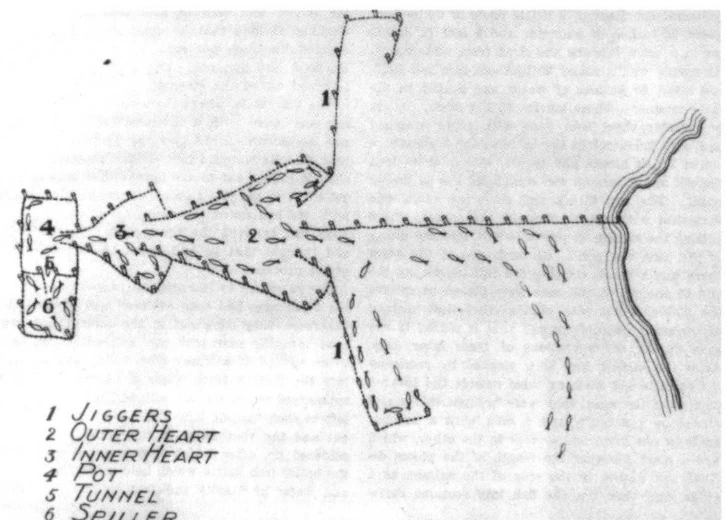

Diagram of a salmon fish trap. From Pacific Fisherman, 1904.

Showing part of the "lead" and "outer heart" of a salmon trap at Lummi Island, Washington. From Pacific Fisherman, 1930.

Island in 1893-1894. By 1900, 15 canneries were operating in the northern area. The rapid expansion of harvests and canneries in Washington State was due to the use of fish traps. In 1893 there were 13 traps and by 1900 there were 163 traps in operation *(ibid.)*. By the 1890s, the Fraser River sockeye fishery was truly an international fishery and the American catch of Canadian-bred sockeye caused concern within the fishing industry of British Columbia—not only because the fish "belonged" to Canadians, but also because of the different regulatory schemes in the two areas. Fishermen in Washington State were under the authority of state officials and their regulations. In Canadian waters, fishermen were bound to comply with federal regulations, not provincial regulations. These two differences would eventually become a stumbling block when discussions between the two countries were held. The complexity of the international aspects of the fisheries would soon be embroiled in national politics. Serious environmental problems, unbeknown to the industry, also began to affect the productivity of the stocks.

Prior to full industry development, sockeye runs were so large that there was a great deal of waste. Richard Rathbun in his report about the American fishery, stated:

> The run of 1897 was one of the largest, if not the largest, in the history of the region… The great body of sockeye first made its appearance about the middle of July and continued until about the end of the first week in August, a relatively short season, but during this period the cannery pack was completed and in addition an immense amount of fish was thrown away, the daily catch being often much larger than could be disposed of. It has, in fact, been claimed, though this is probably an exaggeration, that more fish were caught and wasted than were utilized. (Rathbun 1899)

The subject of waste in both countries' fisheries was mentioned by the British Columbia Commissioner of Fisheries in his report for 1909 as follows:

> The catch that year (1901) was so great that every one of the canneries on both sides of the international line filled every can they had or could obtain; and in addition to the millions of fish which they packed that year, many millions more were captured, from both the Canadian and American waters of the Fraser River District, which could not be used, and were thrown back dead into the water. The waste of sockeye of our own catch and that of the Americans in 1901 is believed to have been greater than the number caught and packed by all the canners on the waters mentioned in any year since, with the exception of 1905 and this year. (Babcock 1902-1932)

Thus, it appeared to many there was an inexhaustible supply of sockeye in excess of industry capacity, at least in the big year. Others, however, were becoming concerned about the status of the runs on the smaller off year cycles.

. . . .
THE START OF DEPLETION

Similar to other new and expanding industries, the initial concerns of the salmon fishing industry in Puget Sound and in British Columbia were related to the day-to-day and seasonal operational problems. The rapid changes in gear technology, plant procedures, and marketing, etc., demanded almost total attention. Little thought or concern was given toward environmental problems. It could readily be argued that the runs had been returning in large numbers for hundreds or thousands of years. Nevertheless, it soon became apparent that the supply of sockeye was limited. The critical relationship between salmon and human activity would play a decisive role in the long-term viability of the international sockeye fishery. Of concern too, was the effect of increased fishing intensity in both countries using traps, purse seines and gillnets.

ENVIRONMENTAL PROBLEMS

The Quesnel-Horsefly River area (Figure 1) became famous in the 1800s because of the 1858 gold rush and the large run of sockeye salmon in dominant years (1877-1881). It is likely that the runs which comprised the Quesnel stock in the big years were in excess of 10,000,000 fish. There was an estimated 4,000,000-plus escapement in 1909 (Babcock 1902-1932).

Prior to 1909 the Quesnel District summer-run sockeye began to suffer depletion as a result of a dam constructed on the Quesnel River in 1896-1897. The dam, placed just below the outlet of Quesnel Lake to impound water for the Golden River Quesnel Company mining operation, was 763 feet long and 18 feet wide. The gates were closed on September 11, 1898 (IPSFC 1961a), and in that year and in 1899, no fish got through. Following protests, a fishway of sorts was installed in 1900 by the mining company in the middle of the dam, but where the river was often dry. The mining operation was subsequently closed down in 1900. This ineffective fishway had an opening for fish passage only 11 inches high and 10 inches wide. In 1901 some fish got through. John Pease Babcock, the first Commissioner of Fisheries for British Columbia, visited the area in 1902 and reported that because of the location of the fishway in the center of the dam few fish were attracted to it; water-flow was inadequate both in the river and through the fishway. Many felt the fishway useless and in 1903 Babcock (the Province) built a fishway into the spillway of the dam, which was more effective in attracting sockeye. It was used in 1905 on the dominant cycle.

Reports on the sockeye migration prior to the existence of an adequate fishway present a staggering picture of the number of sockeye lost. Many thousands were blocked or delayed to the point of energy depletion. Most likely died before spawning. These reports prevailed both before and after the improved facility was in place. Strange as it seems, however, some fish got through to spawn and perpetuate the cycle. Babcock was surprised at the return in 1905. In 1909 he described the run as "so amazing" and "so exceptional." An exceptional run was also produced in 1913 but the escapement was diminished by problems in the Fraser Canyon. The Quesnel fishway, even though inadequate, was probably responsible for the large run in 1909-1913 (Babcock 1902-1932).

The Bullion placer mining operations in the Quesnel River downstream from the dam probably had an adverse impact on the fish as well. Reports of the Ministry of Mines of British Columbia indicate that a total of 12,000,000 cubic yards of earth had been moved by various mining companies up to 1905 (Lay 1936). For many years following, the operations were sold, closed and reopened. The mine apparently was not operated in a serious manner after 1905. The large sockeye runs in 1909 and 1913 could not have been greatly affected by the mine. Although the effects of dumping mine tailings in the river is unknown, it is likely that the dumping was detrimental to some extent to both smolts and adults. It is also impossible to evaluate the impact of the dam on the downstream migration of smolts from Quesnel Lake.

Prior to 1905 there were several placer mining operations in the South Fork of the Quesnel River and in the Horsefly River near the town of Horsefly. In the Horsefly River, the river was diverted to permit work in the dry riverbed. This work would have been in the main spawning area of the lower river population segment of sockeye. Again, it would be impossible to assess the damage the runs suffered, but some loss would have been unavoidable. Cooper (1956) pointed out that:

> The fact none of these placer operations completely destroyed the Quesnel River sockeye run was not because of lack of potential danger to the sockeye, but rather because of the quick rise and decline of the

August 13, 1945. Quesnel River Bullion placer pit. Most of the fine material from placer operations such as this in the Quesnel watershed entered Horsefly or Quesnel River.

August 13, 1945. Deposition of placer mine tailings in Quesnel River.

A dam at the outlet of Quesnel Lake stopped sockeye from entering the lake and reaching spawning areas in Horsefly River, Mitchell River and McKinley Creek. The dam is shown with the spillway gates closed and with very little flow from the totally inadequate fishway (center of picture) which had an opening through the dam of only 11 in. by 10 in.

operations, which resulted in their influence not being exerted over a sufficient number of years to effect complete depletion.

Thompson (1945a) concluded that the first period of poor escapements to the Quesnel-Horsefly system took place from 1899 to 1904 inclusive and he assigned the first period of sockeye depletion in the Fraser River system to the effects of the man-made intrusions as previously mentioned:

> It seems certain that relatively few passed in the other years between 1899 and 1904. The coincidence of this near closure with that of the first period of depletion is very striking. Anything which affected spawning grounds as important as those in the Quesnel district may well have been responsible for the depletion of the Fraser River.

The Quesnel Dam was not removed until 1921.

As we view this matter 90 years later, it is difficult to comprehend how such a valuable renewable resource could have been neglected in such a manner. The lower Adams River late-running sockeye race ranked high with Horsefly River sockeye as one of the great runs in the Fraser River system. It is believed that the Horsefly River-Quesnel District populations were historically the largest. It is astonishing that the Quesnel stock survived the Quesnel Lake Dam and placer mining operations.

Shortly after construction of the Quesnel Lake Dam, another sinister event was taking place on the lower Adams River (Figure 1) just below the outlet of Adams Lake. Around 1908, a logging company built a splash dam at the top end of the spawning grounds just below the outlet of Adams Lake. No doubt the facility was considered by many to be an ingenious device to store water which would then be released to suddenly flash-float logs down the river. This sudden wall of water and logs crashing downstream many times a year would have been far more detrimental than a natural, slower-building spring freshet or flooding caused by fall rains.

Spawning salmon were washed downstream and the scouring effect on eggs and the dislocation of spawners were devastating. At times the artificial floods happened daily for six days each week. Between the floods, the stream bed was nearly dry. Only scant numbers of fish were able to spawn and in the winter the eggs would be exposed and frozen. In 1922, the logging company ceased operation of the dam.

It is remarkable that the lower Adams River run survived the splash dam, but it was not without cost to the resource. A few more years of use probably would have destroyed the race. As reported later, it is more than coincidental that the resurgence of the Adams run began in 1922, the year they stopped using the splash dam. The year 1921 (1913 cycle) was traditionally the dominant run to the Adams River, the same as for the Horsefly River race. But that cycle for both races was destined to be obscure for many years, undoubtedly as a result of the environmentally related problems. The 1922 cycle was quickly established as a major producer for the Adams River race. It is likely that the 1922 Adams River run was originally the sub-dominant cycle and fortunately provided sufficient seeding in 1922.

Concurrent with the cyclic changes and depletion of the late-run lower Adams River population, a disaster was occurring to the large summer-run upper Adams River race spawning at the far end of Adams Lake. Babcock in his 1913 report stated:

> …the run of sockeye to Adams Lake in August and September of 1901, 1905 and 1909 was so great that every tributary of the lake extending to Tumtum Lake at the head of the watershed was crowded with spawning sockeye. I visited the headwaters in 1905 and 1909, and saw countless thousands of dead and spawning fish there. (Babcock 1902-1932)

Thompson (1945a) reported this run was very small in 1913 and 1917 and never reappeared in great numbers. Although a small

The Adams River dam was used to assist in floating logs down lower Adams River to Shuswap Lake. It was a major contributing factor in destroying the upper Adams River sockeye race and in greatly reducing production on the lower Adams sockeye population. The dam was in operation from 1908 to 1921 and it was removed by IPSFC in 1945.

fishway existed in the splash dam, its effectiveness was questionable. However, the splash dam had not completely blocked sockeye in 1909 because large numbers were seen on the up-river spawning grounds. Concerning the ability of upper Adams River sockeye to pass the lower Adams River splash dam, it was stated that no salmon migrated past the dam in 1911 (Edwards 1911). It was pointed out that it was impossible for salmon to go upstream because the fishway was on the wrong side of the river. The comment was made: "What a pity, and the million (sic) of salmon that are trying to go up there to spawn."

Thompson reported that no sockeye destined for Adams Lake were seen below the dam in 1913 and concluded that the demise of this former great race was related primarily to conditions further downriver. Babcock in his 1913 report gave a dismal account of Overseer Fall's records of counts below the dam with a total count of only 1,410 sockeye in August and 670 fish for September 1-24. Indians also reported that few sockeye entered Adams Lake. In his 1917 report, Babcock commented: "The number which entered Adams Lake was so small as to give no promise of assistance to the run of 1921." As far as can be determined, there were no sockeye in the upper Adams River after about 1921, probably due to the combined adverse impacts of the splash dam and problems in the Fraser River. The dam, although not operational after 1921, was finally removed in 1945 by the IPSFC. This action was possible following rejection by provincial authorities in 1944 of a new water licence application to use the facility again. The Commission vigorously opposed the application.

A close parallel between the extermination of the upper Adams River race and sharp decline of the Horsefly River population is evident. Human activity was the major culprit.

The potential of hydroelectric power by damming the Fraser River has been discussed since 1912. One such project was proposed in the early 1900s for a site located about two and one-half miles upstream from Yale, B.C. It never became a reality (Burdis 1912). Thus, Fraser River salmon escaped another disastrous threat to their existence.

$\cdots\cdots\cdots\cdots$
FISHERY RELATED PROBLEMS

In the 1890s, concern was expressed because of the lack of uniform regulations in the fisheries in the Fraser River and in Puget Sound. The general consensus seemed to be that the regulations imposed in United States waters by Washington State officials were more lenient than those imposed in Canada by the federal authorities. In the Fraser River area, the only closures imposed occurred on weekends (Gilhousen 1990 MS). There was a tendency for officials of each side to blame the other while fishermen would fish to maximum gear capacity. Complicating the situation was the fact that the State of Washington and Government of Canada could not legally deal directly with each other in terms of a treaty. At the same time, there were strong feelings between State of Washington and United States' officials concerning the states' right to manage the resource without interference from the federal government. It appeared that the federal government concurred with Washington's interpretation of authority.

There has been considerable speculation and long-term differences of opinion as to whether or not overfishing took place from 1889 to 1913. Unfortunately, accurate escapement information for this period is not available because no systematic accurate method of escapement enumeration was performed. Gilhousen (1990 MS) compared his estimates of the escapements with those calculated by Rounsefell (1949), Killick and Clemens (1963) and those he calculated from Thompson (1945a).

Gilhousen's estimates of escapements from 1897-1934 were 17 percent higher than Thompson's and 52 percent higher than Rounsefell's for the same time period. From 1915-1934 Gilhousen's estimates were fairly similar in total to others; Rounsefell, 4 percent higher, Thompson, 11 percent higher and Killick and Clemens was 12 percent lower compared with Gilhousen.

Escapement estimates for the dominant cycle years up to 1913 (1897-1913) were quite substantial, ranging from 5,100,000 in 1909 to 13,000,000 in 1901 (Gilhousen 1990 MS). The off-year cycle escapements during this same time period were estimated to be from about 215,000 to 865,000 fish. Gilhousen's estimates of off-year escapements up to 1912 in most years did not include late-run escapement. Since all the major runs appeared to be dominant in the 1913 cycle year, the smaller size of the estimated escapements in the off years does not necessarily indicate that overfishing was the primary reason for lower escapement in the off years. The numbers of fishing gear however, did increase substantially after about 1895 (Rounsefell 1949).

Nevertheless, there were strong feelings that overfishing was taking place. British Columbian industry members wrote the Minister of Fisheries on December 29, 1904 to express their concern:

> For several years past the 'sockeye runs' on the Fraser River have shown an alarming decrease in what is called the 'off years', and we strongly feel that unless prompt and drastic measures are adopted for the conservation of the fish supply, the industry will certainly be ruined, with consequent disastrous results to canners, fishermen, merchants and tradesmen interested. (British Columbia Canners 1904)

Examination of sockeye pack data for the off years does indeed show that following the dominant year return of 1897 the three off-cycle years (1898, 1899 and 1900) began to decline substantially by 1902, 1903 and 1904 and in succeeding cycles (Babcock 1902-1932, Rounsefell 1949). The average annual cases packed in both countries for these years are shown in Table 1.

TABLE 1. *Number of sockeye cases packed for the non-dominant year runs from 1894 to 1916.*

Years	Cases Packed
1894-1896	408,141
1898-1900	623,109
1902-1904	412,211
1906-1908	259,959
1910-1912	300,505
1914-1916	272,142

The number of cases packed in these off years from 1906-1912 was less than one-half the annual pack from 1898-1900. While these catch data indicate depletion, the cause of the decline could have been related to a number of factors.

The history of quadrennial dominance in the Fraser sockeye runs revealed that even prior to commercial fishing, the off-year runs were much smaller than the big year cycle runs (Ward and Larkin 1964). These investigators concluded that "dominant populations occurred in upper Fraser sockeye stocks before the development of an intense commercial fishery." A second important conclusion they reached was "that dominance was stable" and maintained "sequence without deviation." They stated that the dominant population belonging to the Adams River race may be as much as a thousand times more numerous than the smallest off-year run. This phenomenon occurred naturally, suggesting that the factor or factors controlling dominance were firmly in place and played a critical role in the maintenance of the stocks. Babcock's surveys of the spawning grounds showed small numbers of sockeye in the lean years adjacent to the big years of 1901-1905. This was to be expected based on the historical nature of the runs,

irrespective of commercial fishing. Babcock, however, believed that all years could be big:

Since, as has already been stated, the catches in both the big and the lean years are the product of the same spawning beds, it is evident that those spawning beds would have produced on the average as great a run in the lean years as they produced in the big years provided they had been as abundantly seeded. It is simply a question of seeding. (Babcock 1918a)

Many others also believed that if the escapements from the off-year runs were increased to the dominant year levels, big runs could be expected every year. We now know that production is not that simple and each cycle has a relationship and bearing with other cycles and to date, no race in the Fraser River system has shown evidence that it is capable of producing a big run every year. For that matter, no sockeye population anywhere has exhibited that capability.

Babcock (1920a) documented his many years of observations on the Fraser spawning grounds (starting in 1901) and was convinced that too much fishing had been allowed: "The catches in recent lean years, have grown less, because they were made at the expense of the fish necessary to seed the beds." So convinced was Babcock that the major problem was overfishing, he stated in his 1920 report that: "The spawning area of the Fraser basin requires no expenditure to bring it into bearing. Appropriations for capital expenditure and upkeep are not required." To emphasize his convictions further he stated: "The inevitable and disastrous trend of events should have been evident to the dullest." He classified the depleted condition as a "monumental record of man's folly and greed" (Babcock 1920b).

A prominent U.S. fisheries scientist, Dr. C.H. Gilbert, from Stanford University, also strongly believed the lean cycle years after 1896 had been overfished (Gilbert 1913-1924). Gilbert later readily acknowledged that the outstanding decline in the 1917 run was directly and entirely due to the blockage at Hells Gate in 1913.

Rounsefell and Kelez (1938) also believed the decline was primarily a result of overfishing. They stated that: "In 1900, while still at a fair level of abundance, this cycle was fished with extreme intensity."

An eminent Canadian scientist, Dr. W.E. Ricker, also maintained that overfishing took place on the 1902-1904 non-dominant cycles (Ricker 1987). He stated: "Thus it was inevitable that Fraser sockeye should be seriously overfished in the "off" years from about 1900 —." With regard to the escapements, he said: "Qualitative and roughly quantitative estimates of sockeye observed in upriver nursery areas suggest that they could easily have added up to 50 or 60 million in the big years."

Estimates of escapement by Gilhousen (1990 MS) were much less (maximum of 13,000,000 for 1901) than those projected by Ricker. All estimates of early year escapements will be questioned and perhaps disputed. It is considered that Gilhousen's estimates are the best estimates available at the present time.

In all likelihood the initial decline of off-year runs was due to increased fishing effort in both countries starting in about 1895 and impacts from the dam at Quesnel Lake. Thompson (1945a) concluded that the dam was the principal cause of the first period of depletion. Others, as mentioned earlier, pointed to overfishing. The dam was probably not responsible for the decline on more than one of the three off years, because the Quesnel system probably did not have large dominant-sized runs in all years but could have had a large subdominant run. In addition, it is possible that the period in question was low in productivity due to natural fluctuations in survival rates, either marine or freshwater induced. However, it is difficult to speculate that survival conditions from 1897 up to 1913 were poor only for the off-year runs when the dominant-year runs of 1897, 1901, 1905, 1909 and 1913 were immense.

Regardless of the reasons, serious concern was expressed in both countries about the impacts of increasing fishing pressure. Some U.S. industry participants who were concerned about the declining catches on several occasions made representation to Canadian federal government officials with an offer to finance the construction of sockeye hatcheries in British Columbia (Pacific Coast Fisheries 1903). This was done as a goodwill gesture of repayment for taking Fraser River sockeye. These offers were subsequently bitterly opposed by the Canadian government *(Pacific Fisherman 1905)*. There is a report of a U.S. industry person who towed a live box with 250 marked Fraser sockeye taken by traps from Lummi Island in Puget Sound to the Skagit River for release. This was to assist the Baker Lake sockeye run in Washington State *(Pacific Fisherman 1904)*. None of the fish showed up at the Baker Lake hatchery (State of Washington Department of Fisheries and Game 1903-1904). Sockeye hatcheries were located on the Skagit and Nooksack rivers.

. . . .

INITIAL DISCUSSIONS FOR COOPERATIVE MANAGEMENT

The first formal discussion between the United States and Canada on the international aspects of the sockeye fishery problems related to the Fraser River system took place at Washington, D.C. on February 15, 1892. Following this, Senator Foster of the United States, on October 4, 1892, proposed to the Canadian government that two experts, one representing each country, should be appointed to enquire into these and other questions:

(a) The limitation or prevention of exhaustive or destructive methods of taking fish or shellfish in the territorial and contiguous waters of the United States and her Majesty's possessions in North America, respectively, and also in the waters of the open seas outside the territorial limits of either country to which the inhabitants of the respective countries may habitually resort for the purpose of fishing.

(c) The close seasons expedient to be enforced or observed in such contiguous waters by the inhabitants of both countries as respects the taking of the several kinds of fish and shellfish.

(d) The adoption of practical methods of restoring and replenishing such contiguous and territorial waters with fish and shellfish, and means by which such fish life may be therein preserved and increased. (Report Joint Fisheries Commission 1897)

The Canadian government quickly agreed to the proposals and an exchange of notes bringing this agreement into effect was completed on December 6, 1892 (Hertslet's Commercial Treaties 1895). The agreement covered a period of four years and a report was submitted on December 31, 1896. The report covered certain mesh-size limitations for purse seines and length of lead nets for traps in United States waters. The findings of the Commissioners indicated that there was no serious problem in the Fraser River District. The international agreement under which the Commission was appointed, provided that as soon as the reports of the Commissioners were placed before the respective governments, an exchange of views would take place and if agreed to, the two governments would undertake a treaty of concurrent legislation. As a result of the report, no action was taken because it was deemed that no serious problem existed.

Westward Development

RAILROAD CONSTRUCTION

The Fraser River flows seaward through one of the largest river canyons in North America. It seems inevitable that this location would be chosen as the best route for a major transportation link between eastern and western Canada. Two transcontinental railroads were built close to the river through this narrow, rugged canyon. The Canadian Pacific was built first and was completed on the west (right) bank of the river in 1885.

Construction for the railroad line along the east (left) bank for the Canadian Northern Railway, later known as Canadian National, in the canyon between Yale and Lytton commenced about June 1911 and was completed near the end of 1912. The track was completed to the east at Ashcroft in early 1915. There is no evidence as to possible damage that may have occurred to the salmon migration during 1911-1912. It is possible that nothing serious was observed because these two years were off years and large numbers of sockeye would not have been present. A photograph of conditions at Hells Gate in 1897 shows that the flow was smooth and placid along the right bank at a gauge height of about 30 feet, suggesting no problem for migrating sockeye. In later years, flow conditions at gauge 30 feet would be very turbulent and obstructive to migrating sockeye.

HELLS GATE ROCK DUMPING/SLIDE

Hells Gate is the name of an extremely narrow gorge in the Fraser River Canyon in southern British Columbia 130 miles from the mouth of the Fraser River (Figure 1). The drainage from about 84,000 square miles of central British Columbia passes through Hells Gate—a constricted area only 110 feet wide at the narrowest point. The variation in the height of the Fraser River at Hells Gate is estimated to have been in excess of 100 vertical feet between 1894 and the present time.

Prior to 1885, Hells Gate was considered to be in a completely natural state (Jackson 1950). However, because of the narrow width of the river, it was generally acknowledged that salmon encountered difficulties here during their migration.

It is not known how much rock was dumped into the Fraser River during construction of the Canadian Pacific Railway along the right bank in the 1880s, but it is likely that the quantity was large. The effect such rock deposits had on sockeye migration is unknown. Sockeye may have been able to migrate up the left bank of the river even though the conditions for upstream passage had been altered along the right bank. In any event, there is no basis to suggest that this railroad construction along the right bank seriously affected the migration or production of sockeye from 1880 to 1912.

New railroad construction in 1911-1912 started a chain of events which were to have a catastrophic impact on the salmon runs to the Fraser River system. Large amounts of rock dumped into the river during these years eventually affected the flow pattern along the river bank. In July 1913, the first sockeye observed at Hells Gate were from a large run which produced a record commercial catch of over 32,000,000 fish. By mid-summer, observers reported "incredible numbers" of sockeye massed below Hells Gate. A photograph of conditions at Hells Gate at gauge 28 on October 13, 1913 indicates turbulent conditions.

Large numbers of sockeye were observed in creeks and rivers many miles downstream from Hells Gate in 1913 (Thompson 1945a). Included among the sockeye were large numbers of pink salmon. While some sockeye were able to pass upstream in 1913, the vast majority failed in their struggle. The stage was set for a staggering decline in future production.

To compound the problem, there was a large rock slide along the left bank at Hells Gate on February 23, 1914 *(ibid.)*, consisting of a huge rock cliff through which the Canadian North-

Hells Gate in 1897 prior to construction of the Canadian National Railway. The relatively smooth flow along the right bank (left side of picture) suggests that sockeye were not delayed in migrating through the canyon at this time. The Cariboo gold rush trail can be seen along the left bank.

In this photo of Hells Gate taken on August 20, 1941, the water level was at gauge 30 feet, the same as in the above photo. The severe turbulence evident on both banks shows that the extensive removal of large rock pieces following the 1913 obstruction and the major slide in 1914 had not been adequate to restore the flow pattern to its previous condition.

ern Railway had driven a cut and tunnel. An estimated 100,000 cubic yards of massive granite rock filled the eddy immediately above the left bank bedrock outcrop which formed Hells Gate. The channel was narrowed to 75 feet and the river was partially dammed, so that it fell 15 feet in a length of only 75 feet along the face of the slide. The significant drop and turbulent conditions at Hells Gate are seen clearly in a April 23, 1914 photograph. Another view of the slide at ground level illustrates the narrow constriction and turbulent flow. Flow of the river was measured at 12.9 ft./sec. at gauge 10 to 17.5 ft./sec. at gauge 60 (Jackson 1950). He reported that even higher velocities prevailed below the metering section.

Babcock reported that sockeye were always delayed at Hells Gate, Scuzzy Rapids, and China Bar in years of big runs, even in the years prior to the slide. However, he felt that the fish eventually passed successfully upriver (Babcock 1902-1932). These observations of pre-slide conditions could have added to the later confusion and misinterpretation about the significance of fish accumulations below Hells Gate in 1913 and in following years. It is very likely that the sheer volume of spawners in those years caused congestion at these locations. Genetically, this was probably advantageous to some races as this would sustain later timing—an important biological point for some summer races to be discussed later. Babcock described conditions at Hells Gate before and after the slide. He stated:

> Previously to the construction of the Canadian Northern Railway a bay of some considerable extent existed on the left bank immediately above the Gate. Here such fish that succeeded in passing on that side found shelter. That bay has now been filled with immense masses of granite, and but little space remains in which salmon can maintain themselves. The filling of this bay has also greatly altered the currents through the Gate. Formerly the waters from the left bank were forced across toward the right bank, so that along that bank, and close to it, a backeddy was formed, which enabled the fish attempting to pass on that side to reach the smoother water above. In former years, the main portion of the ascending fish passed through on the right side. Now (1913) the currents have been so changed that no fish were seen passing through, although many thousands were constantly attempting to do so. No sockeye were observed in the eddies immediately above. Limited as was the space above the left wall of the Gate, it was filled with sockeye struggling to maintain their position. Many of those which had passed through the Gate and gained shelter immediately above, on attempting further ascent, were swept into the channel and downstream. That countless thousands were successful in passing was evidenced by the number noted in the river between the Gate and Scuzzy Rapids. After watching for some hours their efforts to ascend at Hells Gate, it became apparent that the number passing was limited to those which could find resting places in the pockets and eddies just above the Gate, and that those attempting the passage largely exceeded the capacity of the passage-way. *(ibid.)*

In this same time period, huge numbers of sockeye salmon were seen in the eddies and creeks up to 10 miles below Hells Gate during July, August, and September 1913. It has been speculated that millions of carcasses lined the river's edge below Hope. Tributary streams were reportedly plugged with salmon. Babcock wrote that "from Hope to Ruby Creek, and for some miles below the latter, the air was foul with the stench arising from the dead fish that covered the exposed parts of the river."

Hells Gate on October 13, 1913 at about gauge 28. Rock dumped in the river during railway construction from 1911-13 caused increased velocities and turbulence that impeded fish passage but migration conditions were made much worse by a major rock slide that occurred February 23, 1914. From the J.P. Babcock photo collection.

Another view of Hells Gate following the 1914 slide looking upstream from the left jutting rock (from Babcock 1918a).

EMERGENCY MEASURES

Provincial and Dominion authorities began to remove debris at Hells Gate on September 28, 1913 and 2,000 cubic yards of rock was excavated at four sites—China Bar, Scuzzy Rapids, Hells Gate and Whites Creek (Jackson, 1950). This was too late to benefit many races of sockeye. They died below the obstruction because of earlier migration-timing into the river and limited life-span. The passageways that were blasted were temporary because the river was constantly changing in height. This required almost daily surveillance and blasting to ensure that some fish got through. That this difficult work was, to a certain extent, successful explains the presence of remnants of a big year in 1917, although it is likely that some fish were able to get through on their own in 1913.

Work on the huge February 1914 slide began on March 23, 1914 (McHugh 1915). Smaller rocks were dumped in the river and carried away by the currents. The larger rocks, some of which were more than 100 cubic yards, were placed, after excavation, high on the right bank. Before this work was completed, sockeye arrived on July 3, 1914 and on July 15 it was determined that rock removal had succeeded in enabling some fish to get past the slide. The majority, however, were not successful and they found resting areas immediately downstream from Hells Gate. Platforms were therefore erected in this area and three Indians were hired to dipnet fish and transfer them by means of a timber chute to the upstream side of the left jutting rock. This effort was not very effective as only about 1,000 sockeye and chinook were captured. More success in aiding fish passage was achieved by creating fish passage channels through the slide material by blasting and excavating the larger rocks. On August 14, 1914, by which time the number of sockeye reaching Hells Gate had greatly increased, it was recognized that only a very small proportion of the fish were able to get past a 30-foot high-velocity reach around the left jutting rock. A two-foot-deep by four-foot-wide timber flume was built around the outside face of this rock. The flume was 350 feet long and the total drop was 15 feet. Very few fish entered this fishway because the entrance could not be located at the farthest upstream point that the fish could reach. Indians were employed to dipnet sockeye into the flume but in eight days of operation only 16,500 sockeye and 850 chinook were captured. Operation of this fishway became very difficult in the last week of August as a result of rapidly dropping water levels. Since some fish were able to swim upstream through the high velocities around the left jutting rock at this time, operation of the fishway was discontinued and more effort was directed toward aiding the passage of these fish by removing rocks to create fish passage channels through the slide material immediately upstream from the left jutting rock (McHugh 1915).

By March 9, 1915 the work at Hells Gate ended. J. McHugh, the engineer in charge, estimated that 60,000 cubic yards had been removed, 40,000 cubic yards dumped in the river, and 20,000 cubic yards moved to the high ground of the right bank. There were no further man-made changes at Hells Gate for twenty-six years. The drop in the water surface through the Gate had been decreased to 9 feet from 15 feet. The general feeling among Canadian engineers, scientists and industry representatives was that the river conditions had been restored. Although it was thought that all had been done that possibly could be, further study was approved and a comprehensive engineering investigation to outline possible improvements was made from 1926 to 1928. The final report contained several recommendations but the proposals were not carried out by the Canadian government (Board of Engineers 1928).

A Canadian view of conditions in the Fraser River Canyon was given by H.B. Bell-Irving (1930). He believed that Hells Gate itself was not the problem. He pointed to the elimination of resting places above the Gate as the "real cause of the trouble." He did not comment on the large increase in head built upriver of Hells Gate as a result of the rocks deposited along the bank. In support of the discussions about an international treaty, he said: "Nothing has been done, however, as it has been felt that it would be hardly fair to use Canadian taxpayers' money to restore fish, the majority of which were being caught by the traps of Puget Sound."

OBSERVATIONS ON THE RESOURCE

By 1918 Babcock apparently believed that conditions at Hells Gate were as satisfactory as before the slide and that no blockade had occurred since 1913. It seems incredible that these conditions were dismissed as having been corrected. In 1918 Babcock (Babcock

A 350-foot-long temporary fishway was built around the left jutting rock at Hells Gate in 1914. From Canada Department of Naval Service, Fisheries Branch, 1915.

After it became apparent that sockeye were blocked at Hells Gate in 1913 by rock dumped during railway construction, some of the rock was lifted by means of an overhead cable and dumped on the opposite bank above high water level. However, much of the slide material remained on the bank and in the river below the water surface.

April 23, 1914. There was a water surface drop of 15 feet after the huge 1914 slide. Removal of much of the slide rock reduced the drop to 8 to 9 feet by 1915. IPSFC tagging studies showed that this drop was still too great to permit sockeye passage at certain water levels.

1902-1932) commented on the status of the spawning beds: "The only reason they do not produce salmon as abundantly as formerly is because the fish have not been permitted to reach them—because the spawning beds have been made barren by overfishing."

Mitchell (1925) observed large cycle runs to the Shuswap Lake area since at least 1887. He was reportedly an experienced and knowledgeable observer of the salmon runs in several areas. In summing up the conditions in 1925 he stated: "There are now only priceless museum specimens of the rich coloured Fraser River sockeye left. These are worth millions of dollars for breeding purposes."

Mitchell *(ibid.)* described his observations of the Salmon River sockeye runs at Shuswap Lake before and after the Hells Gate rock slides:

> Then a stampede or panic occurred, and salmon came surging down, but the river was so full of ascending fish that they blockaded and made a great flat wriggling dam. So jammed were they that they crowded out, and were rushed up the sloping banks out of water. Where the banks steepened, these struggling, flapping fish were rolled down onto the backs of the fish in the river bed below, into the mass of which they would again sink. The boat was on fish, on a red, flapping squirming mass.

The Salmon River in 1913 was described by Mitchell as "a complete blank" with a return of only 1,000 sockeye, all of which were speared by settlers: "At Eagle River about six hundred sockeye entered, but of these about one hundred would be females. Anstey River had been red with sockeye in 1909." An old settler and his sons rowed 86 miles to the river and back with a barrel of salt and found only eight fish in the whole river. In 1913 Mitchell found only one dead sockeye (male) on the beach of Adams Lake. Scotch Creek (Shuswap Lake) had some sockeye in 1913 but many of the females died before spawning. Mitchell estimated the quantity of salmon along the lake shores in 1913 near Adams River as only 1/200th of previous big years and only one female sockeye for each six males. Observations below Hells Gate in 1913 of blocked fish indicated a sex ratio of twenty females to one male (Einarsen and Mayhall 1928).

Mitchell *(ibid.),* in reference to the dual problems at Hells Gate and the dam below the outlet of Adams Lake, stated: "The females were the first to fail in getting through, but what was worse, the females that I examined hadn't spawned, no doubt owing to the operation of the logging companies' sluice dam."

For several years starting in about 1928, biological and administrative people from the Washington State Department of Fisheries were invited by Canadian authorities to visit the Fraser River spawning grounds. The focus of attention was the condition of the river at Hells Gate although each year the survey team visited many of the spawning areas.

In a memorandum of November 12, 1928, Arthur S. Einarsen and L.E. Mayhall, biologists with the Washington State Department of Fisheries, discussed the variance of opinion that still existed over whether or not the slide at Hells Gate was still a serious problem. They apparently believed it was and stated: "It is lamentable that such an important problem, both to the United States and Canada, should pass by without an attempt at solution." So convinced was Einarsen that he wrote in a further report on September 3, 1929: "Volumes might be written concerning this area but the only point to be gathered from this trip of vital concern to the conservation agencies is that Hell's Gate (sic) is still running as a handicap to the salmon in the Fraser" (Einarsen 1929b).

This also was the opinion of others who observed the migration conditions at Hells Gate. Hill (1929) of the Bellingham Canning Co., following a tour of the spawning grounds, Bridge River Rapids and Hells Gate, wrote: "That Hell's Gate (sic) remains a menace to salmon is unquestionable." He also noted that:

"All salmon seen at Bridge Rapids were still bright in color and quite similar in appearance to salmon taken in salt water, while those seen at Hell's Gate (sic) were often bright red in color, indicating that they had been delayed there for a considerable period of time."

These observations suggested that upstream passage had been possible at certain water levels—a fact that was subsequently established by IPSFC investigations.

While Hill strongly believed that Hells Gate was a problem, he also said: "Overfishing on the Sound and in the mouth of the River is responsible for the depletion of salmon of the Fraser." Hill stated that the Quesnel area escapement in 1929 was only 1,500 sockeye. This was the residual left from the dominant cycle year of 1909 when more than 4,000,000 sockeye were estimated in the escapement. Einarsen (1930) appeared to have accurately evaluated the problem stating: "Investigations indicate that the progress of the sockeye schools past the Gate is entirely dependent upon water stages. If the Gate can be changed to remain at all times easy to surmount, then all sockeye will reach their ultimate goal under normal conditions."

Further, Einarsen (1932) remarked:
As stated in earlier reports, the most precarious stage seems to be when the water ranges between 30 and 40 feet on the gauge and that when the water either drops or raises above these points fish can more readily pass through; however, no rehabilitating program on the Fraser could be complete unless Hell's Gate (sic) were forever removed as a handicap to migrating salmon, as the first step in any such program.

This information, derived from visual observations and from interviews during his surveys was later corroborated by IPSFC investigations. In recent conversations, L.A. Royal, former Director of IPSFC, (personal communication) described Einarsen as a competent and experienced biologist having worked under him at the Washington Department of Fisheries.

It was apparent to some that Hells Gate was still a serious problem for migrating sockeye almost twenty years after the problem started.

Babcock in 1913 referred to the problem pink salmon experienced at Hells Gate and commented: "I am convinced that none of the species was able to get through that Canyon" (Babcock, 1914a). In 1923 Babcock apparently observed pink salmon above Hells Gate (Babcock, 1924a) stating: "For the first time since the fatal blockade of 1913 a few pink salmon were noted at Hells Gate; several were found in Lake Creek at the outlet of Seton Lake, and others in tributaries of the Thompson."

Einarsen (1929a) observed large numbers of pink salmon above the Fraser Canyon in one year stating:
The run of pinks on the tributaries of the Fraser above Hells Gate in 1927 is conceded to have been phenomenal. Areas as far as 200 miles above the Gate having been heavily seeded and conditions seem to indicate a return to runs of a magnitude of the days before the disaster.

Apparently, in some isolated instances, pink salmon did move upriver past Hells Gate after 1913. Pink salmon migration past Hells Gate is usually from September 25 to October 15. During these dates in 1927, the gauge ranged from 22 to 29 feet. In 1913 during the same period the river height was from 23 to 34 feet and in 1917, from 22 to 34 feet. Therefore, the river height in 1927 was not unusual. Motherwell (1927/28) reported that Babcock's observations in 1927 suggested there had been a problem in late 1927 for sockeye migrating in the Fraser Canyon which makes Einarsen's observations even more perplexing. Perhaps pink salmon were of exceptional size in 1927. The importance of size to migration success is discussed later.

The problem of salmon migration past Bridge River Rapids, located upriver from Hells

Gate, had been reported upon as early as 1912 by John Pease Babcock in a letter to F.H. Cunningham, Inspector of Fisheries, New Westminster, B.C. Babcock (1912a) observed great numbers of sockeye blocked at low water levels. He said "the work should be done at once," in reference to blasting rocks. He further stated that: "The matter is of so great importance that I have to urge that you come here at once and view the situation."

Investigations many years later would substantiate the observations made by Babcock and that natural conditions at Bridge River Rapids, at certain times, presented serious obstacles to migrating sockeye.

Based on modern-day knowledge of the tolerance limits of Fraser River sockeye due to life span limitations, stress, and specific energy requirements, many of those sockeye which did reach the spawning grounds in the years after 1912 were, in all likelihood, inferior in quality and less productive as spawners. Energy sources and expenditures in Fraser River sockeye salmon during spawning migration up the Fraser River were studied in 1956 on Chilko and Stuart Lake stocks (Idler and Clemens 1959). Further investigations were made of these and other runs in 1957, 1958 and 1959 (Gilhousen 1980). These studies indicated that sockeye have adapted to make the migration with the least expenditure of energy and their energy stores are just sufficient, with little excess, to make the migration. In addition, the altered sex ratios noticed on the spawning grounds (males are stronger swimmers and could more readily pass Hells Gate) would have significantly lowered potential egg deposition with respect to normal expectations. The poor returns that followed would possibly be attributed by many to escapements being lowered by overfishing. Of the sockeye getting through, many spawned or attempted to spawn in other locations and not in their natal streams. No production could have been expected from

that non-home-stream spawning. It is likely that the cumulative effects of Hells Gate, the Quesnel Dam and the Adams River splash-dam contributed to many poor quality spawners or prevented many of those fish which did overcome the lower-river obstructions from reaching their spawning grounds.

De Witt Gilbert (1988) presented his view of the problem:

Here was the Great Fallacy into which men fell so readily:

1. that the block at Hells Gate had been removed by subsequent excavation;
2. that pre-1913 conditions had been restored;
3. that such conditions were wholly satisfactory for the migration of sockeye;
4. that all salmon which escaped the commercial fishery spawned effectively;
5. that overfishing was the sole cause of the continued low level of sockeye abundance;
6. that the situation could be corrected by controlling men and their fishing;
7. that assumption was knowledge and there was no need for research.

Following the huge decline in the run in 1917, Babcock pointed to the Canadian Northern Railway as the builder who caused the problem. He said: "Some person is responsible for this irreparable loss—and any one of a number of government officials should have foreseen the trouble that would result from this indifferent filling up of the river" (Babcock 1917a).

Compensation for the great economic loss to fishing industries of both countries never seemed to be an issue, at least not publicly. In this regard, an interesting statement was made by Mitchell in 1925: "The white men have destroyed the Indians salmon food supply and should compensate him until they have rebuilt it…the C.N.R. should be made to pay for it."

Depletion of
the Resource

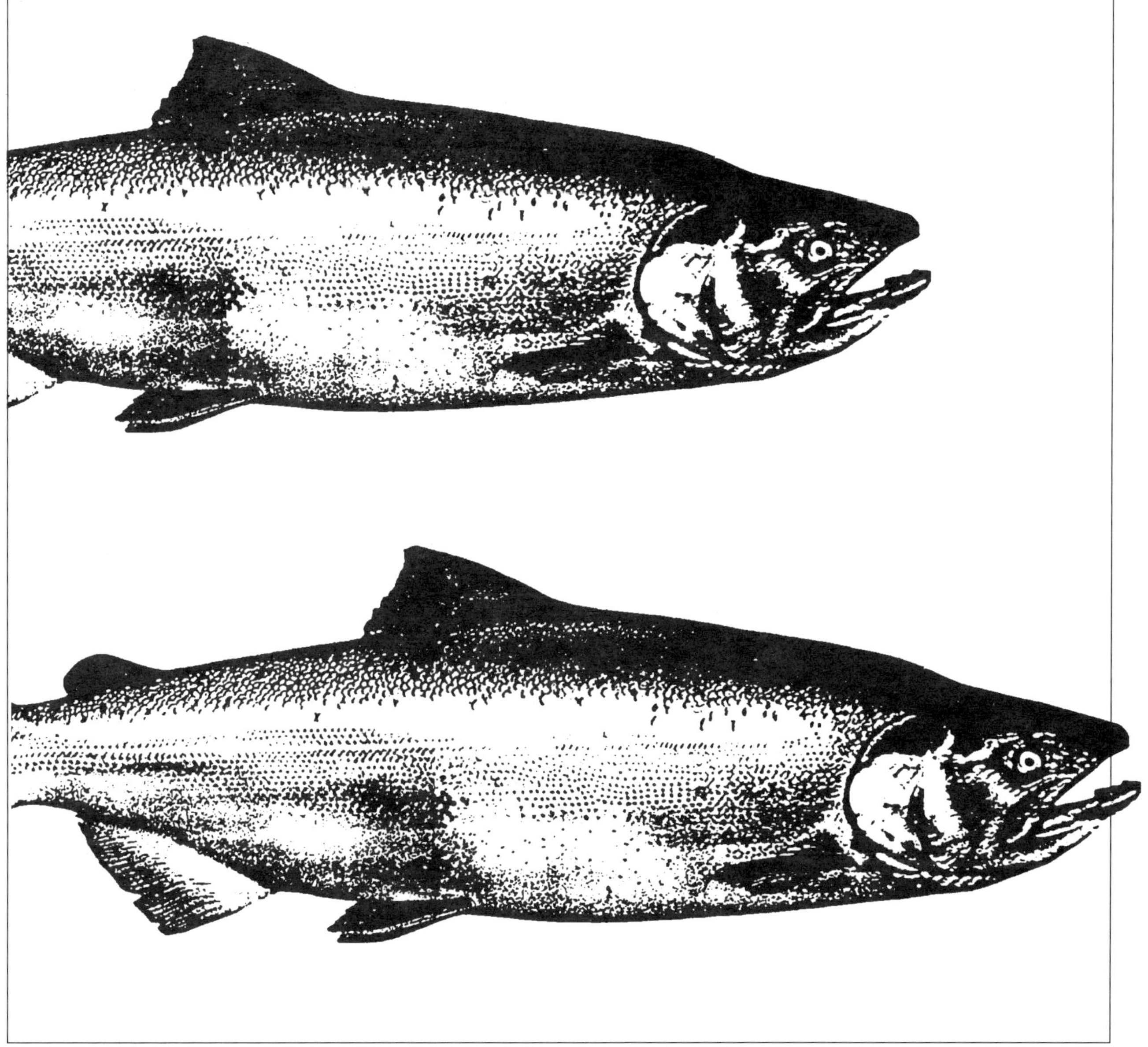

DECLINE OF FRASER SOCKEYE
AND PINK SALMON

The first major effort at documenting the decline of Fraser River sockeye and pink salmon runs was the thorough report of Rounsefell and Kelez (1938). From 1893 through 1916 the Fraser River sockeye run produced an annual average commercial catch of 8,532,000 sockeye (Gilhousen 1990 MS, Rounsefell and Kelez 1938 and PSC). For the period from 1917 through 1932, the average annual catch had declined to 2,018,000 fish, only 24 percent of the catch recorded in the earlier period.

The 1913 record catch of 32,035,000 sockeye was followed by a sharp decline to only 7,030,000 sockeye taken in the return year of 1917. By 1921, the cycle-year catch had dropped further to only 1,685,000 fish. The U.S. trap average catch "index-of-abundance" of sockeye from 1896 to 1916 was 188, whereas from 1917 through 1934 it declined to 48, only 26 percent of the index in the earlier period.

Available data on Fraser River pink salmon also indicated a sharp decline in abundance after 1913. Indices of abundance as derived from trap catches were again examined. The northern index represents the Fraser pink salmon run fairly accurately (compared with Washington State pink salmon) because of the location of the traps and the timing of the catches. The index from 1907-1913 (four cycles) averaged 282, and in 1913 was 285 (Rounsefell and Kelez 1938). In 1915, the index was 50 and for the 20-year period following 1913, the average was only 68 or less than one fourth (24 percent) the pre-1913 index-of-abundance level.

The southern trap index represents Puget Sound pink salmon stocks and was very different from the northern index. The index for the years 1911 and 1913 were 108 and 115, respectively (average of 112). No decline was noted in the index during the period from 1915 through 1933, the average yearly index actually increasing to 130. There was no evidence to indicate that overfishing occurred on Puget Sound pink salmon during this period and many of these fish would have been subjected to fishing in Convention Waters.

Analyses of commercial catch data for both Fraser River sockeye and pink salmon clearly show that the runs declined dramatically after 1913. Based on catch data for sockeye and pink salmon, both species incurred similar depletion in the years 1913 up to 1933, suggesting that a common factor was implicated in the declines.

REASONS GIVEN FOR
THE DECLINE

The abundance of the cycle years following the dominant-year sockeye runs in 1898, 1902, etc. declined 39 percent from the 1898 level during the period 1898-1914 *(ibid.)*. The inference that was accepted by the majority was that overfishing was the principal cause of the decline. Rounsefell and Kelez stated: "…the intensive fishery of 1914 was doubtless instrumental in causing this low escapement… " However, as pointed out earlier, there were other factors—the splash dam in the lower Adams River could have had a greater adverse impact on the lower Adams River run than the Hells Gate slide. The cycle year 1914-1958 developed into the major return year by the mid-1930s, probably mainly due to shutting down the splash dam.

With regard to the big cycle year (1901-1913), the sockeye runs in the commercial fishery were very abundant, based on information from early commercial records in 1877 and from Hudson Bay records in the early 1800s. There was no evidence of overfishing in the early years of the commercial fishery and reports indicate that escapements were very large. The catches averaged almost 26,000,000 sockeye from 1901 to 1913 on the big years, yet huge numbers apparently escaped to the spawning grounds. The escapement past the fishway at the Quesnel dam in 1909 for this stock alone was estimated to be over 4,000,000 fish.

Rounsefell and Kelez speculated that:
The fact that tremendous numbers of sockeye escaped to the spawning grounds on the big years, despite the huge catches, may have occurred because of the presence in all of the big-year cycles from 1901-13 of very abundant late runs, appearing after most of the fishing had ceased.

These fish were not late runs but were most likely predominately lower Adams River sockeye—commonly referred to as "late-run sockeye" which had passed through the outside fisheries in August. After delaying for up to a month off the mouth of the Fraser, they began to move upriver. Lower Adams sockeye were then available to the Fraser River gillnet fishery as lower quality sockeye after mid-September. The lower Adams River sockeye are taken as prime quality sockeye in fisheries away from the mouth of the Fraser River, primarily after mid-August, mixed in with several other less abundant races.

J. Crawford visited the lower Shuswap River area in 1902 and was told that a great many sockeye were there in 1901. Local Indians described the cyclic characteristic of the runs and he reported: "that every fourth year there was a good run of Sockeyes (sic) in this river, but during other years there were very few" (State of Washington Department of Fisheries and Game 1902).

With respect to the other two cycle years, Rounsefell and Kelez did not present definite evidence of the cause of the decline of the runs in these cycles. These authors concluded that the big cycle year was severely reduced by overfishing and the Hells Gate slide.

The history of the increasing fishing effort that developed in U.S. and Canadian waters from 1877 through 1934 is depicted in Figure 2. These data, obtained from Rounsefell and Kelez, show that gillnets were the first gear to exploit Fraser sockeye in the Fraser River area; traps followed in the 1890s in the United States and purse seines started toward the end of the century in the United States. Significantly increased effort was evident, starting in about 1897, with greatest effort from 1913-1917. Thereafter there was a pronounced decline in effort into the 1920s, followed by a gradual increase in gillnets and seines to the mid 1930s. In 1935, salmon traps were prohibited in Washington State. Even though decline in effort took place after 1913, technological improvements were made to gear and therefore fishing efficiency per unit of gear increased.

The early history of the Fraser River commercial fishery for the dominant cycle years (1877, 1881, etc.) and the off-years (1878-80, 1882-84, etc.) shows gradual increase for the dominant years reaching a peak in 1913 (Figure 3). A period of large catches occurred from 1897-1913. The off-year runs also showed the same gradual increase in catch with the annual average maximum catch occurring during the 1898-1900 years (Figure 3). Both the off-year and dominant-year runs as depicted in Figure 3 show a buildup in catch in the late 1800s and early 1900s that closely parallels the increasing trend of fishing licences issued during the same time period (Figure 2). The off-year maximum production peaked about fourteen years earlier than the big dominant year run in 1913. The increased effort likely resulted in a larger percentage catch of the smaller off-year runs. It is possible that the returns in 1905, 1909 and 1913 would have been even larger if better fish passage conditions had existed at Quesnel Dam for the Horsefly River race in 1901, 1905 and 1909.

Maximum catches, totalled for the three off-years, were closely associated with the period of first significant increase in all gear types. From 1902-1916 the level of the three-year total catch declined to about one-half that of the peak three-year catch from 1898-1900. The reason for the high level of catch for three years (1898-1900) is not known. The significance of

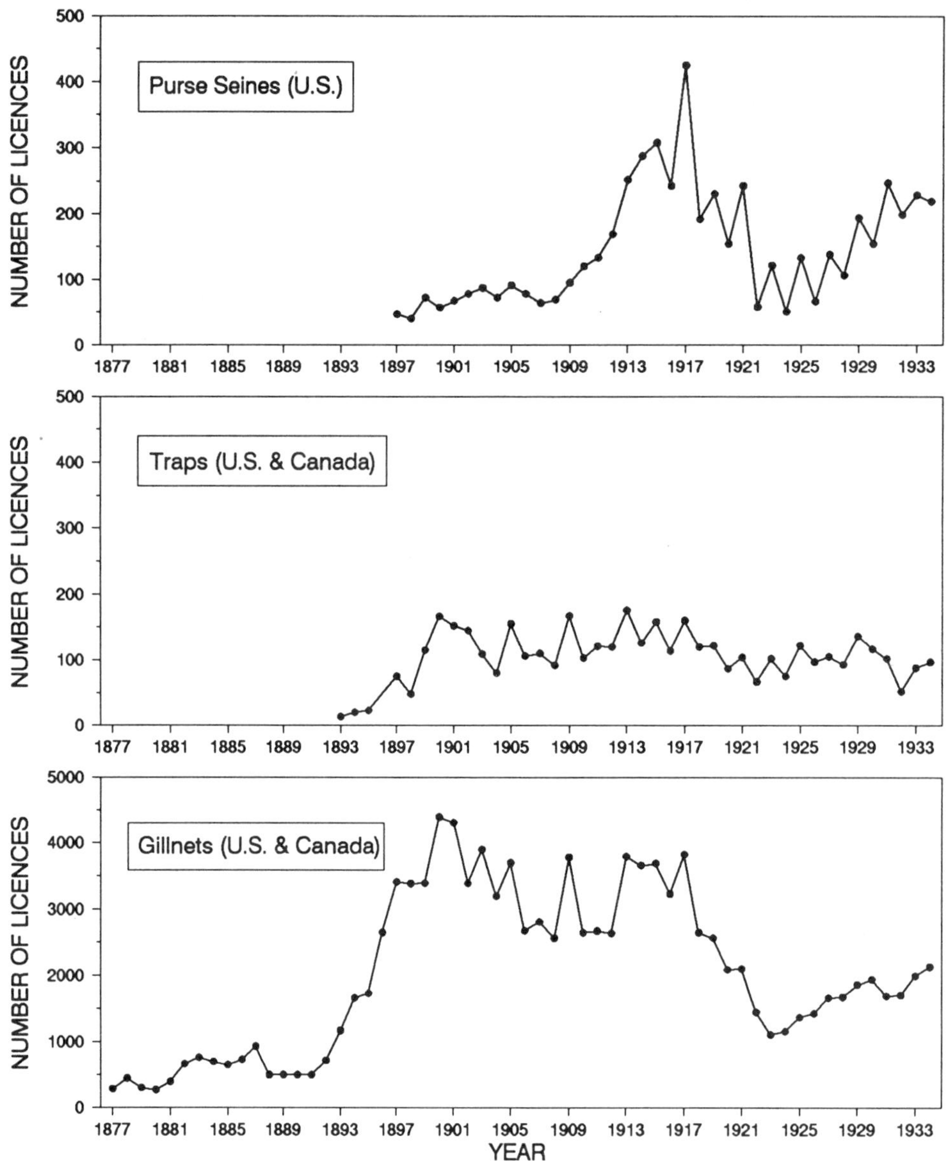

Figure 2. Number and type of gear licences, 1877-1934. From Rounsefell and Kelez 1938.

the decline that followed is difficult to assess since the total three-year catches from 1906-1916 were not much lower than the 1902-1904 summed catch, which in turn, was similar to the catch from 1894-1896. Since freshwater and ocean survival rates from 1894-1916 are unknown, and because commercial fishery exploitation rates were not known, reasons for changes in catch levels for any of the time periods cannot be established with certainty. Overfishing was generally suspected as the cause. However, the principal cause for the decline from 1918-1928 would most likely be attributed to the rock dumping and slide at Hells Gate.

The resurgence of the off-years as they developed in the dominant year (1922 cycle) is seen clearly in the buildup starting in 1926 and continuing on to 1946. This increase occurred without any Canadian government or IPSFC assistance or reduction in fishery exploitation. Based on spawning ground observations, most

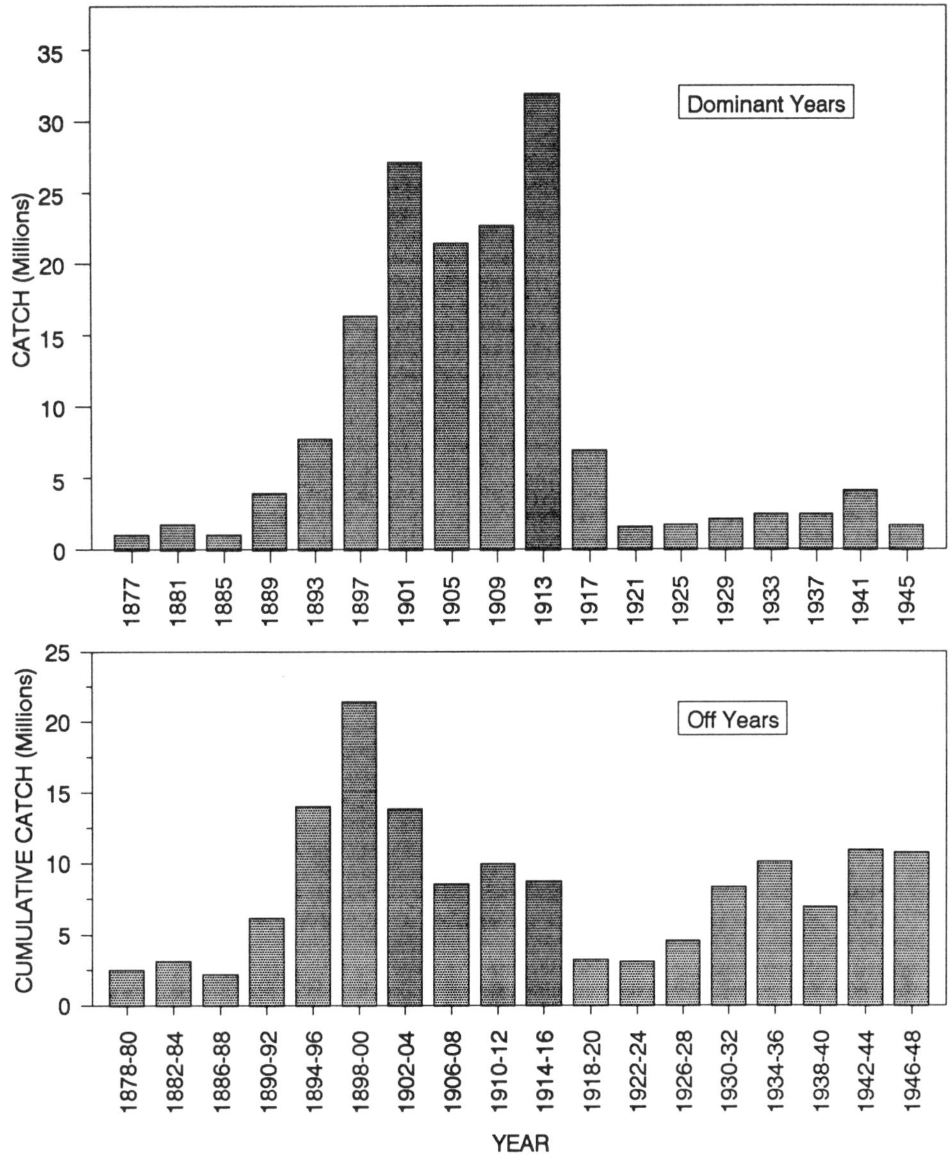

Figure 3. Total commercial catch of Fraser River sockeye salmon in "dominant" years and in the three "off" years of each cycle, 1877-1948. From Rounsefell and Kelez 1938, Gilhousen 1990 MS and Pacific Salmon Commission (PSC).

of the buildup of the off-years from 1926 to 1946 was attributed to one population—the lower Adams River race.

It would appear that there is little evidence of overfishing on the lower Adams River race, except perhaps from 1917-1922. The original dominant cycle (1913) from 1921-1945 (without the upper and lower Adams stocks) continued at a very low level of abundance. The water levels at Hells Gate and resultant blockage conditions in association with high levels of fishing intensity are believed responsible for the sustained low levels of abundance of the big-year runs comprised of summer-run stocks (Thompson 1945a). The off-year cycle runs increased because of the later timing of the lower Adams River race and possibly somewhat lower fishing intensity on that population. The later timing also would provide easier passage through Hells Gate with the river below gauge 30 for most of the migration period. The catch trends shown in Figure 3 after 1917 for the off-year and dominant-year runs by themselves, basically describe and identify passage conditions at Hells Gate as the most significant factor causing and perpetuating the decline of early and summer-run stocks. The biological answer was contained to a great extent in the differing fish passage conditions affecting early and summer-run sockeye compared with late-running races, since fishing exploitation rates (except as noted earlier) on the various races were likely not significantly different. With respect to pink salmon, Rounsefell and Kelez were convinced that the drop in the Fraser River pink salmon run was due to conditions at Hells Gate.

There were two lower Fraser River sockeye races which historically returned to the river very early (May) and spawned in the Alouette and Coquitlam River systems. By the 1930s both of these races were destroyed due to dams on those tributaries. Because of comparatively small abundance, the production loss had little impact on the total fishery.

Formation of
the Convention

I t is interesting to review the setting upon which the Sockeye Salmon Convention was eventually concluded. There was general agreement and awareness between the two governments and the fishing community prior to 1913 that management by a single agency was required to control the industry and manage the resource effectively. The big years were still big or even bigger than in the late 1800s and while there had been some lower production in the off-cycle years, this was generally not considered too serious. The Americans were content to fish, almost at will, but the Canadian industry was increasingly alarmed at the high percentage catch taken in United States waters. Concern about overfishing, however, was increasing. Canada was eager to consummate a treaty, whereas the American industry and Washington State government officials were stalling and fighting the proposed change in control of Washington's authority. Also, the United States federal government was reluctant to invade the constitutional rights of the state. In addition to all this were strong feelings in Canada that the sockeye were of Canadian origin and that the fish belonged to Canadian fishermen.

The decline in runs after 1913 and the increasing intensity of fishing effort probably had more to do with the eventual agreement than anything else. In addition, the 1934 abolishment of traps in the United States (effective in 1935) immediately changed the catch distribution between the two countries. It was then in the best interest of the United States to agree to equal sharing of the resource.

There were political and biological reasons for an agreement between Canada and the United States. Difficult times were on the horizon.

EARLY NEGOTIATIONS

From 1891 to 1900 Canada was taking about 70 percent of the total sockeye catch. However, during the next decade the catch distribution shifted sharply with the U.S. taking 56 percent of the catch. Through 1901, no great concern had been evident concerning the size of the annual sockeye runs, and limitation of the total catch had not been the major concern. The real issue to that point had been the distribution of catch between United States and Canadian interests. In an attempt to counter the superiority of the United States trap-dominated fishery, the Canadian government authorized the use of traps in Boundary Bay in Canadian waters east of Point Roberts. Since this location was not in the primary migration path of Fraser River sockeye, catching success was poor and the action was ineffective. This was probably the first direct involvement of government to affect international catch distribution. There would be more such efforts in subsequent years.

In a 1903 report of the British Columbia Salmon Commission to the Honorable Raymond Prefontaine, Minister of Marine and Fisheries, the Commissioners recommended the use of purse seines be permitted in Canada "so that the Canadian fishermen would be given increased opportunities of taking the Fraser River salmon before they reached the United States limits" (Smith et al. 1903). The plan was to start a Canadian fishery in the Strait of Juan de Fuca. Fearing for their livelihood, inshore Canadian gillnet fishermen opposed such a move.

In the same year, consideration was also given to intercept American-bound salmon headed for Alaskan waters. In a letter to the Minister of Fisheries the statement is made:

> It is more than likely too that fish heading for the fresh water streams in American territory - which streams also empty into Hecate Straits-could be caught in Canadian territory, and thus enable us to handicap the American operations in Alaska in a similar manner to their interference with the Fraser River. (Doyle 1903)

The 1902 season was such a failure that concern was expressed as to the adequacy of regulations—regulations which for 1901 appeared to have been entirely adequate. Nowhere in any of the discussions did it appear evident that escapement requirements for the various stocks on the different cycle years would or should be different.

In 1902, the Canadian government appointed a Commission headed by Professor Prince to investigate the British Columbia salmon fishing industry. This Commission found that the problem was much larger in scope than originally perceived and that a much more thorough and extensive inquiry would be required.

By 1903, conditions worsened and the catch was "very poor." Many in the fishing industry who were opposed to regulations believed that future runs should be assured by fish culture rather than by limitations to fishing. For almost five years, discussions took place between the Canadian federal government and Washington State officials. Part of those discussions centered around the ever-increasing conflicts and competition developing between trap owners and purse seine vessel owners in the United States industry.

In 1905, following three years of poor sockeye returns, Professor Prince was again appointed as chair of another Commission, and the Prince Commission of Canada and the Washington State Fish Commission studied the overall problem. It concluded that treaties could not be consummated between a federal and a state government. Therefore, negotiations between the two federal governments commenced and on April 11, 1908, the Bryce-Root Treaty was signed. The paramount need recognized in both countries was for a joint Commission with control of the international waters. Ratifications were exchanged and the treaty was proclaimed on July 1, 1908. It provided for an international commission, assisted by one expert from each government. This commission was empowered to carry out certain investigations. Its regulations were to remain in force for four years and thereafter

until one year from the date upon which either government notified the other of its desire for a change in the regulations. It was unbelievable that regulations were to be fixed before the season began with no possibility of in-season changes. However, this concept reflected man's understanding of the fish and the biological requirements of the species at the time.

The United States Senate ratified the treaty but did not pass the legislation necessary to give effect to the regulations proposed by the International Commission. Intense lobbying on the part of the Washington State legislature protesting the federal government's involvement in assuming or attempting to assume jurisdiction and control of State waters eventually led to the disappearance of the legislation in Congress. This occurred despite the strong support of President Wilson. Canada, by 1914, had become extremely impatient waiting for U.S. approval, and notified the United States it was abrogating the treaty on October 14, 1914. Feelings between the two countries were strained until World War I when Canada and the United States became allies in 1917. At the same time, factors such as a general food shortage and the failure of the expected "big run" of Fraser River sockeye in 1917 renewed a willingness to resume negotiations for an international fisheries treaty.

In December 1917, an investigatory commission was appointed composed of three experts from each country. Dr. W.A. Found was one of the Canadian representatives. He was later to be one of the original Canadian Commissioners of the IPSFC. Conditions on the Fraser River had deteriorated to the point that a $30,000,000 industry had declined to only $3,000,000 (Report of Commissioner of Fisheries 1933). It was also concluded by many that the resource was being threatened with extinction by overfishing. A "new" convention was drawn up, similar to previous conventions. Canada approved the treaty on October 21, 1919 but Washington State officials again prevailed on the United States Senate not to infringe on states' rights. The Senate did not pass enabling legislation and this lack of action nullified the treaty.

Another meeting in December 1921 resulted in both countries agreeing to stop fishing in the Fraser River District for five years. Unfortunately, a satisfactory conclusion was not reached. There was no agreement on what should take place after the five-year period, and the conference broke up abruptly.

Discussions were resumed in 1922, but these too, were unsuccessful. A bizarre concept had been proposed that might give B.C. fishermen "first crack" at Fraser sockeye and perhaps even eliminate the United States fishery. The *B.C. Fisherman* (1931) reported:

> After several years' investigation under the auspices of the Biological Board of Canada, Dr. A.H. Hutchinson, of the University of British Columbia, has reported negatively upon the suggestion that closing Canoe Pass, the most southerly outlet of the Fraser River, would so direct the influence of the river that salmon bound for the spawning grounds in the Fraser system would be led to take a course through Canadian waters near Vancouver Island instead of through United States waters.

Studies began in 1926 and lasted for several years. The effort was directed at attempting to find a way to change migration patterns thereby reducing or eliminating the catch of Fraser sockeye in the U.S. Point Roberts traps.

By 1928, the sockeye runs had diminished, many canneries had been shut down, and the few remaining operators were satisfied because they were able to survive with lower competition. However, negotiations commenced again and a new treaty was proposed, essentially similar in most respects to the earlier versions of 1909 and 1919. One important difference was that the catch would be divided equally between Canadian and United States fishermen. This clause had some local appeal in Canada because United States fishermen had been taking, with effective traps and purse seines, about two-thirds of the catch each year. United States fishermen were not satisfied entirely because this meant a decrease in their catch. On the other hand, many Canadians maintained that the proposed agreement was not just or fair, since the fish were Canadian in origin.

As early as 1911, United States purse seine fishermen began fishing in high-seas waters (outside national waters) west of the entrance to Juan de Fuca Strait. By 1925, significant catches were being made on the high seas. In this area they were not subject to restrictions. Understandably, Canadians were critical of this extended fishery as they felt it was a further intrusion into their resource. Discussions held in 1928 were no doubt influenced by this high-seas fishing. These seiners, however, did not want to be controlled by an international commission and believed the proposed treaty was designed to eliminate them. Seiners made their largest catch of sockeye in high-seas waters in 1930 (144,000 fish) and in 1934 (675,000 sockeye).

· · · · · · · · · · · · · · · · · · · ·
TREATY SIGNING

A new agreement was negotiated during 1928 and 1929, signed in Washington, D.C., on March 27, 1929, and sent by the President to the Senate for approval. This treaty did not cover the waters where the high-seas fishery operated nor was it clear concerning the obligations of the proposed commission with respect to artificial propagation. Therefore the document was unacceptable to Canada. It was also reported that a Canadian objection was that the international commission would be given certain authority in Canadian territory; for example, the Fraser watershed, for hatcheries, etc. whereas no U.S. land was involved (*Canadian Fisherman 1929, Pacific Fisherman 1929*). This was the first rejection by Canada of a proposed treaty.

The Americans were objecting because the document did not specify that two of the

U.S. Commissioners would be citizens of Washington State. U.S. seiners were also upset at the 50-50 catch division.

Subsequently, changes were agreed upon, including control of fishing in high seas. The convention for the protection, preservation and extension of the sockeye salmon fishery of the Fraser River system was finally signed in Washington, D.C. on May 26, 1930 but was not approved by the U.S. Senate. The Canadian House of Commons approved on May 29, 1930.

An interesting participant who strongly supported the proposed treaty was President Hoover. The *Pacific Fisherman* (1931) reported:

Much credit for the successful conclusion to the negotiations is given to President Hoover. During his term as Secretary of Commerce, Mr. Hoover became keenly interested in the Pacific fisheries and was a strong supporter of protective measures for the salmon fisheries of that coast. It is said that he has personally lent his influence to bring about an amicable agreement on the long-discussed sockeye salmon treaty.

A. L. Hager, an American who moved to Canada and later became a Canadian and IPSFC Commissioner, became a close personal friend of the president. Mr. Hoover was the guest of Hager at his Vancouver residence on several occasions (D. Hager, personal communication).

A new treaty was seemingly assured at this time. However, serious opposition arose almost immediately in the United States over the provision to share the catch on a 50-50 basis and the inclusion of the high seas as "Treaty Waters." After all, the American fishermen had been taking about two-thirds of the catch and the special advantage of the purse seiners would be under the control of an international body. Lurking in the background was the strong opposition of Governor Hartley of Washington State on the principle of state rights vs. federal interference in controlling the fishery (*B.C. Fisherman* 1930).

Signing of the treaty in 1930 did not elicit enthusiasm in Canada as:

The history of the fight for adequate international measures for conservation of the Fraser River sockeye system reflects unpleasantly upon some of our neighbors south of the International Boundary. Four times Canada has officially signified her desire to cooperate, and three times, United States legislators, for political reasons, have refused assent of an international agreement prepared and accepted by men on both sides of the line who were not concerned with politics. Is it any wonder that all B.C. fishermen with a thought for the future accept the news of this new treaty without enthusiasm? *(ibid.)*.

Governor Hartley, in his endeavor to squelch the new treaty, spoke to the people of Washington State by radio on December 4, 1930. His speech was printed in the Congressional Record. He referred to the treaty as: "the so-called Sockeye Salmon Treaty." (Congressional Record 1930a). He refuted the claim that the resource was in danger by stating: "We have not only read but heard much untruthful propaganda in the last few years of the rapidly approaching extinction of the salmon industry on Puget Sound."

He proceeded to defend his position by presenting facts prepared by the Washington State Department of Fisheries. He stated that: "Instead of this natural resource being greatly depleted, the reverse is conclusively shown to be true."

In his plea, he further remarked that:
As Governor of this State, in protecting the rights of its citizens, I regard these facts of dominating importance. What is the crisis, or the necessity which suggests that the State of Washington relinquish its sovereignty over the principal fishing waters of the State, a field in which its activities have been most constructive and fruitful? What can a joint commission, or a federal bureau do better than the people of the State of

Washington in the handling of this or any other natural resource of this state?

It appeared that politics was a more important concern than the well-being of a major resource.

On December 10, 1930, Mr. Miller Freeman delivered a speech to the Seattle Bar Association in regard to the sockeye salmon treaty (Congressional Record 1930b). This speech was a rebuttal to Governor Hartley's speech and it, too, was entered into the Congressional Record. Mr. Freeman was for many years a significant figure in fisheries affairs and the publisher of *Pacific Fisherman*. His presentation was a request for a dispassionate and objective review. He pointed out that the sockeye run had declined dramatically: "Its former magnitude is illustrated by the fact that in 1913 the pack on both sides of the line totalled 2,401,488 cases. In 1929, the combined pack was 172,291 cases, or 7.2 percent of the 1913 pack."

He termed the work could be: "one of the greatest reclamation projects in which the two countries can engage."

Concerning loss of jurisdiction over state fisheries, Freeman said: "It would take from the State of Washington only jurisdiction over the sockeye salmon fisheries, with some incidental effect on other salmon that are taken at the same time and by the same gear as sockeye." Freeman further elaborated on perceived rights of the industry by stating:

We maintain it is the right of the people as a whole to expect that this resource will be maintained to them as a permanent source of national wealth, which is all that this treaty is designed to do.

There was apparently no doubt in Freeman's mind as to the reasons for the Fraser River sockeye decline. He stated:

The principal cause of depletion of the great sockeye run following 1913 was a rock slide in the Fraser River that year, although overfishing, before and since, has been an important contributing factor.

Referring to the Fraser sockeye problem, the Hon. Simon F. Tolmie, Premier of British Columbia, said:

I am not in accord with Governor Hartley's view in this matter.... It is not a state or provincial question.... It is an international question and cannot be dealt with except in an international way. (Quoted by Freeman in Congressional Record 1930b).

The Premier further stated:

the restoration of the sockeye salmon of the Fraser River system has been a burning question for years. The fishery can be restored by a treaty. It cannot be restored in any other way.

U.S. purse seiners did not want the high seas included in the treaty and were opposed to giving the Commission immediate regulatory power. Many meetings were held between 1930 and 1934 with both governments and industry blaming the other for the pathetic status of the runs.

From 1930 to 1934, Canadians, for the most part, had given up hope of a treaty. However, in 1934, a conference was arranged between representatives of the two countries in Seattle, Washington at the request of Governor Martin of Washington and Premier Pattullo of British Columbia. Those present were Governor Martin; B.M. Brennan, Director of Washington State Department of Fisheries; Frank T. Bell, United States Commissioner of Fisheries; Miller Freeman, publisher; Dr. W.F. Thompson, Director, International Pacific Halibut Commission; Hon. Geo. S. Pearson, Commissioner of Fisheries for British Columbia; Dr. W.A. Found, Deputy Minister of Fisheries, Federal Fisheries Department; George J. Alexander, Assistant Commissioner of Fisheries for the Province of British Columbia; and J.P. Babcock, former Assistant Commissioner of Fisheries for British Columbia. The entire Fraser sockeye matter was discussed at this meeting and most of the objection of the United States purse seiners was overcome.

Initiative No. 77 in Washington State to

abolish use of traps was another major development. Traps contributed up to 60-75 percent of the annual catch by the Americans. Immediately following the banning of traps in 1935, Canadian fishermen caught the larger percentage of the catch. In 1936, Canada took 82 percent of the Convention Waters sockeye catch compared to 57 percent in 1935. In 1937 Canada took 55 percent. The increased catch in 1936 was not solely related to the absence of traps in U.S. waters but was probably due to a high number of returning sockeye approaching the Fraser River around the north end of Vancouver Island and through Johnstone Strait. It does not appear that the Americans were aware of the real reason why Canada's catch was so large in 1936. This fortuitous event for Canada convinced the Americans that 50-50 was a pretty good deal and it was now considered in the United States' best interest to conclude a treaty that would provide one-half the catch. Who knows what might have happened if the diversion of 1936 had not occurred or if the U.S. had realized what happened?

Several of the participants of the 1934 meeting eventually served as either members of the Commission or as the Director of Investigations of the Commission.

Early reports by fisheries investigators revealed a general acknowledgement that some Fraser River sockeye migrated toward the Fraser through the northern passage via Johnstone Strait. Seldom was the migration pattern considered a major phenomenon, although occasional reference was noted such as in 1913 (Babcock 1914a). With respect to the proposed sockeye treaty, it is noteworthy that some personnel in the Washington State Department of Fisheries were concerned about a possible loophole in the treaty since the catch of Fraser River sockeye in the waters of Johnstone Strait (non-Convention area) would not count against Canada's total catch in the division of catch between the two countries as provided for in treaty waters. These individuals were also very concerned that published results

of tagging studies on chinook salmon off the west coast of Vancouver Island established that more than 90 percent of the Canadian chinook catch in 1926 was of United States origin.

Examination of the proposed treaty prompted a memo from Einarsen (1931) stating that:

> From reading the treaty as now drafted, it can be seen that Canada receives an unfair consideration in that no provision is made to safeguard those sockeye entering the Fraser from the north through Johnstone Straits as this area is not included in the Convention Waters.

Later in his memo, he stated: "Evidently, a treaty should be drawn up on the same basis as the Sockeye Treaty giving the U.S.A. the same favorable consideration in regard to its chinook salmon." In later years both these points became contentious issues and to a great extent were part of treaty negotiations in the 1970s and 1980s, eventually leading to the demise of the IPSFC.

One of the major shortcomings of the treaty being considered in 1934 from a management perspective, evident in later years, was the exclusion of Commission control in the non-Convention Waters north of the Convention boundary.

The B.C. Fishermen's Protective Association met at New Westminster on November 19, 1934 and Dr. Found, Deputy Minister of Fisheries, was present along with Mr. Tom Reid, Member of Parliament (MP). Traps were no longer going to be used in Washington and there was eager anticipation that the U.S. share of the catch would decline, although there was still strong suspicion about the impact of seines in the U.S. and in Canada. Dr. Found reviewed the key new provisions of the present treaty and mentioned that the important provision was on division of catch:

> … because up to this time we have been getting from between twenty and thirty percent of the early run sockeye coming to the river instead of getting our reasonable

proportion, which should be at least one-half, as the fish are all bred in our waters. (B.C. Fishermen's Protective Association 1934)

He further elaborated that Canadian fishermen would get 50 percent of the catches under the treaty now before the U.S. Senate. Fraser River fishermen wanted strong assurance that they would get 50 percent of the catch. At earlier meetings, they had threatened to fish the runs to extermination. Wild schemes such as closing Canoe Pass so smolts would migrate north through the Gulf Island passes and return as adults away from U.S. waters were also talked about. Interestingly, in recent years it has been determined that most Fraser River sockeye smolts migrate to sea via Johnstone Strait (Groot and Cooke 1987).

Following prohibition of traps, the battle between trap owners and purse seiners ended. In addition, the recreational fishermen's groups which were instrumental in removing traps, were now in favor of a treaty with Canada. Governor Martin said in reference to the treaty: "The State of Washington would be extremely shortsighted if it fails to take advantage of this magnificent offer" (*Pacific Coast News* 1936a).

In 1935, Governor Martin solicited the comments of the fishery communities with regard to the sockeye treaty. The University of Washington School of Fisheries responded as follows:

The staff of the School of Fisheries wishes to indicate its great interest in the success of your efforts to bring about the much needed international care of the salmon runs of the Fraser River. It is one of the most important resources of our State, but one of the most urgently in need of rebuilding, and presents an admirable opportunity for a practical application of fisheries science. (*Thompson et al.* 1935)

Discussions carried on between 1930 and 1935 were intensified in 1936. Fishery organizations were in the forefront of heated debates in both countries on issues such as the inclu-

sion of an advisory committee, no control of gear types, area of control on the high seas, regulatory authority, division of catch, make-up of the Commission, and waters to be covered in the Convention. Mr. Swailes (Member of Legislative Assembly, Province of B.C.) contended: "The Commission has to have power and should not be subject to governmental interference or political interferences" (*Pacific Coast News* 1936b). The 1930 treaty was often referred to as the Trap Treaty and the revised treaty in 1936 as the Seiners Treaty (*ibid.*).

Many in the U.S. industry were concerned about the 50/50 split in catch:

… will it be a fair division or are there some taken up here, are there some taken in here or down this way? … The facts are there is not enough scientific data available to determine just what territory should be taken in order to secure a fifty-fifty division (*ibid.*).

Another said:

… we are passing a treaty for all time pretty well. Let's get it right (*ibid.*)

In another meeting it was stated:

During the debate on division of catch the argument was put forth by some United States delegates that Fraser River sockeye were caught in the waters of Johnstone Strait and adjacent areas. This argument was refuted by members of the upper Fraser Fishermen's Protective Association. (*Pacific Coast News* 1936c)

However, as noted earlier, Babcock in 1913 had reported strong evidence of Fraser sockeye migration down Johnstone Strait (Babcock 1914a).

At the 1936 meeting held in New Westminster, B.C. between industry people from both countries, the Canadian side threatened that if the 1930 treaty was thrown out, they would have a very different attitude in negotiating a new treaty. The Canadian side told the Americans that if they were sincere in wanting a treaty they had better take the current treaty with a 50/50 split (even though six years old) as it was by far the best they would ever do. "The American delegation returned to Seattle sharply rebutted and rather tame" (Freeman 1956).

The State of Washington proposed three additional clauses to the original treaty that Canada had ratified in 1930. It is reported that Miller Freeman, in an attempt to make certain that the treaty did not fail, proposed a compromise which would defer Commission regulatory authority until it had carried out comprehensive scientific investigations for eight years. Thus, after several years of discussions, the governments met for additional discussions. The American side had apparently been unofficially promised that two of the three U.S. Commissioners would be residents of Washington State but that provision would not be officially stated in the treaty. The treaty was finally approved by the United States Senate and ratifications were exchanged July 28, 1937 after several previous attempts had failed (Appendix A).

At the time of signing, there were still strong nationalistic feelings and factions with diverging opinions as to possible treaty benefits. In Canada, strong feelings still existed that the fish were Canadian and belonged to Canada's fisheries. In the United States, there were still those who felt historical precedent should dominate and that the State of Washington should have authority in the matter. Nevertheless, most believed the treaty to be a major accomplishment.

The two major issues that ultimately brought the treaty into fruition were the clear understanding in both countries that the most devastating fishery disaster known at the time needed to be fully identified and corrected, and that fishing in both countries required control by one agency represented by Commissioners from both countries. There was almost unanimous agreement that this could not succeed except under the framework of an international fisheries commission. Canada was also reluctant to invest money for research and rehabilitation unless assurance of benefits was provided. Washington State officials would later object very strongly to the taking by Canadian fishermen of large numbers of chinook and coho salmon produced in Washington streams and hatcheries. An extensive review of the background leading up to the signing of the Convention was given by Tomasovich (1943).

TREATY RATIFICATION

On August 4, 1937, the President of the United States proclaimed that a Convention between the United States of America and Canada for the protection, preservation and extension of the sockeye salmon fishery of the Fraser River system was concluded as signed by their respective plenipotentiaries at Washington on May 26, 1930.

The preamble to that Convention stated: The President of the United States of America and His Majesty the King of Great Britain, Ireland and the British dominions beyond the Seas, Emperor of India, in respect of the Dominion of Canada, recognizing that the protection, preservation and extension of the sockeye salmon fisheries in the Fraser River system are of common concern to the United States of America and the Dominion of Canada; that the supply of this fish in recent years has been greatly depleted and that it is of importance in the mutual interest of both countries that this source of wealth should be restored and maintained, have resolved to conclude a Convention.

The Convention contained eleven Articles. The Convention was ratified by the United States of America subject to three understandings, made a part of the ratification, as follows:

(1) that the International Pacific Salmon Fisheries Commission shall have no power to authorize any type of fishing gear contrary to the laws of the State of Washington or the Dominion of Canada;

(2) that the Commission shall not promulgate or enforce regulations until the scientific investigations provided for in the convention have been made, covering two cycles of sockeye salmon runs, or eight years; and

(3) that the Commission shall set up an Advisory Committee composed of five persons from each country who shall be representatives of the various branches of the industry (purse seine, gillnet, troll, sport fishing, and one other), which Advisory Committee shall be invited to all non-executive meetings of the Commission and shall be given full opportunity to examine and to be heard on all proposed orders, regulations or recommendations.

The three understandings that the United States insisted be added were discussed extensively in the Canadian Parliament early in 1937. The first and third understandings were quite acceptable. However, considerable objection was expressed over the eight-year delay in regulatory control. Canadian authorities nevertheless agreed (some very reluctantly) rather than force destruction of the fishery. In view of United States actions from 1930-1936, Canada had reason to balk at signing, but signed after delaying for one year. It was also agreed that the treaty would be in force for a period of at least 16 years subject to cancellation thereafter following one year's notification.

The Secretary of State of the United States of America stated:

> The Convention is ratified on the part of the United States of America subject to the three understandings contained in the resolution of the Senate of the United States of America advising and consenting to ratification, a copy of which resolution was communicated to the Secretary of State for External Affairs of Canada by the Minister of the United States of America at Ottawa in his note of July 7, 1936. (Hull 1937)

The U.S. Senate advised the ratification on June 16, 1936 and presidential ratification was on July 23, 1937. His Majesty in respect to Canada ratified on June 26, 1937 and the ratifications were exchanged in Washington, D.C. on July 28, 1937. Thus, it took another year after 1936 before the Parties finalized a Convention.

Finally, after 45 years of meetings, negotiations and several proposed agreements, a Convention between the two countries was concluded and proclaimed on August 4, 1937. The simplicity of the Convention and its singular purpose of restoration of Fraser River sockeye would eventually become an outstanding example of cooperation between two countries.

Miller Freeman, publisher and owner of *Pacific Fisherman*, was probably the strongest advocate of the treaty in either country and had worked feverishly for its completion. Governor Martin had appointed Freeman as his official spokesman on the matter. Freeman in his memoirs characterized his role in the difficulties of reaching agreement as:

> The thirty year war … "a running fight" extending over an actual period of 30 years, from 1907 until 1937; and with varying intensity from periods of complete quiet between skirmishes to sustained battles. (Freeman 1956)

He did not withhold his thoughts regarding Governor Hartley's speech. He said it was "sheer political demagoguery" and "rabble rousing." His feelings, and perhaps that of countless others, are probably best expressed:

> To obtain the international cooperation necessary for the preservation of this fishery has been a long struggle, and one in which many people interested in the permanence of our fishery resource have participated. Some of them are gone. Perhaps few of us will remain to see the final accomplishment of its purpose; but we may now feel assurance that this purpose will ultimately be accomplished. A leading part in this movement, from its inception, has been

taken by *Pacific Fisherman* as the organ of the industry, and by its publisher personally. (*Pacific Fisherman* 1937a).

Freeman did not want to be a Commissioner because he did not wish to be subject to criticism of seeking selfish gain. Miller Freeman lived until 1955, Dr. Found died in 1940, and John P. Babcock died in 1936. Babcock was a pioneer who established a consistent yearly series of observations on the Fraser sockeye spawning beds starting in 1901. Estimates of the relative magnitude and numbers of fish for the escapement of the different races have continued to this date. Many of the reports by Babcock were used extensively by the Commission to reconstruct the history of the individual runs and to understand the possible reasons for depletion. It was reported that Babcock traveled the whole length of the river and its tributaries on foot and horseback (Hutchison 1950). He continued his annual tour even after he became almost blind. In 1913, he pondered the loss of sockeye at Hells Gate with great sadness—"in tears." It was said of him that he "loved fish of all kinds as he loved human beings, and about equally" *(ibid.)*.

Thus, many of the people involved in the Commission's beginnings were able to watch its initial progress in investigations and rehabilitation efforts. There continued to be the belief by others, however, that restoration of the sockeye runs would primarily be achieved through regulations of the fishery:

… it is hoped that the Commission will be in possession of sufficient factual data by that time to enable it to effectively regulate this fishery with a view to rehabilitating the runs…. (Report of the Provincial Fisheries Department 1937)

With the elimination of traps and the large diversion through Johnstone Strait in 1936, Canada took 73.4 percent of the Convention Waters sockeye catch from 1935-1936. If the treaty had not delayed harvest regulations for eight years but had required the Commission to divide the catch equally during the nine years from 1937-1945, Canada would have taken about 2,336,000 fewer sockeye. The Canadian catch in every year from 1937-1945 exceeded the United States catch. Therefore, it was profitable for Canadian industry that regulatory control was postponed until 1946. However, this short-term gain by Canada must be considered in light of lost future production and slower rehabilitation because of fewer spawners obtained for the previously large runs of early and summer-run sockeye, particularly for the once-famous Horsefly River race. Based on the biological needs of the resource, there should have been immediate cessation of fishing on many of the sockeye races starting immediately in 1937 (in fact, protection was needed many years earlier), but that was not possible because the IPSFC did not have regulatory control until 1946.

Establishing IPSFC

Canada and the United States had been participants in the Halibut Convention of 1923, a treaty for the conservation of the Pacific halibut resource and management of the fishery. Therefore, cooperation between the two countries in respect to research and management of fisheries was not entirely new, nor was the negotiating of conventions on such matters. Another convention concerning Pacific halibut was signed on May 9, 1930 and a new Halibut Convention was agreed to on June 29, 1937 (Bell 1981). During the many years required to reach agreement on Fraser River sockeye salmon, three conventions covering Pacific halibut had been successfully concluded.

Some of the original members of the first Joint Commission on Pacific Halibut in 1923 and the second in 1930 were John P. Babcock and William A. Found for Canada, and Miller Freeman and Henry O'Malley for the United States. In the 1937 Convention, A.J. Whitmore for Canada and Edward W. Allen for the United States were original members. These individuals had intimate knowledge of the Fraser River sockeye salmon fishery, participated in or supported the negotiations for a treaty, or were original or early members of the IPSFC. The effective but simple format used in the Halibut Conventions was used as a guideline for the Sockeye Convention.

MAKE-UP OF THE COMMISSION

The Sockeye Convention (hereafter referred to as the Convention) did not stipulate who the participants or members of the Commission would be or from where they would originate. Article II of the Convention established that the Commission shall consist of six members, three from the United States and three from the Dominion of Canada. The commissioners from the United States were to be appointed by the President of the United States and those for Canada by His Majesty on the recommendation of the Governor General in Council. The commissioners appointed by each government would hold office at the pleasure of the government by which they were appointed.

Canada appointed its commissioners on April 14, 1937, United States members were not announced until August 24. Because of the late ratification date and the further delay in appointment of the U.S. members, little was accomplished in 1937. A list of all commissioners, directors, assistant directors, and Advisory Committee members from 1937-1985 is presented in Appendix B of this book.

The first Commission meeting was in Vancouver, B.C. on October 28-29, 1937. The Commission consisted of:

Canada:

Wm. A. Found, Deputy Minister of Fisheries, Ottawa, Ontario.

A.L. Hager, Canadian Fishing Company, Vancouver, B.C.

Tom Reid, Member of Parliament, New Westminster, B.C.

United States:

Edward W. Allen, Attorney, Seattle, Washington.

B.M. Brennan, Washington State Director of Fisheries, Seattle, Washington.

Charles E. Jackson, Deputy Commissioner of Fisheries, Washington, D.C.

Mr. A.L. Hager was elected Chairman and Mr. B.M. Brennan, Secretary. The offices of the Commission were established at New Westminster, B.C. The Convention did not specify the location of the Commission, and as might be expected, this was a subject of considerable discussion. Dr. W.F. Thompson, as the new Director of IPSFC, had reportedly suggested that the new Commission be a section of the existing International fisheries Commission (Halibut) in Seattle, Washington. Tom Reid, who was a Member of Parliament for New Westminster opposed this, wanting something visible in his riding.* Many others also insisted

---

* The designation of a Member of Parliament's voting area or district.

that the office be in Canada, though not necessarily in New Westminster (W.E. Ricker, personal communication).

Rules and procedures were adopted at the first meeting. Since the Commission was international in character, it was agreed that the positions of chairman and secretary should rotate between the two countries every second year and for accessibility these positions would, in the future, be filled by Pacific Coast residents. Some of the first commissioners were headquartered in Washington, D.C. or Ottawa. In the ensuing years, the United States membership consisted of the Washington State Fisheries Department Director, a representative of a Federal fishery agency, and a representative from industry. In Canada, a Federal fisheries person was invariably a member of the Commission. The other two members were less rigid appointments than for the United States and were generally industry representatives, politicians, fishermen, or conservationists. In each country there was effort to have one commissioner with close ties to Ottawa or Washington, D.C. The Commission itself had no legal authority or standing.

From 1937-1985, there were 15 commissioners for Canada and 25 for the United States. For the most part, appointments to the Commission were not politically inspired.

The Convention also provided for the appointment of an Advisory Committee membership of five persons from each country to advise on conditions on the fishing grounds. This later grew to six and then seven persons from each country. Advisors to the commission were recommended by industry organizations such as gear groups and processors and Indians and sport fishermen. Over the years there were 39 Canadian and 34 American advisors.

Permanent professional and technical staff was selected by the director of the Commission with the concurrence of the chairman.

The Commission's first Director of Investigations was announced at the October 1937 meeting and Dr. W.F. Thompson, who was at the same time Director of the Halibut Commission, was chosen to fill the position. Dr. Thompson was asked by the Commission to resign his position as Director of the Halibut Commission so he could devote all of his attention to the Fraser River investigations; however, he didn't resign until September 1940.

It was later agreed that the assistant director be of a nationality opposite to the director.

Biographical sketches of the major participants in the Commission's formation and early years of investigations are given in Appendix C.

SIGNIFICANCE OF THE SOCKEYE CONVENTION

Previously, no agreed-upon or binding method was available whereby the two countries could cooperatively conduct scientific investigations, make recommendations for corrective action, and initiate regulatory decisions to improve the status of the sockeye salmon. The Convention provided a means by which systematic, well-conceived investigations could be carried out to determine why the runs were depleted. The mandate of the Convention did not stop here but, as provided for in Article III: "The Commission shall also have authority to recommend to the Governments of the High Contracting Parties removing or otherwise overcoming obstructions to the ascent of sockeye salmon,. . ."

The Commission was also authorized and required to regulate the fisheries after an eight-year investigation. It was extremely important that the Commission not be authorized or required to enforce its own regulations and that this responsibility be undertaken by regulatory bodies in each country. The provision to delay regulations for eight years was considered ridiculous by some Canadians because of the perceived absolute need for immediate and intensive restrictions in fishing, especially on the early-timed races (Ricker 1987).

Two cycles* of possible rehabilitation efforts were lost. On the other hand, the United States industry argued that the Commission knew nothing about regulations and therefore needed time to develop rationale for action. There was a certain amount of truth to this argument but it was really a smoke screen because the seiners did not want to be regulated on the high seas and probably believed they could take more fish than the Canadians. In this regard many in the Canadian industry wanted to delay Commission regulations for eight years since Canada was currently getting the greater percentage of the catch. In reality, the resource was in desperate need of stringent conservation measures such as total fishery closure on some early races even prior to 1937.

Important to the future success of the Commission was the fact it did not have authority to specify gear types. Involvement in politically sensitive issues was minimized as much as possible.

The Convention required that two members from each Party must agree before action or recommendations could be approved.

One aspect of the Convention that received constant attention was Article VII, which required the Commission to "regulate the fishery with a view to allowing, when practical, an equal portion of the fish caught each year to be taken by the fishermen of each High Contracting Party." Fishermen and industry representatives of both countries insisted they receive their national share of the allowable catch no matter what the circumstances.

The 1930 Convention clearly established the intent of the equal sharing of the catch as stated in Article VII:

> Inasmuch as the purpose of this Convention is to establish for the High Contracting Parties, by their joint effort and expense, a fishery that is now largely nonexistent, it is agreed by the High Contracting Parties that they should share equally in the

fishery. The Commission shall, consequently, regulate the fishery with a view to allowing, as nearly as may be practicable, an equal portion of the fish that may be caught each year to be taken by the fishermen of each High Contracting Party.

Article II stipulated that "joint expenses incurred by the Commission shall be paid by the two High Contracting Parties in equal moieties." To many Americans, this was construed as a long-term right to receive an equal share of benefits coming from the resource. Others viewed these contributions as a reflection of a true partnership relationship. In later years, this was a source of friction as long-term investment rights were not acceptable to the Canadian government.

The Commission was required by Article III to make an annual report to the two governments. This accountability was important and the Commission's actions and results were reviewed and scrutinized by both governments, its scientific peers, and by fishermen and industry members of both countries. Accountability also required the Commission to find the facts necessary for it to carry out its purpose and function. Assessment of the Commission's achievements would be based on the status of the resource, not on its ability to position politically.

The Commission also appointed an honorary scientific council made up of three scientists from both countries as provided for in the Convention. L.A. Royal of the Washington Department of Fisheries was one of the appointees who would later become the Commission's third director. The tenure of the Council was short-lived and was not utilized to any great extent.

························

INITIAL EMPHASIS

After the first meeting, the Commission Chair A.L. Hager remarked at the Association of Pacific Fisheries Annual Convention that the Fraser River watershed "is today in as good

---

* Because of the late date (July 28, 1937) the Convention came into effect, little if any regulatory action could have been taken in 1937.

condition as it ever was." (*Pacific Fisherman* 1937b). His remarks were probably directed toward the potential for rehabilitation.

Hager described the new undertaking as "starting forth on a great enterprise." He further stated, concerning the Commissioners, that "they are serious-minded, determined, and men of vision. They are giving of their own time, without salary…. We will be governed by the facts, carefully gathered, and these facts will be laid before the public, so the reasons for their decisions will be apparent" (*ibid.*).

Chairman Hager later reaffirmed the Commission's intentions:

By starting on a non-existent sockeye fishery we hope to create the most valuable of assets—a food asset, a perpetual asset for future generations. …The Fraser River basin, the largest sockeye salmon plant in the world, has been idle for 25 years, the Salmon Commission proposes to put it into operation at the earliest opportunity. (*Pacific Coast News* 1938)

Action was taken in 1938 to protect the watershed. It was done through a request to the Canadian government that the Commission be notified of any proposed project that would affect the salmon-production capabilities. The text of the resolution was:

Resolved that in view of the provisions of Article III of the Sockeye Salmon Fisheries Convention of May 26th, 1930, the International Pacific Salmon Fisheries Commission respectfully requests the Canadian Government to take such action as will ensure the Commission being notified and consulted before authority is given by the government concerned for carrying out any project in the Fraser River watershed that will result in modifying any spawning area therein or in the damming, pollution, or diversion, of any waters thereof that are used by migrating fish either in the adult or young stages. (IPSFC 1939)

In retrospect, this resolution was of such significance that it probably should have been a part of the 1930 Convention.

It was generally recognized that there were many distinct sockeye races to the Fraser system, and identification and timing of each was important. Emphasis was immediately directed to establish methods of accurately enumerating sockeye numbers on the individual spawning grounds.

For many years the industry expressed interest in artificial propagation (hatchery) methods as a quick fix approach to restoring the runs. The Commission believed other investigations were of greater importance and staff investigations were directed elsewhere.

From the beginning, a strong commitment was made to protect natural propagation from the detrimental effects of dams, logging, erosion, dredging, pollution and other human-induced changes in the environment.

The first season of investigation was in 1938. To gain some knowledge of the races involved, the Commission tagged 980 sockeye in Juan de Fuca Strait. This would lead to more extensive tagging/recovery programs. A very high percentage—44 percent—of the fish were recovered (IPSFC 1939). A total of 2,587 sockeye was also tagged off the mouth of the Fraser River and 48 percent of these fish were later recovered (*ibid.*). Another tagging party was established at Hells Gate, and of 2,128 fish tagged below the Gate, 30 percent were recovered in upstream migratory routes or spawning grounds (*ibid.*).

Observers were placed in canneries at Anacortes, Friday Harbor, Bellingham and Steveston to recover tagged sockeye and to gather data on age, size and sex of the salmon delivered there. They also looked for fin-clipped sockeye from a homing experiment conducted in 1936 on the Cultus Lake race by the Fisheries Research Board of Canada. These and other studies were directed at establishing a racial identification program and a statistical system for catch information. The Fisheries Research Board in 1938 informed the Commission that since the Commission was now responsible for

Fraser River sockeye investigations, the Board would no longer be conducting studies on Fraser sockeye.

Drs. W.E. Ricker and R.E. Foerster, Canadian government research fisheries scientists, joined the Commission staff in 1938. For many years both had studied the life history and biological requirements of sockeye salmon at Cultus Lake in the lower Fraser River watershed. This resulted in several publications and findings of significant importance. On many occasions the Commission and staff were guided in their deliberations by these investigations. One of the principal findings was that artificial propagation of sockeye, by the methods known at the time, was not an improvement over natural propagation. A summary of this and other sockeye research findings was published (Foerster 1968). Ricker and Foerster worked for the Commission for a short period then returned to work for the Canadian government.

The Commission conducted an experiment at Cultus Lake to enumerate the escapement by tagging techniques.

The Cultus Lake predator-control and smolt-counting programs performed by the Canadian government were continued. Additional staff was employed to gather and review all available historical data concerning the Fraser sockeye runs, as well as information relating to mining, geological surveys, aerial mapping, water rights, forestry, and Indian harvests of salmon.

By 1939 the Commission was well on its way to establishing the basis for an historical and statistical analysis of the runs. Detailed biological and engineering studies of the obstacles facing migrating sockeye were also under way. The Commission began a study to determine the relative productivity of Fraser tributary systems to produce sockeye salmon. Studies were initiated to define factors affecting first-year survival in various areas of the watershed. Within its first two years the Commission had established procedures and programs to provide biological and engineering information for decisions that would have significant long-term impacts on the sockeye salmon of the Fraser system.

The Early
Research Program

igration patterns and timing of the races needed to be identified if complete resource management was to be achieved. Other studies were discussed and identified relating to the life history of adults, fry, fingerlings and smolts during their freshwater phase. Foremost on the agenda for complete investigation was the obstruction at Hells Gate. Had Hells Gate been primarily responsible for the great decline in the runs? Was Hells Gate still a problem? Facts were urgently required. With so many possible directions for investigation and so many answers needed, research was carried out on a priority basis.

MARINE TAGGING PROGRAM

It was generally believed that Fraser River sockeye races were present at specific times in the marine fishing areas. Fishermen often remarked that different types of sockeye (weight, length, etc.) appeared consistently each year at specific times of the season. It was also recognized that sockeye from river systems other than the Fraser were probably caught in Convention area fisheries outside the river, but where and when and to what extent was unknown.

To provide answers, the Commission continued the sockeye tagging program that began at Sooke in 1938 and fish were tagged each year through 1948 (Figures 4 and 5). Tagging was also performed in many other locations including Salmon Banks, Point Roberts, Sands Heads, and Johnstone Strait. During these eleven years, 33,334 tags were applied and 17,219 or 51.7 percent were later recovered (Verhoeven and Davidoff 1962).

Previous to the Commission's tagging program, O'Malley and Rich reported on the results of tags placed on 4,494 sockeye at Sooke

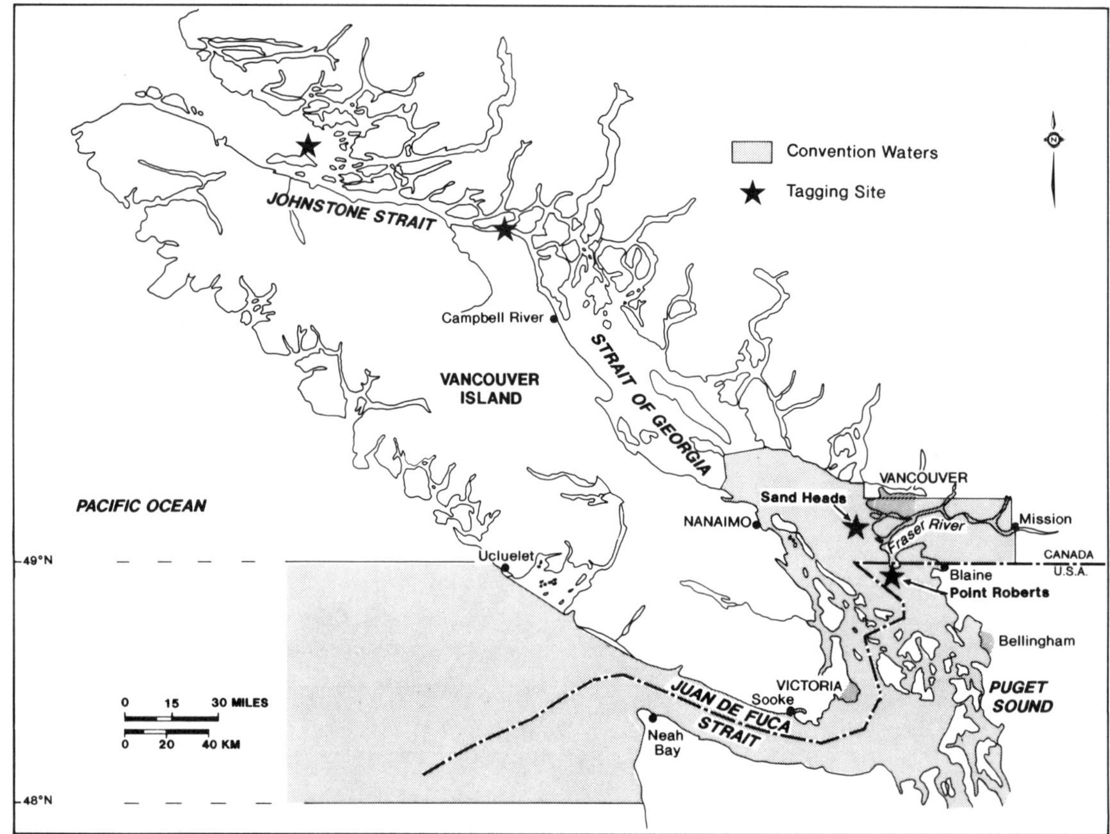

Figure 4. Convention Waters fishing area and locations of sockeye tagging from 1938-1948. From Verhoeven and Davidoff 1962.

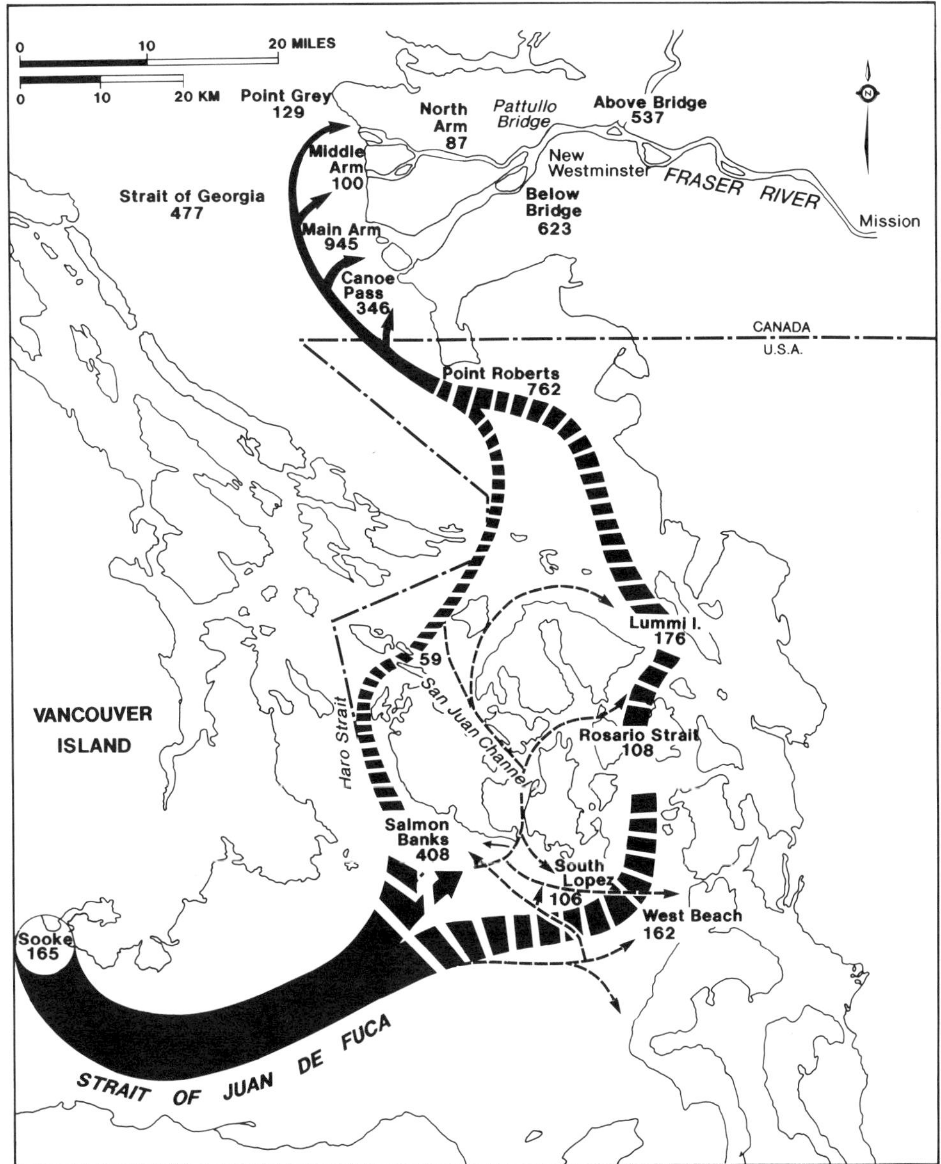

Figure 5. Migration routes of Fraser River sockeye salmon as indicated by tagging at Sooke, 1938-1948, and by information from tagging in other areas. The numbers of Sooke tags recovered in each area are shown. Broken portions indicate that the relative importance of the routes could not be determined precisely. From Verhoeven and Davidoff 1962.

and in Puget Sound in 1918 (O'Malley and Rich 1919). The 1,199 recovered tags indicated that most of the fish were headed towards the Fraser River after first approaching the Salmon Banks area and then going eastward to Rosario Strait. It appeared that almost all the sockeye tagged from Sooke to Point Roberts were Fraser River sockeye. The migration routes determined from the 1938 tagging program at Sooke showed fish reaching the Salmon Banks and migrating northward through Rosario Strait or Haro Strait (Figrue 5). It was essentially the same migration pattern as determined from the 1918 program.

On some occasions fishermen reported large numbers of sockeye moving through the San Juan Channel between Haro and Rosario Straits. The general consensus was that early-migrating stocks moved mostly through Rosario Strait and that later migrating sockeye such as the Adams River race used Haro Strait. In fact, in 1954 and 1971, a very high percentage of Adams River sockeye moved through Haro Strait and were not readily available to United States fishermen except later in the season at Point Roberts.

Sockeye were tagged in Johnstone Strait (Figure 4) in 1940 and 1941 and the fish appeared to move directly toward the Fraser River. A total of 2,746 sockeye were tagged and 1,554 (56.6 percent) tagged fish were subsequently recovered, almost exclusively in Canadian fisheries with 27 percent in Johnstone Strait and 62 percent in the Fraser River area. Only 10 recoveries (one percent) were reported in United States fishing areas, mostly at Point Roberts (Verhoeven and Davidoff 1962). The results of this tagging at the north end of Vancouver Island indicated that sockeye tagged in Johnstone Strait were more abundant at the northerly entrances to the Fraser River (Point Grey and the North Arm) than were sockeye tagged at Sooke, and that most of the sockeye moving through Johnstone Strait were Fraser fish.

Tagging at Sand Heads at the mouth of the Fraser River substantiated fishermen's belief that in September and October sockeye had delayed in the area for several weeks and were subject to significant movement due to wind and tidal action. On many occasions, large numbers of Fraser sockeye were captured in the most northerly U.S. waters near Point Roberts during or following periods of strong westerly winds. The tagging in 1938, 1939 and 1941 resulted in 92 recoveries (five percent of all recoveries) at Point Roberts. More than half (55) of the recoveries were made in 1938, the dominant year of the Adams River race. The fish captured under these conditions have been commonly termed "blowbacks" by fishermen and result from delaying late-run (fall) races that hold off the mouth of the Fraser River for up to one month. The majority were of lower commercial quality because of the influence of freshwater from the Fraser River.

Tagging results showing the migration routes of Fraser River sockeye were incorporated into the Commission's management regime starting in 1946. In subsequent years migration routes were determined each year by where the greatest commercial catches were made.

Tagging also established the time of passage of non-Fraser sockeye stocks. Recoveries of 417 tagged sockeye—2.4 percent of recoveries—were made in spawning areas on Vancouver Island, in the Skagit River system, and in the Lake Washington drainage system during the tagging period (1938-1948). The majority of these stocks migrate earlier than present Fraser sockeye. They were generally found in the Strait of Juan de Fuca and Puget Sound from June to mid-July.

Tagging studies also provided information on the migration rates from one area to another, information critical for effective management. After making corrections to the recovery data to allow for the effects of tagging, erroneous tag recovery dates, etc., Verhoeven and Davidoff concluded that Fraser sockeye tagged at Sooke took an average of two days to reach the Salmon Banks in United States waters. The estimated migration time from Sooke to Point Roberts was three to four days and to the mouth of the Fraser River, four to five days. The late-run migrating races moved slightly slower in reaching the Fraser mouth and delayed in that area much longer than summer-run sockeye. Some sockeye delayed from 19 to 34 days off the mouth. On occasion, summer-run sockeye delay several days before moving up the Fraser. In some instances, retrograde movement occurs in the lower reaches of the river.

Sockeye tagged in Johnstone Strait in 1940-1941 took an average of eight days to reach the mouth of the Fraser River. In general, it appeared that sockeye tagged in the northern approach area required from one to two more days migration time to reach the mouth of the Fraser than did sockeye tagged at Sooke.

Results from fish tagged in the Sand

Heads area were more difficult to interpret because the area at the mouth of the Fraser River is an area of delay depending on the races involved, time of season, and water conditions. Therefore, when fish were tagged, it was not known how long they had already been in the area before capture. The best estimate available from the early tagging studies suggested a migration time of three to seven days from Sand Heads to the "Above Bridge" area between Mission and New Westminster, B.C.

HELLS GATE STUDIES

Passage conditions for sockeye migration at Hells Gate had generally been conceded as favorable since a large amount of the rock and debris from the slide had been removed between 1913 and 1915 (McHugh 1915; Babcock 1902-1932; Napier 1914). While there were some doubters, such as Einarsen, apparently most industry individuals, engineers and scientists believed the problem was solved and that overfishing was the major factor responsible for the poor returns. The opinion of some Americans was that a problem still existed at Hells Gate, whereas most Canadians associated with the industry believed it had been corrected. W.H. Pugsley however, wrote:

> It is the general consensus of opinion of people that live in this section that the river has never been cleaned out properly after the slide, that the upward migration of the fish is considerably hampered yet by this. (Pugsley 1924)

There were also some interesting developments taking place that required explanation. Why did some races return in good numbers (Chilko, Birkenhead and lower Adams) and others continue to decline to extinction or near extinction (upper Adams and Horsefly)? Few seemed to question why the Chilko race, which migrated during mid-summer, continued to have relatively good numbers of sockeye on the spawning grounds each year whereas Horsefly River sockeye migrating at a similar time declined to 200 spawners in 1937 from over 4,000,000 in 1909 (Kew 1937). Timing and movement of all races was not clearly identified at that time. The lower Adams River sockeye race was increasing significantly on a different cycle (1922) and was establishing dominance on that cycle rather than the original cycle of 1913. If the Hells Gate passageway was still the serious problem, how could these and several other seemingly contradictory developments be related to Hells Gate? On the other hand, if Hells Gate was not a serious problem, why did Seton Creek and Thompson River pink salmon disappear? There were many unanswered questions.

If overfishing during the period of depletion had been the principal factor responsible for the decline of Fraser River sockeye and pink salmon following 1913, then other salmonids should have exhibited severe depletion. This did not occur, not even for all sockeye races.

One of the most intriguing features of sockeye response from 1913 to 1945 was the selective depletion and rebuilding of certain races. The upper Adams River spawning population, a summer-run race, was immense in 1905 and 1909 but had practically disappeared by 1921 and was extinct before the Commission came into being in 1937. Other summer runs to Shuswap Lake, such as the Anstey, Eagle and Salmon River races, had very large escapements before 1913. These disappeared after 1913. In recent years small numbers of sockeye have been observed in these streams.

The lower Adams River race had very large returns in 1905, 1909 and 1913, only to diminish to a few spawners by 1921. This decline commenced with the 1913 escapement. From an escapement estimated at only 20,000 fish in 1922, the fish of that cycle increased rapidly (an estimated escapement of 300,000 in 1926) to an estimated total return in excess of eight million fish in 1942. The 1942 escapement had increased to two million fish.

This occurred to the same lake systems (Shuswap-Adams) where the summer-run races previously referenced became extinct or extremely endangered.

The Quesnel Lake (Horsefly River) sockeye spawning populations, estimated (based on extrapolation of counts) at more than than four million fish in 1909, declined rapidly and by 1929 only 1,500 spawners were reported. The Stuart Lake sockeye runs also showed similar sharp declines.

The Chilko Lake-River populations, reported by Babcock to be extremely scarce after the Hells Gate slides up to 1928, began to increase. Escapement estimates made by the Canadian Department of Fisheries list escapements of 70,000, 100,000 and 110,000 fish for the years 1929, 1933 and 1937, respectively. Sizeable escapements at Chilko were also recorded for the years 1928, 1932 and 1936 of 20,000, 70,000 and 74,000 fish, respectively. By 1940 and 1941 net escapements were about 300,000 fish each year, even with substantial loss in 1941 due to blockage at Hells Gate. Chilko sockeye migrated upriver at about the same time as the Horsefly and late Stuart populations. The importance of Hells Gate water levels at specific times during the migration of different stocks is presented later in this section.

Lower Fraser River races, such as the Birkenhead River population, spawn in tributaries below Hells Gate and therefore would not be affected by migration conditions at Hells Gate. Records of the annual collection of sockeye eggs at the Owl Creek hatchery on the Birkenhead River were used by Thompson in support of the conclusion that this race was one of the most consistent producers both before and after Hells Gate. Interestingly, data showed the lowest egg take on record in 1917, the first year of adult returns after the block conditions prevalent in 1913. The egg take in 1918 was also extremely low. This suggests* low survival

---

* Birkenhead sockeye contain fair numbers of five-year-olds so brood year production is not known.

factors, unrelated to the poor migration conditions in the Fraser River Canyon in 1913 and 1914.

Most of the upriver races were greatly reduced in 1917 and 1918. It appears possible, based on the great reduction at Birkenhead, that in addition to the Hells Gate blockage, some feature of the environment—either freshwater or marine, that was common to many Fraser stocks—significantly reduced productivity. The rate of return per spawner for many races in the returns of 1917 and 1918 appeared to be very low.

Babcock commented on the inconsistent pattern of returns in the upriver and lower river sockeye populations. In reference to the Harrison-Lillooet Lake runs his 1920 report stated:

> The number of sockeye that spawn in that section has not declined; on the contrary, there is evidence that it has increased. The question then naturally arises, why has the sockeye run to that section been maintained while the runs to all other sections have declined? (Babcock 1902-1932)

In his 1927 report, Commissioner of Fisheries Babcock said again: "The Birkenhead is the only section in the Fraser River basin where the run of sockeye shows no sign of diminishing" (Babcock *ibid.*).

It appears that he was cognizant of a very important biological distinction but did not fully comprehend the significance of the observation.

Babcock speculated that either the hatchery was very successful or that these runs passed through the fishing areas early in May and were not heavily fished. Neither explanation, as we now know, was plausible since sockeye hatcheries were later shown to be no more effective than natural propagation.

Fraser River pink salmon abundance also declined substantially following 1913, whereas Washington State pink salmon returns remained steady.

Fraser River chum salmon (*Oncorhynchus*

keta) do not spawn above Hells Gate. Therefore, a catch analysis of this species prior to and after the Hells Gate slide might shed some additional light on whether Hells Gate or overfishing was the primary problem causing the sockeye and pink salmon declines.

Abundance data for chum salmon are available for Puget Sound purse seines from 1910-1934. Fraser chum salmon migrate through the fishing areas earlier than Puget Sound chum stocks and later than Fraser sockeye and pinks. The data from Table 55 of Rounsefell and Kelez's 1938 report shows that the trap catch index from 1917-1934 was 51 percent of the catch from 1913-1916. If all other factors are assumed equal, this would suggest that some overfishing had taken place; however, the decline in chum catch was not as pronounced as for sockeye and pink salmon. The sockeye index of abundance from 1917-1934 was only 26 percent of the 1902-1916 index and the pink salmon index of abundance from 1915-1933 was only 24 percent of the index from 1907-1913.

For coho salmon *(Oncorhynchus kisutch)*, the trap index of catches north of Sandy Point (mainly Fraser stocks) showed that coho (from 1916 to 1934) were 53 percent of the level of catch from 1906 to 1915 (Table 39 Rounsefell and Kelez 1938). Since coho migrate later in the season and were probably fished less extensively, they would find water-level conditions at Hells Gate more favorable than at the time sockeye move through the Gate (similar to lower Adams River race), hence less decline in the status of this species after 1913 compared with sockeye and pinks would be expected.

Chinook salmon *(Oncorhynchus tshawytscha)* index of abundance from the traps north of Sandy Point from 1918-1934 showed no evidence of overfishing and increased 18 percent over the catch level from 1910-1917 (Table 46).

Both Fraser sockeye and pink salmon runs showed a greater decline following 1913-1914 compared with chum salmon which spawn below the Gate, and coho which migrate past Hells Gate late in the year (September-November). Chinook, which are very strong swimmers, increased in abundance. This limited analysis, based solely on catch data, suggests that overfishing might have been responsible for some of the decline of most of these Fraser River species following the Hells Gate slide, but major responsibilities lie with the environmental disruptions at Hells Gate, lower Adams River and Quesnel River.

Because of later migration, it is likely that harvest rates on the lower Adams River race were lower than on the summer run races. However, the Birkenhead and Chilko Lake-River races migrate earlier and would have been exposed to fishing intensities similar to Quesnel and Stuart Lake stocks. Yet while the former races produced well, with Chilko even increasing, the Quesnel and Stuart sockeye were endangered and those of the upper Adams River and Salmon River races exterminated. The principal cause for decline could not have been overfishing, but the number of high quality spawners may have been reduced following 1913.

One of the more interesting declines and subsequent rehabilitation of a sockeye run took place in the Shuswap Lake area. The lower Shuswap River race had been very large on the 1909 cycle. The river was reported to be "red" with sockeye from Mabel Lake downriver (personal communication from local resident). There were a great many sockeye spawning in the river below Mabel Lake in 1901 (State of Washington Department of Fisheries and Game 1902). Babcock (1902-1932) and others noted that in years of big runs the river was filled with sockeye. The returns were possibly equal to that produced by lower Adams River spawners; however, a local resident reported the 1913 escapement at lower Shuswap River to be less than fifty fish compared with thousands in 1909 *(ibid.)*.

The history of the lower Shuswap River

sockeye run has not been extensively investigated and escapement records are fragmentary. Several hundred fish were reported in 1921 and a medium-sized run was reported in 1922 (Anonymous 1969). No record of fish on the dominant 1909-1921 cycle is available after 1921 but the 1922 cycle increased only slightly to 2,000 fish by 1942 and in 1946 1,200 sockeye reached the river. In contrast, the lower Adams River escapement increased significantly on the 1922 cycle to over two million sockeye in both 1942 and 1946 from an estimated 20,000 fish in 1922.

The splash dam on the lower Adams River could not have had any direct influence on the failure of the 1913 escapement to the lower Shuswap River. The Hells Gate obstruction is the only logical primary reason for reduced escapement of South Thompson/ Shuswap runs. Yet even though the lower Adams escapement in 1913 was below that of previous cycles, considerable numbers of sockeye did spawn in the lower Adams and Little rivers in 1913 (Babcock 1902-1932).

While the Hells Gate slide is clearly implicated, no satisfactory explanation has been given as to why the lower Adams River race increased significantly from 1922-1942, whereas the lower Shuswap River population did not, nor why the Adams escapement in 1913 was fairly good but the lower Shuswap was very poor. The Adams total run in 1942 was estimated in excess of eight million fish compared with an estimated total run of less than 10,000 sockeye of the lower Shuswap River race.

It is significant from a cyclic-dominance perspective that the lower Shuswap River run also switched to be dominant on the same cycle on which the Adams run re-established dominance in 1922. Productivity of the nursery lake environment would appear responsible.

Possibly, the longer migration to lower Shuswap River increased the mortality rate of fish already weakened at Hells Gate and decreased the spawning efficiency of those fish, comparatively more so than the Adams River

sockeye. Also, it was possible some straying of lower Shuswap sockeye to the Adams River occurred during this period, perhaps because of energy requirements, while some straying of Adams River fish to lower Shuswap River occurred in 1942 and 1946 when the Adams escapements were large. During this period, spawning began to take place in several scattered locations along the shoreline of Shuswap Lake in areas adjacent to and distant from the Adams River. As recent as 1982 more than 30,000 late-run sockeye from lower Adams River or, more likely, lower Shuswap River were observed spawning in the Eagle River, traditionally a summer-run spawning area with river entrance in close proximity to the lower Shuswap River.

The decline and continuing low abundance of the lower Shuswap River race from 1913-1946 could hardly be directly related to overfishing since this race migrates at the same time as lower Adams fish. Perhaps escapements were below a threshold level and resurgence was not possible until Commission regulations starting in 1946 provided the necessary impetus.

In 1926, 1927, 1930 and again in 1934, substantial numbers of presumably lower Adams River sockeye were observed spawning or attempting to spawn in streams below Hells Gate. J.A. Motherwell's description of the late season migration problems at Hells Gate in 1926 and particularly in 1927 is informative. In his report he said:

> It has been suggested that the lack of male fish in the Kawkawa Lake spawning area for instance, which is tributary to the Coquihalla system, is evidence that they were probably able to pass Hell's Gate (sic), but that the female, being weaker, were obliged to turn back and passed up the Kawkawa Lake spawning grounds. It is hoped that investigations will divulge the facts in this matter in the very near future. (Motherwell 1927/28)

In 1930, *B.C. Fisherman* reported that every stream for sixty-five miles below Hells

Gate had large numbers of spawning sockeye salmon. In 1926 and 1927 Kawkawa Lake and River contained sockeye and in 1930 and 1934 the area was again well seeded (Motherwell 1931 and 1934). In those years, sockeye were observed below Hells Gate in December. These were most likely late-run sockeye destined for Shuswap Lake that were blocked or delayed at Hells Gate. These observations indicated that all was not right in the river and that many sockeye did not reach their natal spawning area. What seems so obvious now was not understood then. In spite of environmental problems, the resilience of the lower Adams race prevailed. How many sockeye from this race and other races were lost each year from 1913 on is not known. These losses, however, certainly "fueled the flame" that overfishing was the principal factor responsible for the poor runs.

Cultus Lake sockeye have migration timing similar to that of lower Adams River sockeye. Data on sockeye counted at a fence below Cultus Lake since 1925 shows that large numbers of sockeye were present in 1927 (56,376 females) at a time when other upriver runs were low. Average annual female escapement from 1925-1936 was 11,921. For the next 12 years the average annual escapement of female sockeye was 16,794. There was no evidence of overfishing on this race.

It was shown that early season sockeye for the years 1938 to 1942 were less affected by the Hells Gate slide due to the higher water levels which made fish passage easier (Thompson 1945a). It follows, then, that the early migrating portion of races such as the Horsefly would have had easier passage at Hells Gate than the peak or later migratory segments, such as sockeye coming from the lower Horsefly River spawning area. These more recent conclusions, based on tagging results, had been noted earlier by Babcock, who said:

> The first division of the great run of sockeye that escaped capture in 1913 reached the canyon at Hell's Gate (sic) late in July, and successfully passed over the obstruction because of high water. Many thousands passed over, in late July and early August, during the stage of high water. The bulk of those reaching the canyon after the high stage of water had passed were unable to get through the canyon. (Babcock 1919a)

Most of the summer-run sockeye that tried to migrate past Hells Gate in 1913 after mid-August would have been Horsefly River sockeye. It is likely that the only remnants of the once-great Quesnel escapements to the Quesnel system (552,000 in 1913) (Thompson 1945a) came from the early portion of the run—primarily upper Horsefly River-McKinley Creek fish. When the Commission began its surveys in the late 1930s and the early 1940s, these were the only locations where spawning sockeye were found in the Horsefly area. No salmon were observed in the once-productive and vast spawning grounds of the lower river near the town of Horsefly. Little wonder then, that the Commission was faced with serious problems relating to genetically-influenced sockeye adaptations, with an early-migrating stock not suited to the timing of environmental conditions of the spawning grounds. It took many years for transplanted Skagit River sockeye from northern Washington State to adapt and be productive in the Lake Washington system in Seattle (Royal and Seymour 1940).

Thompson (1945a) concluded that the reason why the Chilko run survived at a much greater level than the Horsefly/Quesnel stock was because Horsefly sockeye passed up the Fraser River one month earlier than the Chilko run. He termed this "a point of great significance in explaining the survival of the run to the Chilcotin."

He believed that the improvement in the Chilko run while the Horsefly run was decreasing steadily: "may have been due to the passage of the Chilcotin fish through the Fraser Canyon at a date later than that of the Quesnel fish so that it was not subject to the same difficulties in passage."

Tag recoveries in 1939-1942 (Thompson 1945a), showed that the first Chilko sockeye entered the Fraser River after mid-July. This corresponds closely to other IPSFC data which showed the first Chilko sockeye entering the Fraser on July 13 (Henry 1961). Thompson believed the first Horsefly sockeye arrived on July 1. Henry determined that the Horsefly race arrived on July 20, slightly later than the Chilko race. Thus, both Chilko and Horsefly populations migrated up the Fraser Canyon at similar times. There is no reason to believe the pre-Hells Gate Horsefly population was favored with a much earlier migration, and therefore easier passage.

Since 1973 the dominant Horsefly run (1973-1985 cycle) has been much later in its migration—about two weeks later and probably quite similar to the pre-Hells Gate timing. It is likely that the Horsefly run in 1913 and in earlier years migrated up the Fraser River somewhat later than the Chilko population.

Therefore, some other explanation must account for the fact that Chilko escapements were maintained at high levels following the slide years and consisted of about 300,000 sockeye in both 1940 and 1941, whereas the historically much larger Horsefly run was at the point of near extinction—200 fish—by 1937 and 1,000 fish in 1941.

It is worth noting that another upriver race migrating at a similar time to the Chilko and Horsefly was also diminished. The late Stuart population in 1941 had an escapement of only about 10,000 fish. On the other hand, the Stellako River race migrates at a similar time as Chilko, Horsefly and late Stuart sockeye, yet it was in fairly strong condition by 1942. Escapement in that year was 48,000 fish and in 1946 about 250,000 sockeye.

Why were the Chilko River and Stellako River populations able to withstand the effect of the Hells Gate barrier better than other summer-run populations? It is postulated that the answer possibly lies in energy reserves and spawning time of these races.

The Horsefly and late Stuart populations spawn soon after arrival at the spawning grounds. Peak spawning for the Horsefly population in the 1940-1970 period occurred around September 1 or earlier. Peak spawning period of the late Stuart race was later because of the greater distance to the spawning grounds (650 miles compared to 470 miles) and some delay in the lakes. The Stellako (610 miles), Chilko (400 miles) and late Stuart populations peak spawning periods are about three weeks later than Horsefly. These two stocks may have been able to withstand longer periods of delay* at the Gate waiting for an opening and still reach the spawning grounds with sufficient time and energy reserves. Also, the Chilko race was adapted to migrate to a height of about 4,000 feet, the highest of any Fraser sockeye. These sockeye as they pass Hells Gate may have the highest stamina. The Horsefly River population spawns very soon after arriving at the spawning grounds whereas the other races delay in the lakes. It is clear, however, that not all Chilko sockeye reaching Hells Gate were able to wait out the block conditions. In 1941 for example, when only 1,000 Horsefly sockeye reached Horsefly River, there was a large loss of several hundred thousand Chilko sockeye at Hells Gate, even though an estimated 280,000 fish eventually reached the Chilko River spawning grounds. It should be noted, however, that the relative magnitude of these two races differed as the Horsefly run totalled about 5,000-10,000 fish compared with a Chilko total run of several million.

Subsequent studies have shown why escapement to Adams River increased on the 1926-1942 cycle; the passage conditions at Hells Gate from mid-September to mid-October were, in most years, favorable because of the low river discharge and because the Adams River splash dam was not in operation. It appears that most of the differential production of the various races can be attributed to the genetic

---

* See results of tagging on page 75.

August 20, 1941. Blocked sockeye accumulated in an eddy below Hells Gate.

September 1, 1941. Sockeye attempting to migrate through the high velocities and extreme turbulence at the left jutting rock at Hells Gate during a block stage.

make-up of the races and their timing at Hells Gate associated with existing water levels.

The year of the dominant run for several races changed as a result of environmental conditions at Hells Gate and lower Adams River. As mentioned previously, Adams River sockeye became dominant on the 1922 cycle instead of the 1913-1921 cycle. The Seymour and Stellako River runs did likewise. The Chilko run had also been dominant on the 1913 cycle but following the 1941 blockage, the subdominant run of 1940 became the dominant cycle. The earlier mentioned three races changed dominance during the period 1922-1942 while the Chilko change was more recent, starting in 1940.

> In every case where major runs have shifted their dominant year it has been to the year of the subdominant run regardless of whether the subdominant run occurred before or after the dominant one. A new subdominant run has formed in its original relationship to the dominant runs. In every case where dominance has shifted, the run in the year of the original dominant run has declined significantly in size. (IPSFC 1964)

The Stuart and Quesnel system runs have maintained their dominance on the original big year cycle of 1913.

Examination of the field records of sockeye tagging data at Gorge Camp at Hells Gate for 1939 and 1941 determined if any pink salmon were taken incidental with the sockeye. Records of all salmonids captured by dip nets were taken, and pink salmon were apparently trying to pass upriver past Hells Gate in both years. In 1939, 383 pinks were captured (September 1 to October 17), along with 325 chinooks and 1,516 coho salmon. In 1941, only 23 pink salmon were taken. The timing of migration in those years, based on the more extensive data of recent year returns, was consistent with modern-day pink salmon migration past Hells Gate. Data suggests that some early run Fraser River pink salmon probably attempted to migrate upriver in each odd-numbered year from 1913-1943. Whether these fish which in most years did not pass Hells Gate, eventually spawned further downstream in the tributary streams or in the main Fraser River after being blocked at Hells Gate is not known.

The most logical place for the Commission to have focused its research was at Hells Gate. Visual observations had been made for years and while these provided subjective conclusions, the murky and turbulent water conditions prevented collecting conclusive data on the ability of sockeye to migrate past the Gate. The sockeye tagging method that had proved successful in marine areas was used in the Fraser River, starting in 1938 below Yale, B.C. Peterson type tags were used, consisting of two celluloid disks about one-half inch in diameter, one of which bore a serial number. The tags were attached, one on each side of the fish, with a nickel pin inserted through the upper part of the body just below the dorsal fin.

Favorable sites to catch, tag and release sockeye were found at Hells Gate and from 1939 through 1942, 31,490 sockeye were tagged and released. Most of these, 28,408, were released below the Gate. The results were reported by Thompson (1945a). An additional 24,864 tagged sockeye were released below Hells Gate from 1943 through 1947. Of these, 13,963 were tagged in 1945, 1946 and 1947 in years after the fishways were completed. Analysis of these studies was given by Talbot (1950).

Fish were captured by gillnets during 1939-1941; however, due to the size selectivity associated with gillnets and the poor condition of captured fish, dip nets were used part of the time in 1940 and 1941 and exclusively from 1942-1947.

The results of the 1939-1942 tagging program, based on analysis of 10,098 (32.1 percent) recoveries, (6,441 above Hells Gate and 3,657 below), provided a great amount of data on racial migration timing. The tag recoveries were made primarily by Indian fishermen throughout the Fraser River watershed and

Adult sockeye salmon with a "bullseye" Peterson tag attached by means of a nickel pin inserted through cartilage and muscle tissue just below the dorsal fin.

July 22, 1947. Indian dip nets and fishing techniques were used for capturing sockeye for tagging at Hells Gate.

August 26, 1945. The biological staff also adopted techniques used by Indians for capturing fish for tagging at Bridge River Rapids.

The rock slide on the left bank caused by railway construction at Hells Gate increased the water velocity to such an extent that, even after most of the rock had been removed, sockeye were unable to proceed past the bedrock protrusions on both banks. This photo of the right bank was taken during fishway construction in late 1944.

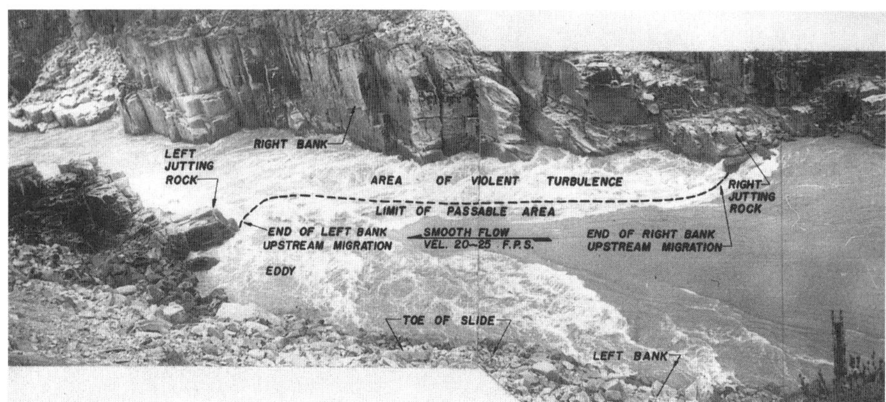

The high velocities and steep drop in water surface at the right and left jutting rocks prevented sockeye migration over a wide range of water levels. Fish migration was also severely impeded by violent turbulence, primarily upwelling, along the right bank below the right jutting rock.

Even in 1944 when the fishway construction was started, rock material from the railway slide was still evident on the left bank. Although this rock could have been excavated, there was an unknown quantity of submerged rock that would have been difficult or impossible to excavate. Fishways were considered to be the best solution since removal of the submerged rock could have dropped the water level and created block conditions at many upstream sites where rock had been dumped in the river during early railway construction.

also by Commission research staff at various spawning grounds.

These investigations clearly established that recoveries of sockeye tagged at Hells Gate were proportionately much higher for the lower Adams River and Little River races than for other races (Thompson 1945a). Tag-recovery data also identified year-to-year consistency in the arrival and timing of various races at Hells Gate. Some races experienced great difficulty in passing Hells Gate between a gauge height of 26 to 50 feet, the major block occurring between 25 and 40 feet. There were brief periods of open passage between water levels of 40 and 50 feet. The lower Adams/Little River races do not reach Hells Gate until after mid-September and at that time water levels are usually declining and average about 20 feet. Tagging information quickly supplied evidence indicating why the Adams River race was steadily increasing on the 1922 cycle and why most of the summer-run races passing Hells Gate in July and August were depleted. The recoveries upriver from Hells Gate of fish tagged below the Gate were not proportional to the daily numbers tagged. There was no doubt that factors other than the variable Indian fishing rate and variable tag recovery rate at the various spawning grounds were responsible for differential rates of tagged salmon appearing upriver of Hells Gate. Careful analysis of the data showed conclusively that periods of difficult passage and complete blocks occurred at certain water levels. This explained, in part, why certain races were diminishing while others increased.

In 1941, evidence of the block was most vividly seen as hundreds of thousands of sockeye were prevented from moving upstream. An unusual flow pattern was the cause and resulted in an extended period of blockage, the first blockage an estimated 36 days, and after an opening of one or two days, a second blockage of 31 days. It is likely that the sockeye blocked at Hells Gate in 1941 was the largest accumulation since 1913. It was speculated that more sockeye died below Hells Gate than passed through. Adams River sockeye were also blocked, a most unusual occurrence, because of high water levels late in the summer. The 1941 visual evidence of the blockage conditions at the Gate confirmed research findings that were being conducted by Commission staff.

Thompson (1945a) reported on the research findings and in one phase of his study he examined in detail the return success of the annual Fraser runs from 1912-1939. He compared the returns with the days of impassable or passable conditions at Hells Gate from July to the end of the season during each of the relevant years. He found that correlations were high, sufficiently so to indicate the major cause of return variations following the slide was related to the varying number of days that difficult passage conditions existed. It was remarkable that the data illustrated this so clearly, as later studies showed that yearly marine survival rates of Chilko River sockeye varied at least twenty fold. If these varying survival rates had been known for the period analyzed by Thompson, his results could have been even more striking. Of course, if freshwater survival rates had also been known, an even more complete and accurate assessment of the effect of conditions at Hells Gate could have been made.

Thompson's conclusions indicated the primary cause of decline was due to selective depletion, not overfishing. The data also showed nearly complete absence of recoveries of tagged fish below Hells Gate and a high recovery rate above Hells Gate during periods when river levels were determined to be passable. He concluded:

> The removal of the obstruction is necessary, because it is the principal cause of the present depleted condition of the Fraser River sockeye run, although heavy fishing may have contributed. *(ibid.)*

Prior to the publication of the findings from the tagging program conducted at Hells Gate from 1938-1942, the Commission met on several occasions to review progress of the studies. Tentative conclusions were reached as to what would be required to ensure safe passage for sockeye at all water levels. The Advisory Committee was also given reports by the Commission and staff on the progress of the investigations at Hells Gate. The Commission meeting of June 13, 1941 took an interesting turn:

> During the discussion with the Advisory Board the question of whether the time was now ripe for the Commission to assume all responsibilities of the treaty instead of waiting until the end of the two-four year cycle run were up was debated at considerable length. The question arose when the matter of regulation was mentioned... (IPSFC 1941a)

The Commission considered a resolution put forth by the advisors regarding assumption of regulatory control four years earlier (1942) than provided for in the treaty, but the Commission concluded as follows:

> ... after careful consideration decided for the present not to take any steps or make any proposals affecting changes in either the treaty itself or the understandings attached thereto. . . *(ibid.)*

Early in 1943 the Commission's Chief Engineer, Milo Bell, presented to the Commission plans that he and his staff had prepared for one fishway on each bank at Hells Gate. Total projected cost was $1,207,000.

Why spend large sums of money on fishways when rock abutments could have been blasted and removed? Between 1926 and 1928 an Inquiry into the Fraser River Conditions by a board of engineers made extensive hydraulic studies, mapped the contours of bed and bank, and measured velocities and drop through the reach (Board of Engineers 1928). Model tests in 1943 by the Salmon Commission, using this data, showed that the loss in head at the Gate would be only slightly reduced by rock removal. It also was clear that no alterations should be made to river margins since this would increase velocity. It was concluded that

blasting and rock removal would not improve the runs (Bell 1945).

The Commission had provided only slight improvement to sockeye passage by blasting through rock at the left bank of Hells Gate in the fall of 1942, allowing a small temporary fish-ladder facility for limited numbers of sockeye to use and providing a brailer to capture sockeye for transportation above the Gate. Operation continued into 1943 and 1944. In 1943, 1,097 sockeye were brailed and in 1944, 12,466 sockeye were assisted upriver (Talbot 1950). It was obvious these measures would not resolve the fish passage problems.

Late in 1943, the Commission met again with its Advisory Committee to discuss in detail the scientific and engineering findings at Hells Gate. The Commission's engineering consultant, Professor C.W. Harris, also reported on the Hells Gate model studies reproduced on a scale of 1:50 at the University of Washington. A permanent solution for correcting salmon migration problems at Hells Gate was outlined and proposed. Based on all the findings, the Commission recommended to the two governments that known areas where blockage conditions existed, most at Hells Gate and Bridge River Rapids, be corrected. The Advisory Committee endorsed the proposal. Two million dollars ($16,500,000 at 1985 value) was requested for projects to be done over a five-year period. By 1943 the Commission had identified 37 obstructions or points of difficult passage. Support for the Commission's programs and its investigative methods was clearly evident. Additionally, Commission minutes for November 26, 1943 record that the "canners" were 100 percent in favor of the Commission being responsible for the management of all species.

The report submitted to the two governments on January 11, 1944 included both biological and engineering data. Approval was given in 1944 for construction, a contract was signed on August 24, 1944, and work began shortly thereafter. The scope of this undertaking was enormous, risky and, to some, controversial. The fishway design was new and untested for a project of this magnitude. This was a gigantic decision for the Commission and the two Parties.

Investigations by IPSFC at Hells Gate continued even though basic conclusions had been reached regarding construction efforts that were required to facilitate fish passage. Since tagging had been performed at the right bank for only one year (1942), more information was needed. Tagging after the fishways were operational would evaluate their effectiveness. The results of studies from 1943 to 1947, inclusive, were reported (Talbot 1950). Tagging in 1943 and 1944 gave similar results to studies from 1938-1942, even though a temporary fishway was in place in 1943-1944 along the left bank.

. . . .

RACIAL STUDIES

Knowledge about the time of passage of specific Fraser River sockeye races was taken from the marine tagging programs in the late 1930s and 1940s and from tagging at Harrison Bay near Harrison Mills in the lower Fraser River area, Hells Gate and Bridge River Rapids. Tagging in freshwater areas such as Hells Gate and Bridge River Rapids was done in many years when conditions were not favorable for migration; therefore, data on the time and duration of migration in many instances are not completely representative of undisturbed populations. In most instances, the date when sockeye first appeared is most likely the most accurate of the data obtained, as the last date of appearance could have been affected by delays at the Gate. It could not be determined however, how long delayed fish had been at the Gate before the tag was applied; therefore, the tendency would be to have dates of first arrival too late.

Early Stuart sockeye passed Sooke between June 18 and July 18 and Bowron sockeye, between July 10 and August 4 (Verhoeven and Davidoff 1962). Upper Pitt River sockeye passed through Sooke between July 18 and August 8

followed closely by Chilko River sockeye from July 18 to August 14. Stellako River sockeye were somewhat later in migration and moved through the Sooke area from July 15 to August 22. The Birkenhead race appeared on July 31 and was past the area by September 4. The migration timing of individual lower Adams River sockeye extended from July 23 to September 23, whereas the late-run Weaver Creek race migrated from August 5 to September 23. The above information was based on the return of only 348 tagged fish in the 11 years of tagging. The limits are broad and give only a general impression of the migration timing. Examination of the run timing in later years showed considerable annual variations; nevertheless, the information was valuable in increasing the understanding of the migration habits and timing of the various races.

Tagging at Hells Gate showed that the early Stuart race passed the area from June 28 to August 6 based on 944 recoveries over 10 years (*ibid.*). Bowron River sockeye (n = 463)* were present at the Gate from July 6 to August 26 (9 years of data). Because of the small size of the Horsefly run and only 7 years data (n = 8), little information on the timing of the Horsefly race was obtained. The first Horsefly fish tagged and recovered was tagged on July 25 and the last on August 3. Raft River sockeye with 10 years data (n = 145) migrated past Hells Gate from July 23 to September 4. Seymour River sockeye with six years data (n = 49) were passing the Gate from July 26 to September 4. Numerous Chilko River fish were tagged and recovered in 11 years of tagging (n = 1919) and migration dates were very broad ranging from July 13 to September 24. Late Stuart sockeye, based on four years of data (n = 84), appeared at Hells Gate from July 28 to September 4. Many Stellako sockeye were intercepted in the 11 years (n = 1823) of data and their time of passage was extended from July 26 to October 10.

---

*n = number of tagged sockeye recovered

Results of the early-year tagging programs in both salt and fresh water areas were less than complete, but did provide information used later for setting fishing regulations. Tagging information was supplemented by more precise racial identification obtained from scale analyses, an identification method which began in the late 1950s. Killick (1955) used the tagging data to show chronological consistency of the races during spawning migration.

SPAWNING GROUND ENUMERATION

The Commission immediately recognized that information on the number of spawning sockeye in each spawning system was necessary to manage the runs on a scientific basis. Thompson (1939) stated: "Foremost among the necessities is an accurate system of estimating the runs, and the numbers which escape to the spawning grounds of each race."

The population-estimation program based on tag-recovery results is derived from the following relationship:

$$N = \frac{nt}{s}$$

where N = Total number in population

t = Number of fish tagged

n = Number of fish sampled on spawning grounds

s = Number of tagged fish recovered

The results of the Cultus Lake experiments in 1938 were reported by Howard (1948). The Cultus Lake sockeye run was ideal in the respect that for this experiment escapement counts could be obtained for verification from a weir or fence located just below the outlet of the lake.

In 1938, 4,416 fish were tagged, one-third of the 13,342 sockeye counted passing the Cultus weir. In 1939, 3,660 sockeye, one-twentieth of the 73,189 fish entering the lake, were tagged. Dead fish examined in 1938 for tag ratio determination totalled 4,735 and in 1939 totalled 9,832 sockeye. In 1938, 1,519 tagged fish were recovered and 477 in 1939.

Population was estimated at 13,765 sockeye in 1938 and 75,441 in 1939. The 95 percent confidence limits for the population estimate in 1938 was 13,090 to 14,475 and for 1939, 68,966 to 85,523.

Results from the Cultus Lake programs were promising and therefore the method was applied throughout the Fraser system. Minor modifications depending on each area's physical characteristics and the size of the run were necessary. Carcass sampling was done in a random manner with respect to both tagged and untagged fish.

In 1939, 1940 and 1941 similar tagging techniques were applied in the Harrison River system (Schaefer 1951). All sockeye spawning in the Harrison system are in rivers or streams, many of which are subject to flooding during the spawning season. Most of the spawning takes place in the earlier Birkenhead River run and late runs in Weaver Creek and the rapids of Harrison River (Figure 1).

Tagging was done at the mouth of Harrison River, Skookumchuck, Birkenhead River and Weaver Creek. A trap was used in the Harrison River to capture migrating sockeye. In the Skookumchuck area, sockeye for tagging were purchased from Indian fishermen using dipnets. Fences were used to capture sockeye in the Birkenhead River and at Weaver Creek.

Tagging and recoveries consisted of three methods: live fish capture by trap or dip net, live tagged and untagged salmon counts on the redds, and dead fish examinations on the spawning grounds.

Schaefer made an exhaustive examination of the data obtained from his experiments and concluded that: "the serious loss of tagged

September 11, 1948. Jim Jensen and
Roy Jackson, who became Assistant
Director before leaving to become
Executive Director of the
International North Pacific Fisheries
Commission, show their catch of
Tetachuk River trout.

August 18, 1950. Miss Isla Davies of the
U.S. State Department was shown the
Commission's field operation at Chilko
River by staff librarian Nina Wesson
Hughes and two members of the Chilko
crew, James Hickey and Tom Harvey, Jr.

Three of the early members of the IPSFC biological staff: Roy Jackson, Dr. Earl Foerster and Ed Whitesel (left to right). Identity of the prankster is unknown.

October 3, 1947. Biologist Stan Killick (later to become Chief of Operations) holding a male sockeye at the tagging site at the mouth of Adams River.

July 20, 1945. Biologist Gerry Talbot and his assistants Rod MacLeod and Bruce Krug building a weir for sockeye enumeration on Kynoch Creek tributary to Middle River.

August 25, 1947. A sockeye weir on Bowron River being attended by biologist Jim Mason and his assistants.

In 1961, the Bowron enumeration weir was built at the outlet of Bowron Lake where a long fence was required. The use of weirs for enumerating early and summer sockeye races was discontinued when the adverse effects of interfering with the migration were recognized.

May 21, 1947. The waters of the turbid Fraser River and the clear, beautiful Thompson River create an interesting scene as they mix at the Fraser-Thompson confluence.

November 1, 1958. High velocities and turbulence in Thompson Canyon, seven miles above the Fraser-Thompson confluence, create very difficult passage conditions for both sockeye and pink salmon in low-water years. The severity of the problem was reduced by minor rock excavation in 1946 but fishways may be required if the spawning populations increase. The 1958 Adams River run entered the Fraser very late and was further delayed at Thompson Canyon and at other locations because of the very large size of the run.

October 20, 1946. A typical commercial fish trap was built on South Thompson River near Pritchard for capturing sockeye to permit studies of the tag-and-recovery method of enumeration for Thompson River populations and to establish arrival timing of the many sockeye populations spawning in the Thompson system. After one year of operation, the trap was removed and all of the runs were enumerated on their spawning grounds by the tag-and-recovery method for the large runs and live counting for the smaller runs.

fish between the Harrison trap and the Birkenhead River is probably the combined results of actual mortality during migration and of a differential mortality of tagged fish." The loss was so great that enumeration was hopeless. Tagging can have an adverse impact on the fish. Howard (1948) found a greater number of tagged fish immediately above the tagging area at Cultus Lake than in other areas of the system.

Schaefer conducted a tagging-versus-fin-clipping (one ventral fin) experiment at Weaver Creek throughout the 1940 migration using a 1:2 ratio. Data indicated that tagged and fin-clipped fish remained distributed throughout the stream and throughout the recovery periods in the same ratio as when released. The two groups did not suffer any differential mortality. The greater handling of tagged fish did not cause increased mortality; however, these fish were tagged at the spawning grounds. The Birkenhead River sockeye had an additional 80 miles to migrate through river and lake after tagging before they reached the spawning area.

Enumeration of the Weaver Creek run, based on tagged fish releases at the Harrison traps and recoveries at the Weaver Creek spawning grounds, also showed serious discrepancies: 24 percent in 1940 and about 61 percent in 1941. In both instances, the Harrison trap data overestimated the population.

Tagging at Skookumchuck resulted in a Birkenhead River population estimate about 25 percent higher than that obtained at the spawning grounds. Schaefer concluded: "that an estimate of high precision of a given spawning population can be obtained only by tagging very close to the spawning grounds, preferably at the mouth of the spawning stream." He emphasized, however, that careful consideration needed to be given to the interpretation of data in each instance for a high degree of accuracy.

Conclusions based on critical examination of the investigations established criteria to provide the most accurate population estimate. These were:

(1) tagging a minimum of two percent of the total population,

(2) tagging done immediately below the spawning area,

(3) tagging done throughout the duration of sockeye arrival,

(4) numbers tagged over time to be proportionate to arrival abundance,

(5) dead recovery to be continuous during the dying period and with uniform effort, and

(6) all recoveries lumped together regardless of sex or size unless other evidence of varying tag ratios is available.

Adaptations of the method, such as using live-counts rather than tagging or weir counts to establish indexes for various smaller streams were used where tagging was not considered necessary or practical. Accurate escapement estimates could be achieved by tagging as low as one percent of the total population provided other requirements were met, such as tagging and dead recovery throughout the arrival and dying periods. Most of the tagging in subsequent years was done by using small beach seines. Sockeye escapements by area and years (1938-1985) are summarized in Appendix D.

· · · · · · · · · · · · · · · · · · · ·
COMMERCIAL CATCH STATISTICS

Even though the terms of the Convention prevented commercial fishery management until 1946, it was early recognized that an accurate statistical system would be required to determine commercial catches. Daily catch by gear and area for each country on a timely basis would be very important for management and for fulfilling the Convention requirements to divide the catch equally between the two countries.

In 1939 Mr. F.H. Bell of the International Fisheries Commission (Halibut) was asked to advise on the methods of collecting catch sta-

August 4, 1945. Stellako River, one of the most stable sockeye spawning streams in the Fraser watershed. With a stable flow of clear, silt-free water, it is a very valuable sockeye and trout producer.

August 1950. A weir was operated for several years in Stellako River to enumerate sockeye populations spawning in Francois Lake tributaries and to determine the arrival timing, especially of the early and late Nadina runs. After the run timing had been established, the fence was no longer used.

September 11, 1950. Chuck Walters, Milo Bell, Jim Mason and Roy Jackson (left to right) on Stellako River. This river was extensively studied because of its great value as a sockeye producer.

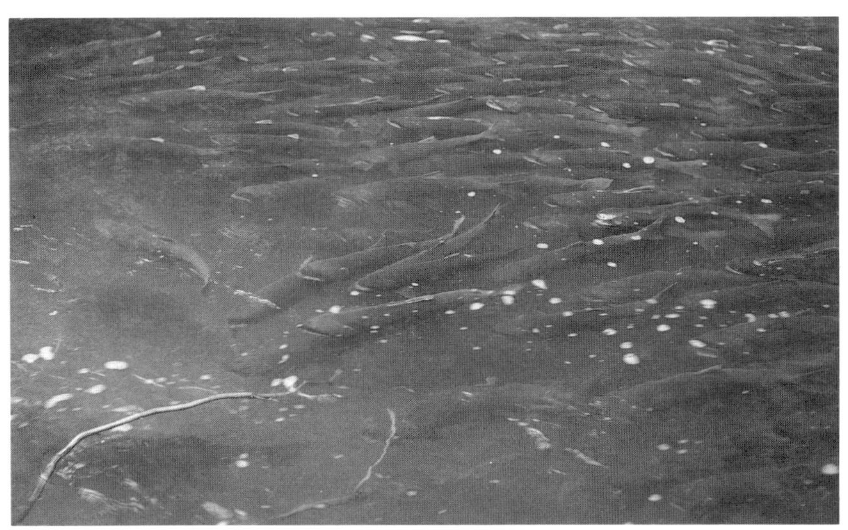

August 12, 1949. A school of sockeye resting in a pool above a counting fence in Forfar Creek, tributary to Middle River. The number of sockeye spawning in this stream increased in one cycle (1945-1949) from 7,081 to 80,484.

August 12, 1949. Sockeye spawning in Forfar Creek. When they have matured and are ready to spawn, sockeye spread out and defend their redd territory.

October 18, 1946. Sockeye spawning in lower Adams River.

August 24, 1947. The upper Bowron River sockeye spawning grounds across from Sus Creek, an area where grizzly bears were frequently seen.

A part of the Driftwood River considered to have potential for sockeye spawning. A biological survey was completed in August 1947 and many miles of excellent spawning gravel were found.

Developing techniques for scientifically enumerating populations of spawning sockeye required considerable effort in the Commission's early years. In this 1938 photo, a crew was engaged in the very difficult task of attempting to use dip nets and gaffs to catch sufficient sockeye for tagging on Adams River. It was later learned that beach seining in Shuswap Lake at the mouth of Adams River was a much more effective capture method.

Commissioner Donald R. Johnson (1971 - 1980) worked as a biological staff member (1939-1941) exploring sockeye spawning areas in the upper Fraser watershed.

July 21, 1945. Three staff biologists - Gerry Talbot, Roy Jackson and Bruce Krug (left to right) stop for lunch on Middle River in the Stuart watershed.

April 1, 1949. Biologist John Weir, who was in charge of IPSFC activities at Chilko Lake for many years, frequently travelled the trail between Redstone and Chilko Lake. His vehicle became bogged down so often in one particulary soft area that it became known as "Weir's Wallow". Travelling to the upper parts of the Fraser watershed was often very difficult. IPSFC vehicles were easily recognized in the early years of IPSFC activities because dual licence plates were used on both the front and the rear.

March 31, 1949. Even the main Chilcotin highway at Riske Creek sometimes became impassable during spring break-up. Road improvements in later years made travel much easier for IPSFC staff.

May 2, 1956. Hew Dunlop required a 4-wheel drive-vehicle with chains on all wheels for transportation from Riske Creek to Farwell Canyon where he supervised construction and maintenance of the fishways. Having grown up in the Lillooet area, Dunlop had a broad knowledge of the Fraser watershed and the historical sockeye and pink salmon runs. For this reason, he was hired as a guide by Dr. W. F. Thompson in 1937, one of the first members of the IPSFC field staff.

August 14, 1947. Dr. Dick Van Cleve after a long portage on the Driftwood River system, tributary to Takla Lake. Transportation to far-flung sockeye spawning areas was often a major problem.

July 23, 1945. The outboard motor provided the easiest and most common mode of transportation. This photo was taken from Gerry Talbot's Middle River cabin, which was used by IPSFC staff working on Takla and Trembleur Lakes.

September 1, 1949. Funds were sometimes obtained to use aircraft for transporting men and equipment to explore possible sockeye spawning areas in the northern part of the Fraser watershed.

tistics. He was retained again in 1940 and later was appointed Assistant Director of the Commission in 1943. Bell soon returned to the Halibut Commission and served for many years. To collect statistics, observers were stationed at Steveston, Bellingham, and Anacortes. A system of records or logs on catch numbers and locations was kept by fishermen of purse seine and gillnet fleets. Accurate information on numbers landed at each cannery was also required.

Because of great diversity in record keeping at each cannery, different methods were looked at. Photographing cannery records was one method investigated. It was to provide inexpensive information on total catch by day, location, gear and fishery (nationality). As early as 1938, the Commission established criteria for obtaining usable and accurate commercial catch information on a timely basis.

Some data were collected from the canneries by mail in 1939, other information gathered by personal visits by Commission staff. In 1940, all information was collected at landing points by Commission staff. Records from fish tickets were copied by hand and the data telephoned daily to the Commission office. Buyers in Canada were usually contacted by telephone from the Commission's headquarters. Such a method meant preliminary landing figures for both countries were available soon after the fish were landed. This system was in place by 1944.

Carbon copies of buyers' tally sheets in Canada were made available to the Commission, but the system was not immediately universal in Canada. Reports from the Washington State Department of Fisheries and the Canadian Department of Fisheries were used to verify Commission data. The Commission also secured copies of available historical data from various sources to describe catches and runs of previous years.

These early efforts laid the groundwork for a sound statistical system important to the success of the Commission's overall program.

C H A P T E R   7

As discussed earlier, study results showed a migratory block at Hells Gate at certain water levels—a block that had a serious impact on the large escapement destined for the Chilko River and other smaller races (Horsefly) in 1941. In that year, hundreds of thousands of sockeye failed to pass Hells Gate.

Engineering investigations for a fishway started in a small way in 1941. By 1942, a special appropriation from the governments was made for engineering work, and a complete survey of the Hells Gate situation was undertaken. Bell (1945) reported on the findings and explained why fish could pass at certain water levels but not at others. He found that on the left bank at a gauge reading of 25 feet and lower: "the central flow which moved at high velocity and impinged on the 'left jutting rock' in the Gate, shifted out and upstream thus removing the surface drop of 6 feet at this point."

When river level increased above the 25-foot gauge mark, the major portion of the flow poured around or over the rock on the left bank and the drop, velocity and turbulence created an obstacle that prevented fish from migrating upstream.

Slide material had been removed in 1914 by Canada; however, by 1942, as photographs revealed, new slide material had completely filled the left bank excavations and conditions appeared to be very similar to those which existed in 1913. It is not known how many years passed until the 1913 bank conditions returned.

The conditions on the right bank showed that the block occurred about 240 feet upstream from the Gate at the narrowest point. Below the gauge level of 40 feet, a drop of six to eight feet occurred at the right bank jutting rock, preventing upstream migration. When the water level exceeded the 40-foot level, the drop decreased so that fish could pass.

The problem at Hells Gate was initially one of correcting conditions on the left bank when the gauge reading was above 25 feet and on the right bank below the 40-foot gauge mark. With velocities in the center of the river in excess of 20 feet per second, the drop of eight to nine feet along the banks had to be overcome. Prior to the 1914 slide, the drop was five feet. Following the slide, it had increased to 15 feet. By the end of 1914, following removal of rock, the head was nine feet. It varied from eight to ten feet from 1915 into the 1940s (Jackson 1950).

Two fishways were built into the hydraulic model as previously described in Chapter 6. The one on the right bank was to be 20 feet wide and 220 feet long, and on the left bank 20 feet wide for 160 feet and 12 feet wide for the remaining 300 feet. The left bank fishway was to function from elevation 23 to 55, a range of 32 feet. The right bank fishway was operational from elevation 20 to 55 or for a range of 35 feet.

Fish entrances at both fishways were located where upstream passage became impossible.

In order for fish to pass safely through the fishways, and to reduce the velocities, pools were developed by using vertical baffles that would create uniform conditions inside the fishways over all water levels. Studies showed that safe passage for sockeye could be provided by installing fishways along both banks of the river without major alterations to either bank structure.

· · · · · · · · · · · · · · · · · · · ·

HELLS GATE

Construction of the fishways began along both banks of the river in the fall of 1944. It was necessary to work during the low water period from late fall to early spring before the annual snow-melt freshet. By the spring of 1945, the right bank fishway and lower end section of the left bank fishway were almost completed. During the 1945 season, these fish ladders provided improved sockeye passage past Hells Gate.

The first fishways on both banks at Hells Gate were completed on May 8, 1946, providing fish passage from gauge levels of 23 to 54

September 12, 1947. An operating model of Hells Gate was built at the University of Washington to a scale of 1:50. It was used from 1943-1949 to assist in fishway design, especially for studying potential locations of fish entrances and exits. This work was directed by Milo Bell, Chief Engineer and Professor C.W. Harris, Director of the Hydraulics Laboratory.

September 12, 1947. Jim Pyper worked on the design and construction of the model and performed most of the testing. Here he is examining the flow at the fish entrance of a proposed low-level right bank fishway.

September 12, 1947. The fish entrances to three Hells Gate right bank fishways are shown in this view of the model - the high-level fishway, the main 20 ft. wide twin-jet vertical-slot fishway and the proposed low-level single-jet vertical-slot fishway. The latter wasn't built because the late-run populations migrating on the right bank at that time were too small to justify the cost. With increasing pink salmon populations in recent years this fishway may be required in the near future.

feet. Early in the season, water levels were quite high—more than 50 feet, and late in the season the river level dropped below 17 feet. The high and low river levels gave Commission staff the opportunity to examine fish passage conditions outside the designed operating range of the fishways. Observations showed that salmon migration was difficult between gauge levels of 50 and 65 feet. In addition, a block developed on the right bank when the water level fell below 17 feet and was maintained throughout the 11-17 gauge level. These water levels usually exist at the time of the Adams River-South Thompson late runs. A serious delay in 1946 at Hells Gate and in the Thompson River rapids occurred for the Adams River run.

The numbers of tagged sockeye recaptured below Hells Gate declined sharply, indicating successful upstream movement. In addition, the percentage of sockeye tagged each day below Hells Gate and recovered above Hells Gate the same day increased significantly. A dramatic decline in the number of scarred or injured sockeye was immediately noticed in the spawning areas above Hells Gate. The fishways were successful.

Sockeye buildup below the Gate, as noted in other years, was not evident and fish were migrating relatively consistently upriver. A minor problem still existed at gauge height of 52-54 feet because the river surge caused some fish to be stranded on the deck gratings. The problem was studied and plans were formulated to extend fishway capabilities for fish passage at both higher and lower water levels.

Construction of the right bank high-level fishway was started in October 1946 and completed in 1947. The range for safe passage during high water was increased to operate between 54 and 70 feet. As the sockeye runs continued to increase and migration patterns were studied over a wide range of river levels at Hells Gate, it became obvious that further fishways at both higher and lower levels were required. In 1951 the left bank high-level fishway was constructed and finished in De-

cember of that year. It covered an operating range of 54 to 70 feet.

Further high-level fishways were necessary as the early Stuart run began to increase. This race migrates during the high waters of spring freshet. Early Stuart sockeye were about 15 days late in arriving at the spawning grounds in 1960 due to high water levels. Some red-colored sockeye—the redness at this location being an indicator of delays—were noted at Hells Gate. Unspawned fish were found dead many miles below the spawning grounds. An estimated 21,000 sockeye did not reach the spawning grounds. Another migration block existed somewhere at or below Hells Gate and analysis suggested the problem was at Hells Gate. It was recommended that the operating range of the left bank fishway be extended to gauge 92. In early 1965, an upper level fishway was completed to cover water levels from gauge 70 to 92, encompassing the highest flows that had occurred during the period of sockeye migration.

Delays in migration of large Adams River escapements had been observed when the river level dropped below gauge 24, the lower limit of the initial fishways. Thus, in late 1965 and early 1966, construction proceeded on a series of sloping baffles placed against the left bank of the river. This new fishway design, completed in March 1966, had a daily capacity of 250,000 sockeye and pink salmon. A serious delay like that of the Adams River run in 1946, was eliminated.

The total cost of $1,470,333 for all fishways at Hells Gate was shared equally by the United States and Canada.

The first Hells Gate fishways were constructed during World War II and difficulty was experienced by the contractor (Coast Construction Company, Ltd.) in completing the fishways on time and within cost estimates. For a variety of reasons, the cost over-run was large and for several years there were prolonged discussions between the Commission staff and its legal counsel and Coast Construction officers and their legal representatives. The con-

May 16, 1945. Construction of the Hells Gate fishways had progressed sufficiently to permit partial operation for passage of the 1945 sockeye populations. The main right bank fishway was completed and the 20 ft. wide portion of the left bank fishway was also completed. The 12 ft. wide extension of this fishway, which carried the fish further upstream to a less turbulent area, was completed in time for the 1946 runs.

May 26, 1945. The main left bank fishway was overtopped for the first time in 1945 by the rising water level of the spring freshet.

September 13, 1946. Gauge height 22.4 feet. Located at the farthest upstream point that the fish can reach, this entrance to the Hells Gate main right bank fishway appears ideal.

A tunnel was provided on the right bank in an effort to avoid the necessity for fish to swim through part of the violent turbulence downstream from the fishway entrance. The upstream end of the tunnel is in a back eddy immediately downstream from the main fishway.

June 22, 1952. Gauge 56 ft. With rising water level, the two fish entrances of the left bank high-level fishway begin to operate as the main fishway becomes submerged.

September 1981. An aerial view of Hells Gate showing the main fishway and the high-level fishway on the right bank and the low-water open-baffle fishway, the main fishway, the high-level fishway, and the upper level fishway on the left bank. The tunnel on the right bank is not visible.

March 16, 1966. Gauge height 9.5 ft. The four fishways on the left bank at Hells Gate are shown - one open-baffle fishway and three vertical-slot fishways: the 20 ft. wide twin-jet and two single-jet fishways are used by the smaller populations that migrate during high-water periods. The water gauge is painted on a cliff shown at the right side of the picture.

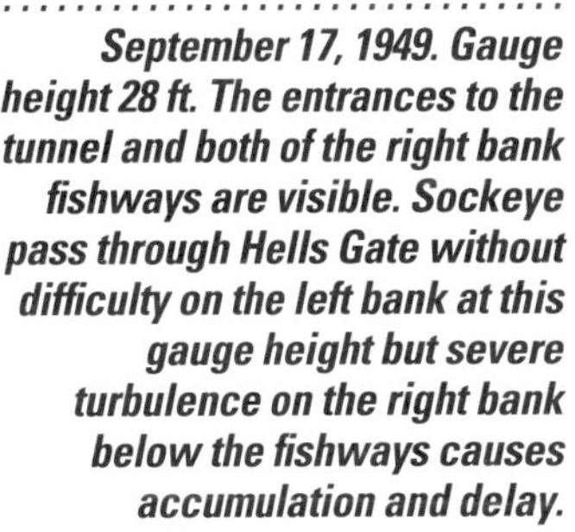

September 17, 1949. Gauge height 28 ft. The entrances to the tunnel and both of the right bank fishways are visible. Sockeye pass through Hells Gate without difficulty on the left bank at this gauge height but severe turbulence on the right bank below the fishways causes accumulation and delay.

September 10, 1965. Gauge 27.5 ft. The different operating levels of the main left bank fishways are apparent. All of the loose rocks near the fishways were carefully concreted into position to prevent erosion. The upper fishway, shown on the right side of the photo, was completed in March 1965. It operates between river levels 70 and 92. The small high-level fishway in the center of the picture, built in 1951, operates between river levels 54 and 70 and the main fishway, which began operation in 1945, operates from river level 18 to 54.

May 26, 1948. A few days after this photo was taken the foot bridge and the warehouse building on the left bank were washed out by the highest discharge since 1894. The water level reached 108 ft. in 1948. The highest water level that occurs during the sockeye migration period is 92 ft.

October 21, 1954. Gauge height 30 ft. The peak o the large Adams River run of over two million sockeye was migrating through Hells Gate when this picture was taken. Passage conditions were excellent on the left bank. Because of the large run and the restricted passageways through the Fraser Canyon, accumulated sockeye were visible in low-velocity areas over 50 miles of river canyon.

March 31, 1966. Gauge height 15.0 ft. Twenty years after completion of the first Hells Gate fishways it was found necessary to build a low-level fishway, primarily for the late-run Shuswap sockeye and the Seton Creek-Thompson River pink salmon populations. An open-baffle design was used but scouring by large pieces of bed load during high water created major maintenance problems.

tractor requested a payment of an additional amount of $414,000 for over-cost at Hells Gate. The Commission eventually did make a sizeable additional payment to the contractor but not equal to that requested. The contractor was never satisfied with the outcome. Thus, a considerable amount of the Commission's attention was diverted at a time when new regulatory responsibilities were soon to be assumed.

BRIDGE RIVER RAPIDS

Bridge River Rapids are about 75 miles above Hells Gate or 205 miles from the mouth of the Fraser River. In the early years of the Commission's investigations, it was noted that a blockage appeared to exist at low water condition. This generally occurred late in the migration season. On numerous occasions sockeye were observed spawning in Seton Creek, downstream from Bridge River Rapids. These fish were believed to be there because of a migratory block at the rapids. Studies began in 1940 and tagging was conducted annually from 1942 through 1945. The effect that the Hells Gate blockage had on arrival of fish at the rapids prior to 1945 could not be clearly ascertained until 1945, the year the Hells Gate fishways became operational. Tagging in 1945 at Bridge River Rapids clearly established that a serious block occurred at certain low water levels. In 1945, 1,627 sockeye were tagged below the rapids between July 7 and September 3, during which water levels were greater than a gauge reading of 651 feet. A total of 31.4 percent of the tags were recovered above the rapids and only 5.2 percent below the rapids. When the river dropped below 651 feet (September 3-October 15), 353 sockeye were tagged but only 5.9 percent were recovered upriver and 37.7 percent below the rapids. These results provided conclusive evidence of a natural low-water blockage at this location.

The Commission decided early in 1945 to construct two fishways, each on the right bank at the upper and lower rapids. Con-struction started in December 1945 and was completed at the end of April 1946. Tagging done in 1946 at gauge levels higher than 651 feet to confirm the effectiveness of the fishways resulted in 34.5 percent of the tags being re-covered above the rapids and only 4.5 percent below. During tagging below 651 feet, 16.8 percent were recovered above the rapids, whereas only 0.9 percent were recaptured or recovered below the rapids. These data, when compared with the 1945 tagging results, clearly indicated that the fishways were effective in passing sockeye upriver at low water levels.

YALE

The early Stuart sockeye run in 1955 did not appear at Hells Gate on its historical schedule. Downriver observations to investigate the cause found a high-water natural blocking condition about 15 miles below Hells Gate. A later-than-normal spring freshet resulted in a six-day block in the upstream migration. The early Stuart escapement of 30-35,000 fish was almost destroyed. Only 2,170 sockeye arrived at the spawning grounds and these fish were late, badly bruised and in general were poor spawners. As during earlier years of Hells Gate blockage, sockeye were found in creeks not normally used, such as Spuzzum, Seton, Texas and Williams Lake. Scale analysis showed these fish were of early Stuart origin. The fish apparently did not have sufficient energy or time to reach their natal spawning grounds.

The 1955 studies indicated the obstruction point near Yale occurred between gauge 70 and 73 (at Hells Gate) and that there were four separate points of obstructions, two on each bank (Cooper and Henry 1962). Fishways were designed for the right bank to provide passage at gauge readings of 62 to 78 feet at Hells Gate. Construction was recommended and both governments approved the plan. The structures were completed in May 1957. It was decided that even though fish would be blocked along the left bank, they would, after a

August 26, 1945. Tagging at Bridge River Rapids showed that many sockeye were blocked by the severe drop during late summer and fall.

May 6, 1946. Two fishways built on the right bank of Bridge River Rapids in 1946 were very effective in expediting fish passage. The lower fishway is shown, looking upstream. Sockeye had no difficulty at this water level but the drop was much greater at lower water levels.

In the winter of 1943, two consultants retained by IPSFC - civil engineer Professor Ted Pretious and geologist Dr. Henry Gunning - examined the proposed fishway construction site on the Fraser River just upstream from the Bridge River confluence along with staff members Jim Pyper and Stan Killick (left to right).

Resident Engineer Larry Bomberger (on the left) and Chief Engineer Milo Bell (on the right) accompany the contractor's superintendent on an inspection of Bridge River fishway construction in 1945.

February 16, 1945. As at Hells Gate, the fishways on the Fraser River at Bridge River Rapids had to be built when water levels were low. This necessitated working under severe winter conditions.

Looking upstream at the lower of two vertical-slot fishways built on the right bank at Bridge River Rapids. Like the Hells Gate fishways, both have an inside width of 20 ft. with vertical baffles or walls spaced at 18 ft. centers. There are two 2 ft.-wide vertical slots in each baffle.

May 6, 1946. The lower fishway at Bridge River Rapids, looking downstream.

The upper fishway at Bridge River Rapids.

After fishways were completed, sockeye were able to pass on the right bank and the accumulation of fish below the rapids was reduced so that sockeye seldom dropped downstream to enter Seton Creek. As shown in this photo , taken on April 26, 1946, the water slope through the fishways is very gentle compared to the river.

In 1940, when the above photo was taken, migration conditions were so poor that sockeye were delayed for days and many moved back down river and entered Seton Creek, which flows into the Fraser about six miles below the rapids. Some of the thousands of sockeye that accumulated below the rapids during the migration period prior to fishway construction may be seen in the above photo. Courtesy Pacific Fisherman.

short delay, find their way to the right bank and proceed upriver. One of the fishways at Yale was a reinforced-concrete structure similar to the left-bank, high-level fishway at Hells Gate. The other was a trapezoidal rock-cut, with three concrete baffles providing flow stabilization.

River discharge analysis and sockeye migration timing data in earlier years indicated that a natural block to early Stuart sockeye migration probably occurred in 1900, 1914, 1933 and 1954 *(ibid.).*

River levels were low during early Stuart migration in 1957, 1958 and 1959 and sockeye did not have to use the new Yale fishways. In 1960 water levels were higher and early Stuart fish were seen using the right-bank, high-level fishway. With a drop in river height, below the operating range of the Yale fishway, some red-colored sockeye were noted both below and above the Yale fishways.

Prior to 1955, most observations of delayed sockeye in 1913, 1926, 1927, 1930 and 1941 were noted below Hells Gate and in tributary streams many miles downriver. In more recent years (1955, 1960) when migratory problems were experienced for the early-timed races, moribund and dead sockeye were observed in the tributary streams above Hells Gate. This probably occurred in the earlier referenced years as well, but was either undetected or unreported. Other recomended improvements at Yale included drilling and blasting passage parallel to the existing right bank fishway to improve fish passage between gauge 53-62. It was also recommended that a fishway be provided along the left bank to operate from gauge 61 to 92. In addition, in April 1961 a rock pinnacle located 1,000 feet above the Yale fishway was removed.

Funds were not available until 1963 to start the fishway construction and rock cut on the left bank at Yale Rapids. This work was completed in March 1964. The operating range on the right bank was extended to lower river levels by a rock cut completed before the spring freshet in 1964. New improvements provided passage for salmon on the right bank covering river levels from gauge 53 to 78 (Hells Gate readings), and along the left bank from gauge 61 to 92. The facilities at Yale covered all the river levels at which fish had encountered migration difficulty. In 1965, three concrete baffles were placed in the right bank rock-cut to improve the hydraulic characteristics in this channel.

The Commission's efforts at Yale and Hells Gate provided access to spawning grounds for the early Stuart race that never existed historically. This race now travels easily along the best migration path ever available.

FARWELL CANYON

Chilko River sockeye migrate 110 miles up the Chilcotin River to their spawning grounds just below Chilko Lake at elevation 4,000 feet. These sockeye undergo one of the most arduous migrations of all Fraser River races, passing through several turbulent areas on the lower Chilcotin River. Biological and engineering surveys were conducted for several years beginning in the early 1940s at Farwell Canyon, located one hundred miles below Chilko Lake and about seven miles upriver from the Fraser River. By 1945, surveys indicated a suspected natural obstruction and delay at the canyon (IPSFC 1946). Tagging investigations indicated that at a certain water level, the canyon formed a partial obstruction because of high marginal river velocities.

The obstruction was about 1,000 feet long where the river dropped 15 feet over that length. The Commission approved construction of five fishways in 1946. Construction started in May 1947 and was 80 percent complete by the end of 1947; however, geological formations in the canyon were less solid than originally believed and one of the fishways had to be rebuilt after a major slide occurred in 1949.

July 22,1964. A vertical-slot fishway and a deeper rock cut assist sockeye passage on the right bank at Yale Rapids.

July 18, 1960. The entrance to the main right bank fishway shown in the upper photo at Yale Rapids was just above the water level on this date and sockeye were migrating through the rock cut.

June 7, 1964. A single-jet vertical-slot fishway on the left bank at Yale Rapids. Five fishways were built at this site three miles above Yale in the Fraser Canyon to assist the passage of early sockeye runs at high water levels.

Spring freshet was extremely high in 1948 and large quantities of rock and sand were carried into the fishways. It was anticipated that further trouble would occur because of the unstable nature of the river banks in the Chilcotin River.

In the spring of 1950, the fifth and final fishway at Farwell Canyon was completed. Further slides occurred in the construction area but the first four fishways were functioning properly and sockeye were observed using them. Later that same spring, the lower right bank fishway was partially closed by a small rock slide into the outlet tunnel and the fishway channel. The material was removed and all five fishways were operative in the 1950 migratory period. In the spring of 1955, a slide occurred in the fractured rock of the canyon wall above the lower fishway on the right bank. Large rocks fell on the deck gratings and partially filled the entrance to the fishways, but they were once again operational during the 1955 season. In 1956 the trash racks at the upper end of the lower and middle left bank fishways had to be fitted with stop logs to prevent rocks and gravel from accumulating in the fishway during freshets.

Following the hydrological events between 1948 and 1955, it was determined that the 1948 flood had changed river flow patterns to such an extent in the area above the fishways that fishways were no longer required (M. Bell personal communication). The fishways have not been maintained and were not operative through 1985. Changes in the flow pattern in recent years have reportedly increased the possibility that the fishways may again be required (F.J. Andrew, personal communication).

. . . .
TOTAL FISHWAY EXPENDITURES

Fishway construction, while very rewarding, was not accomplished without difficulty. Most of the fishways had to be built during winter when river levels were low. Extreme difficulties were encountered at this time of year due to low air temperatures and high snowfall. Special precautions had to be taken, especially in central British Columbia (Bridge River Rapids and Farwell Canyon), to permit pouring of concrete during cold weather. Annual maintenance to repair concrete, remove gravel and other debris, stabilize river banks and repair steelwork such as gates, intake trash racks and deck gratings also had to be done during cold weather. However, the cost of fishway operation and maintenance represented a very small part of the Commission's annual expenditures.

Divisional expenditures for fishway construction from 1944-1966 were as follows:

	Number of Fishways	Total Cost
Hells Gate	6	$1,470,333
Bridge River Rapids	2	234,133
Farwell Canyon	5	211,212
Yale Rapids	4	118,462
Planning and Overhead		363,573
Total	17	$2,397,713

Appropriations are in Canadian dollars. Combined fishery benefits are expressed in dollars which are the sum of the values in both countries (U.S. plus Canadian dollars). When values are given for a specific country they are expressed in that country's currency.

It is not possible to quantify the benefits provided to resource enhancement as a direct result of the fishway construction; however, investments by both countries in fishway construction have already returned values far in excess of the costs. These benefits continue into the future.

The biological and engineering divisions of the Commission joined forces as closely integrated disciplines in the fishway efforts. Scientific data required to show where migration difficulties persisted were used by the engineering division to design and ultimately supervise construction of the many fishways. Following construction, the effectiveness of each facility was assessed by additional biological investigations as well as by engineering studies.

May 19, 1948. High velocities and turbulence through Farwell Canyon on the Chilcotin River about seven miles above the Fraser confluence delayed the migration of sockeye to Chilko and Taseko Lakes. Five fishways were built to correct the problem.

September 11, 1945. Resident engineer Charles Clay, consultant Ted Pretious and Chief Engineer Milo Bell inspected Farwell Canyon during their investigation of the feasibility of fishway construction.

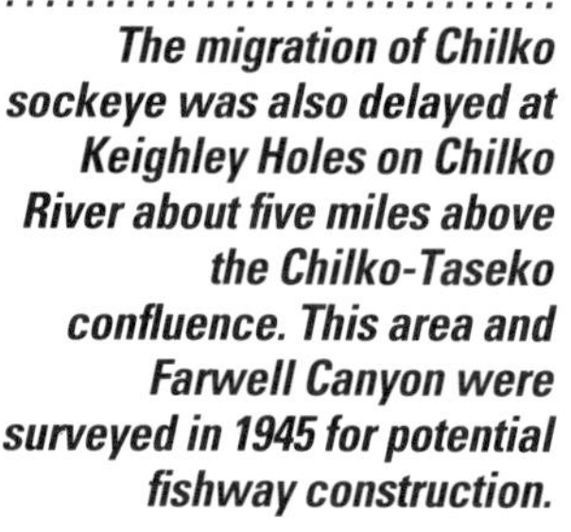

The migration of Chilko sockeye was also delayed at Keighley Holes on Chilko River about five miles above the Chilko-Taseko confluence. This area and Farwell Canyon were surveyed in 1945 for potential fishway construction.

August 18, 1948. The lower fishway on the left bank at Farwell Canyon, completed in time for use by the 1948 sockeye run. Indians fishing at this popular site during the 1945-1950 period, when the canyon was being surveyed and the fishways built, were very pleased with the Salmon Commission's work because the sockeye were in much better condition than in previous years.

July 1948. The upper fishway on the left bank of Farwell Canyon. Steep, unstable banks on both sides of the Chilcotin River were a severe hazard during construction. Slides of rock and debris also caused many maintenance problems after construction was completed.

May 14, 1952. The lower fishway on the right bank in operation. Severe floods in 1948 and 1950 changed the river bed and reduced the water surface drop through the canyon, making the fishways unnecessary. They will again be needed if the riverbed erodes to the previous condition.

CHAPTER 8

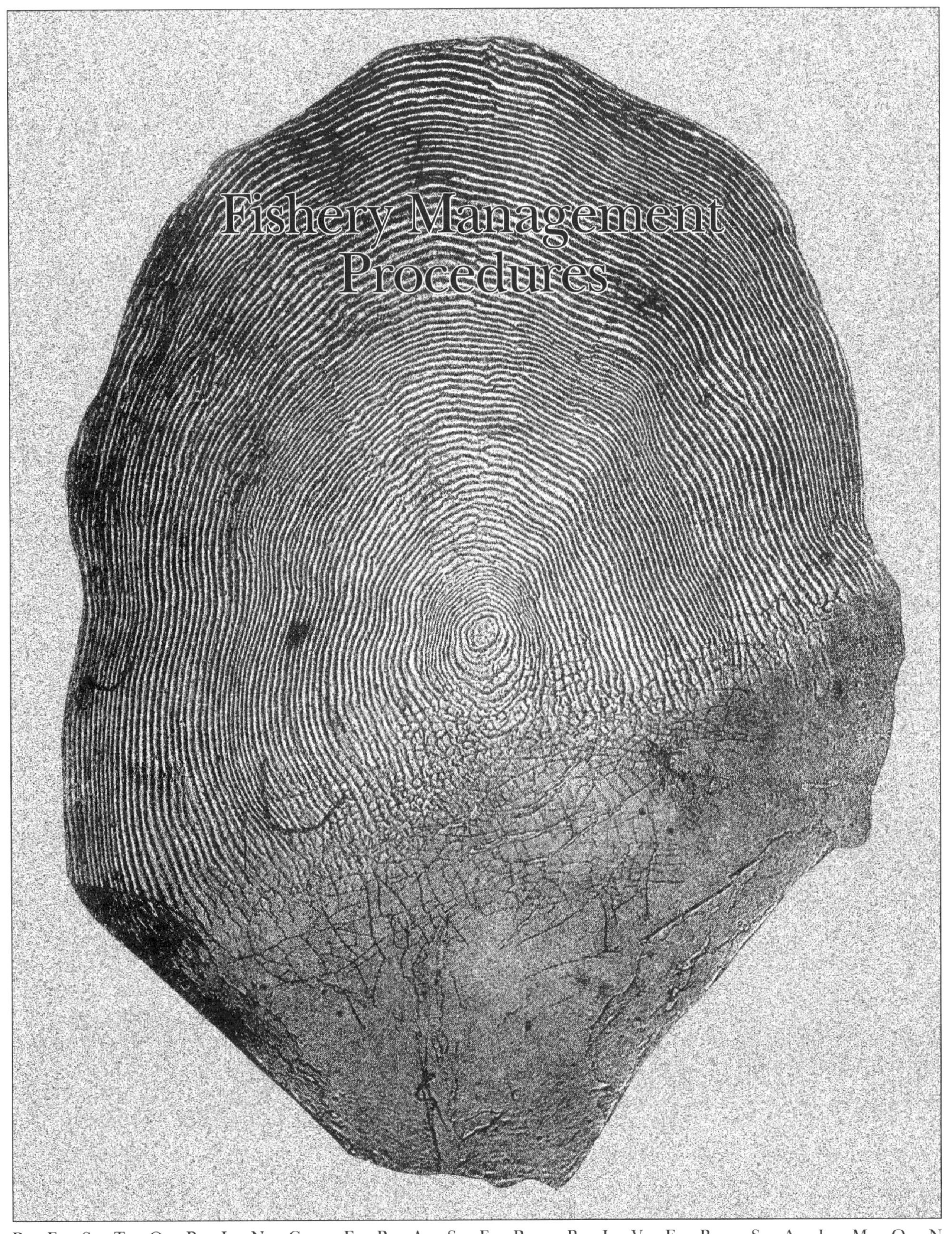

Fishery Management Procedures

The first year of Commission sockeye fisheries regulation in Convention Waters was 1946. Significant information on fisheries had been obtained from 1937 through 1945 for a reasonable understanding of management needs and procedures necessary to implement regulations. Timing of the different spawning races had been gathered from tagging programs and a sound statistical system to gather commercial catch data from both countries was also established prior to 1946. In addition, population estimates at the various spawning grounds provided information on the relative strengths or weaknesses of many races.

Restricting the Commission's initial activities to research was beneficial, it gave the Commission opportunity to focus on the primary problem affecting Fraser River salmon—Hells Gate. A negative aspect of the eight-year delay was that many of the early-run and summer-run races were already in very depleted conditions and would have benefitted from total fishery protection.

PROTECTION AND REBUILDING OF DEPLETED STOCKS

Although actual numbers of fish in the escapements to the spawning areas of separate races were not known prior to the Commission's involvement in Fraser River investigations, Canadian government fisheries personnel and provincial authorities had surveyed most spawning grounds for many years before and after the Hells Gate slide and their escapement estimates were useful. Visual surveys indicated great depletion by the late 1930s to mid-1940s in numbers to many spawning areas, with the exception of some races as noted earlier.

Escapements in 1945 of upriver runs such as early Stuart, Horsefly, late Stuart and Stellako River—all above Hells Gate—were five times greater than those in the 1941 brood year. The Chilko River escapement was less than that in 1941. The size of these increases were the most encouraging since 1913. What is of significance is the fact that the Commission did not have management control of the fisheries in 1945. There were no special closures to protect the stocks and fisheries continued as they had in previous years. Both countries had five to six days fishing each week. In addition, water levels were unfavorable during much of the migration period. Increased escapements reaching the spawning areas were a result of the fishways at Hells Gate. It was reported that Indians were astonished by the fine physical condition of the fish in the canyon above Hells Gate and on the spawning grounds. The timely passage of salmon through the fishways showed conclusively that if the commercial fisheries were restricted sufficiently to allow even minimum escapements, the fishways would provide suitable migration passageways. The net escapement to the spawning grounds would be increased, as would the potential for greater production of future cycles. A great step forward had been taken by the Commission and with no time to spare.

In 1946, answers to how to divide the allowable catch equally and how to obtain optimum escapement of specific races were needed for proper management.

It was yet to be established that an international commission could harmoniously and effectively manage a series of complex fisheries—a harvesting system for Fraser sockeye salmon that had been historically contentious.

The Commission staff examined the 1942 brood-year escapements and recommended that all of the early-run stocks above Hells Gate, including the early Stuart, Nadina River, Bowron River and other early-run races needed complete protection with no harvest by commercial fisheries. Gillnet fishing in the Fraser River in 1946 did not start until August 8 and United States waters were closed to sockeye fishing until July 25. A strong message had been sent to the industry. The benefits from time-specific closures in the commercial fisheries and the fishways at Hells Gate and Bridge River were immediately evident. In 1942, a total of 95,173 sockeye reached the Raft, Seymour and Chilko rivers; the Nechako and Stuart Districts; and the Bowron River. Escapement in 1946 was about 3.5 times greater at 327,019 fish. The only stock that showed a disappointing increase—17 percent—was the early Stuart run. These sockeye encountered water levels well above the 1946 operating range of the Hells Gate fishways and it is surmised they suffered delays, damage and losses. The complete closure of the commercial fisheries during early Stuart sockeye migration must be credited for the 17 percent improvement.

The 1946 closure in the early season was followed even more stringently in 1947. The commercial fishery was not opened until August 18 in both countries and the Fraser River was not opened until September 8. Because of the poor run and low escapements in 1943 maximum protection was given to the 1947 return. Escapements to upriver stocks increased to 158,001 in 1947 from 36,360 in 1943—more than four times. The early Stuart escapement increased to 15,000 from 3,000.

Fishing restrictions were somewhat relaxed in 1948, the year of the dominant Chilko run. Fishing began in the Fraser River on July 28 and in United States waters on July 18. The escapements above Hells Gate in 1948 increased to slightly more than twice those of the 1944 brood year. A significant increase in the early Stuart escapement was noted in 1948, going to 12,000 from the 400 in 1944.

For the fourth consecutive year (1949), fishing in Convention Waters was delayed until later in the season, during the last week of July. Everyone watched with anticipation: 1949 was the first return of fish whose parents were assisted by the fishway. The rate of return for early Stuart sockeye was phenomenal—27,000 spawners in 1945 produced more than 575,000 spawners in 1949. Most reached the spawning grounds. The only harvest was taken by the Indian fishery in the Fraser River and tributary streams. Harvest restrictions in 1949 turned out to be too extensive, for many of the early Stuart spawning areas were overseeded. Large numbers of sockeye spawned in tributary streams where few, if any, had been found in earlier years. The excessive escapement; however, did have long-term benefits. To this day these streams continue to produce a significant part of the early Stuart production.

Other dramatic increases occurred for early runs bound for the Nadina and Nithi rivers and Ormonde Creek. Escapements to these Nechako District spawning areas increased to 25,500 in 1949 from 1,200 sockeye in 1945. While the early-season closure gave almost complete protection for the early Stuart race in the commercial fishery, it also protected the Horsefly River population. The escapement of this race increased to 20,000 in 1949 from 3,000 in 1945. The long-standing position of Canadian fishery scientists such as Dr. W. Ricker, that regulatory authority should have started in 1937, is supported by the improved returns in 1949. The increases in sockeye escapement and production were the result of fishery closures and fishway facilities.

The importance of the Commission's regulatory actions during its first four years of management authority (1946-1949) cannot be overemphasized. The principle of maximum protection of depleted stocks was critical to the restoration and future production of Fraser River sockeye. These populations continued to increase and valuable harvests were again possible in the early 1950s.

While the Commission received the overall support of many in the industry, there were some who bitterly opposed the severity of the restrictions. Benefits accrued in subsequent years from the difficult times of the late 1940s, however, completely justified the restrictions.

The Commission's first year of management of the sockeye fisheries in 1946 was somewhat less than auspicious with regard to international sharing. The catch by Canada (54.4 percent) was 689,000 fish greater than the U.S. catch (45.6 percent), whereas by treaty agreement each country should have had an equal share. While this difference was substantial, total catch for both countries was almost 7.8 million fish.

Other more serious problems occurred in 1946 which would be the forerunner of management problems for the next thirty years. Timing of escapements was critical for the successful reproduction of the stock. Due to natural causes, the large dominant-year escapement to the Adams River was about 12 days late entering the Fraser River. In addition, river discharge was very low (below 18 feet on the gauge at Hells Gate for most of the migration) and the fish encountered difficult migration conditions at Hells Gate, China Bar rapids, Scuzzy rapids and Thompson River rapids. Consequently, an already-late escapement was delayed an additional 9 to 10 days before the fish—some in poor condition to spawn—arrived at Adams River. This produced a total Adams run of only 2.5 million sockeye in 1950 compared with almost 9 million in 1946. Even though poor production was not attributable to 1946 Commission management actions, the first return from Commission control of the fisheries was viewed by some as a disaster. The total return (all races) dropped from an estimated 11.1 million in 1946 to only 4.4 million in 1950. Reasons for the poor return were not known at the time, but the Commission's research results in later years were eventually explained to fishermen and industry. This knowledge gave industry confidence in the Commission.

Escapements of lower Fraser races after the Hells Gate slide were proportionately much greater than those to upriver areas above the Gate. For example, from 1938-1941 total escapements (prior to Commission management and the fishways) of lower Fraser River races were 472,000 fish or 22 percent of the total Fraser escapement. By 1966-1969, prior to any enhancement efforts (other than fishways/ management) total lower Fraser escapements had increased to 630,000 fish, but were only 13 percent of the total Fraser escapement. By the late 1960s, the principle of applying racial fishery management practices to all stocks, along with improved access to upriver spawning areas, reestablished the historical balance of escapements between the two areas. By 1982-1985, total lower Fraser River escapement had been significantly increased by production of the Weaver Creek and Pitt River spawning channels (described in Chapter 13). The four-year total was 816,000 fish, 10 percent of the total escapement. Even though total escapement to the Fraser watershed from 1982 to 1985 increased, primarily by natural methods, to 8,070,000 fish—about four times more than in the 1938-1941 period—and lower Fraser escapements between the same periods had almost doubled because of enhancement efforts, the percentage of total escapement below Hells Gate from 1982-1985 was still less than one-half (10 percent) that of the 1938-1941 period (22 percent). This ratio is further evidence of the debilitating effect of the Hells Gate blockage prior to 1945.

The combination of extensive early-season closures and improved access to the spawning grounds led to large and immediate increases in spawners. Fishery management and fishery technology launched the rehabilitation of the Fraser River sockeye salmon resource.

Two possible consequences might have taken place with regard to fishing regulations and the fishways. First, fishery closures by themselves would have had limited benefits, but would have increased the returns somewhat. The increases, however, would not have been significant. Also, fishways, by themselves, without the added benefits of harvest restrictions, would have been only partially successful in rebuilding the stocks. However, those few sockeye that escaped the fisheries would have been far more effective as spawners, compared with those fish that encountered difficult passage or complete blockages at Hells Gate without the fishways.

It is believed that those sockeye races completely lost, were lost before 1937. Therefore, while increased escapements could have been immediately achieved if the Commission had been given regulatory authority starting in 1937, such action by itself would not have been as beneficial as the combination of fishery closures and the fishways operational in 1946. If the national authorities had reduced harvests in both countries starting in 1937, the acceleration and buildup of the resource starting in 1945—when the first fishway was operating—would most likely have been greater.

CATCH STATISTICS

A basic requirement for effective management of any commercial fishery is accurate catch statistics. This was given high priority by the Commission. The total daily catch by each country plus the daily catches by area and gear were required in order to establish harvesting rates for future management decisions.

The work conducted by the Commission from 1938 to 1945 laid the foundation for the 1946 fishing season. In subsequent years, a special carbon copy sheet of the Washington Department of Fisheries' fish ticket book was provided to the Commission. Each fish buyer or cannery in Washington was required by state law to forward that copy to the Commission's office. The Commission's statistical staff eventually relied almost exclusively on phone calls to each buyer for daily catch information. These daily catch numbers were later verified by checking against the fish tickets as they arrived by mail at the Commission's office.

Collecting catch statistics from the Canadian fisheries was more difficult. Catch records in Canada were provided voluntarily and no IPSFC copy of fish ticket was available in the government fish ticket book. The fact that large numbers of Fraser River sockeye and pink salmon were caught outside Convention Waters complicated things. Only a small portion of the United States catch of Fraser River pink salmon was taken outside Convention Waters and a small catch of Fraser sockeye was occasionally taken in Southeastern Alaska.

In its early years the Commission used the small cabin cruiser, M.V. Statistic, to survey the number of gillnet boats and sockeye landings in the Fraser River area. Since only general estimates of the number of fishing gear were available, catch-per-unit-of-effort data were used to estimate total catch. In more recent years, the Commission and government agencies counted gear from the air. These counts were more accurate and provided better estimates of total catch based on catch-per-unit-of-effort data. Duplicate records mailed after the fishing period directly to the Commission by canners or cash buyers provided actual daily-catch data. Daily phone contact with all major buyers was maintained during fishing periods. As the offshore troll fishery developed in the 1960s, it was necessary to station a Commission observer in the Ucluelet/Tofino area on the west coast of Vancouver Island. During most of the summer, the observer visited each buyer daily to determine troll landings by species and to gather the critical information of where and when the fish were caught.

The Commission's ability to formulate and implement rapid regulatory decisions depended upon getting accurate and timely catch estimates for each fishing day. Such information was also critical for equal catch division between the two countries and to determine in-season estimates of run size. The credibility of Commission statistics was challenged on many occasions and fish buyers would often question catch estimates.

Since there were different methods of counting fish caught and landed between the two countries and between buyers in the same countries, it was not surprising that accuracy would be challenged. Buyers were reluctant to count and pay for the small precocious jack sockeye because "it took three jacks to equal one adult sockeye." In some cases, buyers would weigh the sockeye in the first brail (dip net) that was unloaded, count the fish, calculate an average weight-per fish, and use that number to estimate the number of fish in each weighed brail-load. In other cases, counts were made by applying the average weights calculated to the total weight of fish received at a fish-landing station. Commission statistical staff had to be constantly alert to detect short-cuts in landing and counting procedures and make necessary adjustments to catch estimates. Ultimately, the fishing industries of both countries accepted the accuracy of the Commission's data on the catches by area (country), gear and time. The cooperation of the industry in providing preliminary daily catch data was very important to the Commission's understanding and control of the fishery.

On one occasion, some industry members requested the catch be divided between countries by weight and not by numbers of fish. From a biological point of view, effective management would not be possible without knowledge of numbers. The request was rejected.

During the 1970s and 1980s, some fishermen began to sell their catch off docks or directly from their boats. When sales were conducted in this manner fish tickets were seldom, if ever filled out and Commission staff were required to make spot checks for estimates of the unrecorded catch. These were added to each country's total catch during the season. These add-ons were not popular with industry representatives in either country. The largest share of fish sold off docks took place in British Columbia. Another problem arose when Canadian purse seiners began to brine-chill their fish, thus tendering their own fish with two to three days catch. Records of daily catches were difficult to obtain and hailed estimates were often misleading.

DEVELOPMENT AND APPLICATION OF RACIAL IDENTIFICATION TECHNIQUES

When the Commission began its investigations it recognized the need for methods to identify individual sockeye races in the Fraser system.

Extensive tagging studies in marine areas and at Hells Gate in the 1940s showed that separate spawning races migrated at certain time periods. In the early 1950s as the Commission developed and applied the concept of racial identification and specific racial harvests to management of the resource, a more exact management technique was required. If depleted races were to be protected and restored, an accurate and fast system to determine the location, timing, and need for harvest or protection was needed. Extensive seasonal closures were too broad in scope.

Critical, too, was proper identification in the commercial catches at all times, in all locations and on a daily basis. This meant that the Commission also needed to identify the non-Fraser stocks.

Early studies by Dr. C.H. Gilbert (1913-1924) had shown promise—noting differences in scale characteristics. For example, he noticed that smolts from Quesnel Lake were larger than those from Chilko Lake. The growth patterns on their scales were larger and more consistent during the entire spring and summer growing period than growth patterns on smolts from Chilko Lake. Circuli, or growth ring numbers on the scales of Quesnel smolts, were also greater when compared with Chilko migrants, reflecting the superior growing conditions in Quesnel Lake. Circuli are analogous to the growth rings evident in the trunks of trees.

In 1939, and for all years thereafter, scales and length measurements were gathered from sockeye in the commercial catch and in the spawning grounds and analyzed for age and racial identifications. In the mid-

114

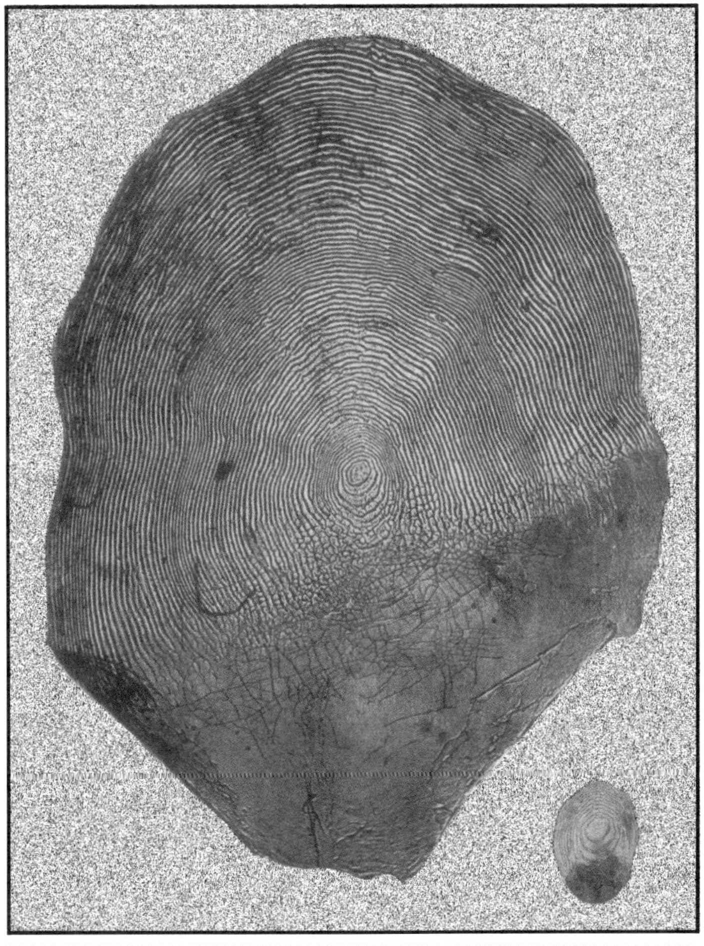

1940s, the Commission began to search for ways in which the sockeye scales could be used to identify individual races. Schaefer (1951) identified and separated Birkenhead River sockeye from other races in the Harrison Lake system on the basis of scale characteristics. These findings held promise that greater application might be possible in identifying other races of Fraser River sockeye.

Clutter and Whitesel (1956) developed the methodological basis for identifying fish to spawning stock by scales and applying that information to Fraser River sockeye management. These procedures were refined by Henry to provide a system for determining reliable estimates of racial composition (Henry 1961).

Differences in numbers of freshwater circuli was used to separate different races of Fraser River sockeye. It was essentially an adaptation of the method originally proposed by C.H. Gilbert. Sockeye salmon scales contain a record of the life history beginning in the early fry stage (Figure 6). The growth patterns displayed in the freshwater and marine zones of the scale, such as circuli counts and spaces between circuli, are unique within particular races. Gilbert showed that the growth patterns on the scales of fry were the same as scale patterns of adult sockeye returning to the same area. This was even more support for the home-stream theory for Fraser River sockeye.

Prior to the return of four-year-old sockeye—about 90 percent or more of each year's return—scale growth information for

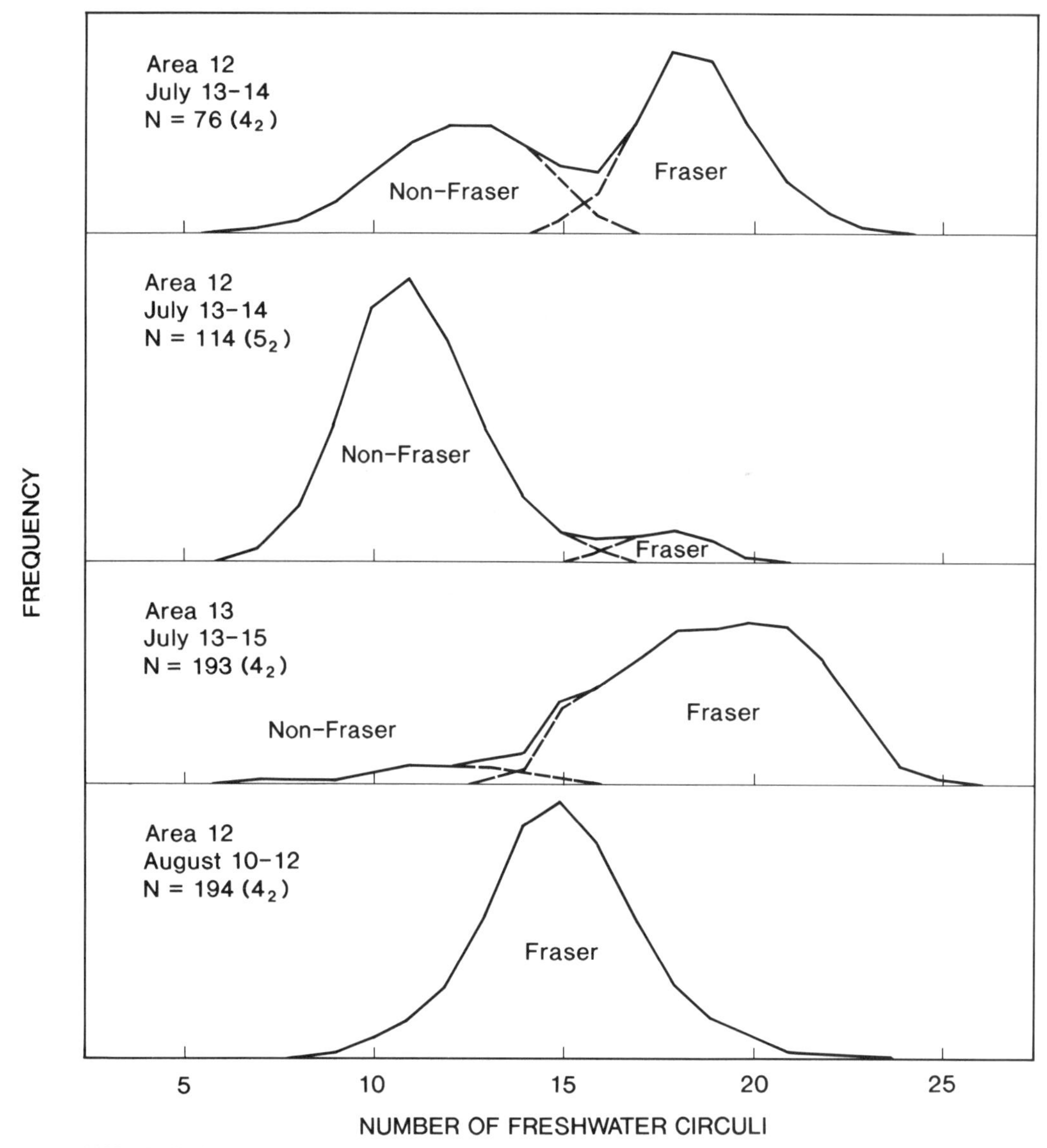

Figure 7. Smoothed freshwater scale circuli frequencies from Johnstone Strait sockeye salmon catches, 1959. The upper line in each graph is the composite frequency for the total sample. From Henry 1961.

major races was available from scale samples taken during the outmigration of smolts, or from scales taken from three-year-old jacks that returned to the spawning grounds one year earlier. The fingerprint-type identification characteristics on scales are created in the freshwater nursery lakes. These lakes have different levels of productivity and varying densities of sockeye fingerling populations from year to year. Some lakes have relatively low productivity resulting in low average numbers of circuli (10) for sockeye fry and fingerlings. Other lakes, relatively productive, can have a low circuli count of 10 as well, but this is due to competition by large numbers of juveniles. Conversely, good growth and high circuli count (17-18) can occur in low-cycle years in less productive lakes when few sockeye fingerlings compete for food.

The identification method developed in the late 1950s by Henry provided estimates of racial composition of different races in the catch. Unknown was the degree of variability. There were accuracy tests using scales from fish of known races, with known numbers of each race in the sample. But there were no tests

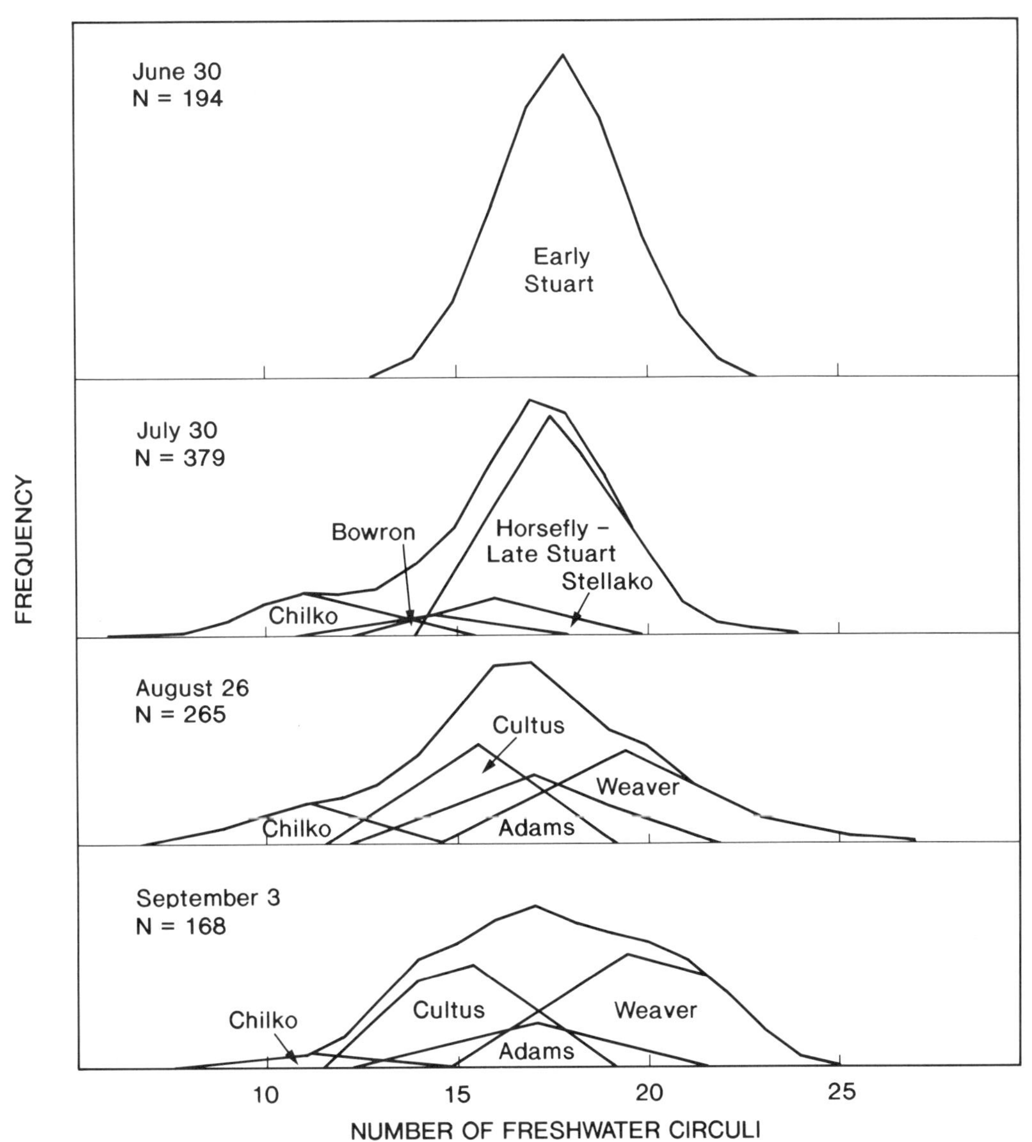

Figure 8. Smoothed freshwater scale circuli frequencies for samples of 4$_2$ sockeye salmon from San Juan Islands catches in 1957, showing estimated racial composition. The upper line in each graph is the composite frequency for the total sample. From Henry 1961.

possible during the fishing season to evaluate the process. Another critical aspect of the methodology was that the determination of which races were present in what abundance in the fisheries during each fishing period, was never absolutely known.

In spite of problems associated with the methods developed (Figures 7, 8) for obtaining racial catch percentages, such as gear selectivity, sample size, representative sampling; the results were successfully applied to resource management. At the same time, there was great value in the method developed, such as early detection of aberrant timing of the runs, unexpected variation in run size, delaying stocks, and unusual migration patterns.

During the 1970s and 1980s attempts were made to reduce subjectivity and thereby improve accuracy. Several races had other distinctive characteristics such as length, spring growth, (scale width following the last freshwater winter-mark), the distance from the scale center to the tenth circulus, and marine growth. By the 1980s, a computer program was developed, using discriminant function analysis to more accurately identify fish to race and to determine racial percentages in catches and escapements. Since then, other technological advances have been explored and developed (Cook and Guthrie 1987).

Even though determination of racial percentages in the catches during the 1958-1985 period was not 100 percent accurate, it provided data sufficiently useful for management of Fraser River sockeye. Continuing research will no doubt provide improved methods for determining racial composition of the catches. However, the representativeness of the collected scale samples may have been a more severe shortcoming in racial composition estimates than the scale-analysis procedure.

TEST FISHING WITH COMMERCIAL FISHING GEAR

During some of the early years of Commission regulation, commercial fisheries in Convention Waters were open five days each week. Under these circumstances, a great amount of catch data was available to determine timing and relative abundance of the runs. As gear numbers and efficiency increased; however, it was necessary to reduce the numbers of fishing days. In some instances, allowable fishing time decreased to one or two days per week. Occasionally it was necessary to impose closures for an entire week to protect a weak stock. In these situations, it was mandatory that the Commission have information to assess run status and racial composition.

Test fishing results were first used in the Commission's management regime in 1946 during a closed period, when the late-run Adams River race was migrating through the fishing area. Some of the Commission's Advisory Committee members doubted the usefulness of test drifts by gillnets near Mission in the Fraser River and the soundness of purse seine test fishing results at the mouth of the Fraser River. The initial reaction of the fishing industry towards test fishing was negative.

By the mid-1950s the Commission was more earnest in its search for methods to gather reliable information during closed periods, and how to estimate the daily escapement past the commercial fishery. It had been suggested in 1946 that an electronic fish counter might be installed in or near the Hells Gate fishways to count the salmon migrating upstream. Investigations showed it was not technically feasible.

In a study of the selective action of Fraser River gillnets on Fraser sockeye salmon it was noted that during the test sets made in 1947 and 1948: "the experimental catches were reasonably representative samples of the catch of the entire gillnet fleet" (Peterson 1954). It was also noted that the abundance of Chilko River sockeye arriving at Farwell Canyon was related to the commercial open (abundance lower) and the closed periods (abundance higher) about 18 days earlier in the lower Fraser River. This gave promise to the value of test fishing to assess escapements during closed periods.

Systematic test fishing began in 1958 and was used each year thereafter. All fishing gears (except traps) were employed at one time or another; however, gillnets and purse seines were more extensively used.

Test fishing operations were established at several locations in Convention Waters. Originally, test fishing was confined to the Fraser River area with a number of different sites extending from the lower reaches of the river to the upper limit of commercial fishing near Mission, B.C. After a number of years of experience, and based on the consistency of results, two sites were selected that provided the most reliable information— the Whonnock drift in the upper area and the Cottonwood drift in the lower river. A test-fishing procedure was established that standardized all catches for any differences in length of net fished or in duration of time fished (Henry 1961). Because of the differences in efficiency between fishermen, the Commission retained each test fisherman as long as possible to maintain consistency in the catch data.

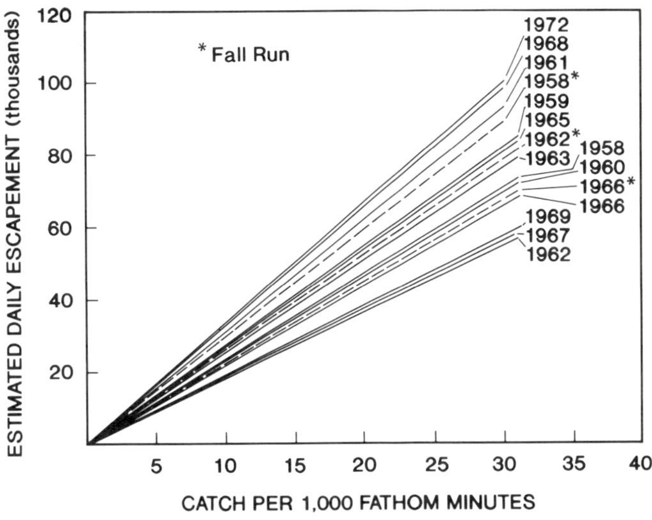

Figure 9. Relationship between test fishing catch at Whonnock and known adult sockeye salmon escapements. From Roos MS 1969.

Based on years of data, a statistically-defined relationship was established between test fishing catches and known adult sockeye escapements determined from individual enumeration programs for the races at the spawning grounds (Figure 9). When test fishing was conducted over the entire season during closed periods, daily escapement estimates were available. As a general rule, only small numbers of fish escaped the commercial fishery during open fishing periods. During the 1960s gillnet test fishing was conducted every season and in some years efficiency of the fishing varied considerably. Starting in the 1970s, acoustical methods were employed to improve the accuracy of the estimates (IPSFC 1973, 1974). Scale samples collected daily from test fishing determined the racial composition of the escapement.

Tide, river volume, size of fish, time of drifts and volume of fish encountered affected the accuracy of the test-fish catches. Despite this, escapement estimates were useful for day-to-day management operations. Auxiliary information, such as total commercial catches in the river on days adjacent to test fishing catches, helped evaluate the methodology and refine escapement estimates. Also, the Commission maintained an observer at Hells Gate (about five days migration time from the lower river for sockeye) and daily estimates of escapement past Hells Gate were examined continuously during the season to verify the test fishing estimates and relative effectiveness of the Fraser commercial fishery. The Indian food fishery catches below Hells Gate were also included in the evaluations.

One of the keys to the Commission's success in establishing and applying the concept of harvest management by racial group, was the fact that reliable estimates of the total number of sockeye escaping upriver could be ascertained in a timely manner. These estimates, combined with the racial composition data from scale analysis, gave information to control harvests and allow the escapement of adequate numbers of sockeye of specific spawning races.

The Commission soon realized that not only information on run status (size) in the Fraser River was essential but also that information from other areas in Convention Waters was needed. In the early 1960s test fishing with purse seines was started in Area 20 of Canadian Convention Waters at the entrance to Juan de Fuca Strait (Figure 4). In most instances, after being confined near the vessel, salmon were permitted to escape over the submerged net corks. The number of fish in the net were estimated by visual counts for each purse-seine set. These indicated the strength or weakness of the runs several days in advance of their arrival in U.S. Convention Waters and the mouth of the Fraser River. This was very helpful in 1961 and 1962 when, for the first time, extended closures were necessary due to poor pink and Adams River sockeye salmon runs (Roos 1969). Although this test fishing was valuable, the information was inconsistent and difficult to interpret due to vessel and crew differences and variable migratory routes in Juan de Fuca Strait. Procedures were improved by standardizing number of sets and location and duration of open net time.

Test fishing also provided information on fish abundance during closed periods—a time when anxiety levels peak within the fishing community. Industry was assured that the Commission was monitoring the status of the stocks and that excessive numbers of undetected fish were not moving past the fishing fleets.

Because of vessel and crew expenses, purse seine test fishing was costly, and many argued that sufficient fish should be retained to pay daily expenses. The Commission was reluctant to do this except on occasions when information needs required it.

Purse seine test fishing in Area 20 was restricted to when Adams River sockeye were migrating through the area. The test fishing also provided information in odd-numbered years when Fraser River pink salmon were returning. Seine vessels monitored the runs in the Strait for about one month during August and early September.

In the mid-1960s, the Commission began gillnet test fishing in Area 20. The costs of this program were considerably less than for chartered purse seines. Experienced fishermen were hired but it was soon apparent that individual fishing procedures carried out by fishermen had to be standardized. Commission staff, always aboard test-fishing boats, monitored a relatively uniform fishing procedure to standardize results from night-to-night and from season-to-season. Fishing experiments involv-ing two test-fishing boats indicated that one vessel could give adequate information. Because of the different migration characteristics of early and mid-summer sockeye races (early Stuart and Chilko) compared with fall races, such as the lower Adams, test fishing with gillnets was more effective and results more consistent on the earlier migrating stocks.

For 20 years the Commission received timely and useful data from Canadian Area 20 gillnet test fishing. The data were compared to known numbers of fish reaching inside areas and were sufficiently accurate that daily estimates of sockeye entering the Strait could be calculated. Gillnet test fishing for pink salmon was never productive, mainly because of significant changes in the size of individual pink salmon from year to year and the poor gilling characteristics of pink salmon because they are slower swimmers and do not "charge" into the net like sockeye.

On many occasions staff recommendations and Commission decisions at in-season regulatory meetings were based to a great extent on the results of test fishing the previous night in Area 20. Frequently, scale samples from the test-fish catch were transported to the laboratory by plane, processed and analyzed within hours.

Although information from gillnet test fishing performed in the Neah Bay area (Figure 4) of United States waters was generally inconsistent, it was useful on those occasions, when sockeye migration had shifted to United States waters; thus helping to explain low catches in Area 20.

Since gillnet test fishermen were paid with a percentage of the catch, there was incentive to fish each night as effectively as possible in the approved area. In the early years of gillnet test fishing there was a standard contract fee with the fishermen. Fairly substantial incomes for test fishermen created resentment and animosity; however, Commission staff wished to maximize consistency in effort and results by using the same fishermen from year to year. Eventually, agreement was reached on a maximum number of years of employment. This provided that test fishing would be shared with other qualified fishermen. Test fishermen were also asked not to fish during commercial fishery openings so as to prevent what was called double-dipping.

Gillnet test fishing was subsequently used in the Salmon Banks area, off the southern tip of San Juan Island in United States waters. Because of the wider area available for fish migration in northern Puget Sound, results were more difficult to interpret and evaluate. This information; however, was useful to confirm run-sizes determined three to four days earlier by Area 20 test fishing. Gillnet test fishing was also performed in Area 17 in the Canadian waters of the Strait of Georgia north of the Fraser River mouth to assess escapement from Johnstone Strait. The consistency of these results was not encouraging. Although in the 1940s the Commission's legal counsel advised that under the Convention it was permissible to gather data in Johnstone Strait (non-Convention Waters), the Commission refrained from test fishing there because of the political sensitivity of the issue.

In many years purse seine test fishing in the Salmon Banks and Point Roberts areas was conducted on late-season sockeye runs and on pink salmon runs, but no systematic procedure or consistent results could be established from year to year. Nevertheless, it did provide a general impression of the migration status and relative abundance of the stocks.

Observations and counts of sockeye passing through reef net gear in the Lummi Island area of United States waters were made during closed periods for several years, and since reef net gear is stationary, the results were entirely dependent on the location of fish migration paths. Wind, tide, etc., affected the migratory paths and behavior of the sockeye. Reef net observations—noting the abundance of salmon migrating through "open" reef nets proved to be of limited value and were often misleading.

Test fishing by troll fishing in the high seas west of Vancouver Island was used occasionally, but results were not used to a great extent in decision-making by the Commission. In the early years, test fishing catch information was not released to the public but this was subsequently reconsidered and a daily hot-line was established which provided regulatory announcements and test fishing results.

ACOUSTICS

In 1972 the Commission started to use echo-sounding techniques to reduce variability in escapement estimates from gillnet test fishing in the Fraser River These, with frequent modifications, were employed each season from 1974 through 1985. The operation consisted of a 20 kHz echo-sounder mounted on a boat that ran repeatedly on consistent transects back and forth across the river at Mission for 16 to 24 hours daily during the major migration periods. Target abundance data were later examined in the office and periodic stationary gear sounding was used to give rate of migration information. About 0.5 percent of the total migration was detected with acoustic gear. Also, echo sounding was done during the commercial fishery to establish a base-line count of resident non-migratory species and smaller fish not taken in the fishery. The efficiency of the Fraser River gillnet fleet inside the river, in most years, approached 100 percent.

Echo sounding was used in the Fraser River at Mission and at the Cottonwood test fishing site near the mouth of the river. The latter was eventually discontinued because of the frequent milling of fish and migration back downriver.

Test fishing provided scales for racial analysis as well as confirming findings from echo sounding. Unfortunately, both echo sounding and test fishing had deficiencies. Echo sounding could not distinguish between different species of fish or other salmonids, jack sockeye could not be differentiated from adult sockeye, and there were times when the downstream movement of sockeye at Mission may not have been identified.

On occasion, discrepancies of up to 15 percent were found between sockeye estimates from echo sounding and the numbers of fish known to have gone upstream (Indian catches above Mission, plus enumerated escapement). Other sources of unaccounted removals could have been poaching, unreported or inaccurate Indian fishery catch, natural mortality and drop-outs from the Indian gillnets. This does not preclude possible errors in enumeration of certain escapements. Even with such discrepancies reasonably accurate escapement estimates were obtained. Echo sounding techniques were not applicable for enumeration of pink salmon escapement.

The Commission also conducted exploratory acoustical investigations for daily estimates of sockeye and pink salmon abundance in outside areas. Efforts were made in Area 20 on migrating sockeye in the 1970s and in the 1980s, off the mouth of the Fraser River on delaying late-run populations. Many difficulties were encountered and the efforts did not result in useful programs.

. . . .

OPTIMUM ESCAPEMENT DETERMINATIONS

Another important challenge facing the Commission was to determine optimum escapement numbers for each race. Historical records such as the Hudson Bay Journals, the Reports of the British Columbia Commissioner of Fisheries and other Canadian government documents, were examined to establish historical cyclical abundances and approximations of spawning escapements and the abundance of returning adults. Since the Commission itself began enumeration of many individual sockeye races in 1938, more accurate data on spawners and returns began accumulating in 1942. These escapement-return data were subsequently augmented by extensive surveys of the physical nature of major spawning grounds (Pyper and Vernon 1958). The wetted perimeter of these spawning grounds was measured and the estimated suitabilities were classified by quality of substrate. Information was available to estimate the capacity of each spawning ground. Observations of spawning sockeye yielded estimates of the spawning area required by each female sockeye.

Through the years, the Commission developed estimates of optimum annual abundance for escapement of the major stocks in the watershed. For example, for the Chilko River population, a net escapement of 450,000 sockeye in the dominant cycle-year was desirable (optimum), while the subdominant cycle-year escapement goal was 300,000. Although a statistical examination of the escapement-return relationship was made, enough information was not available to establish egg-to-fry and fry-to-smolt survival rates or marine survival rates for many other races.

There are factors other than number of spawners which affect adult returns. Timing of escapements and arrivals at spawning grounds, such as for the Adams River race, must be incorporated into the analysis. Late-timed escapements to the Adams River were seldom productive even if an optimum number of spawners reached the river. The necessity to obtain the proper number of spawners at the right time is important to stock productivity and to the understanding of changes in productivity (Royal 1953).

There may be at least a two-fold difference in egg-to-fry survival rates from year-to-year for dominant-year populations. Fry-to-smolt survival rates may also change by that much. Smolt survival from the time they leave the nursery lake, go to sea and return may vary from one percent to twenty-two percent. Even with the same abundance in a dominant-year escapement, considerable variation in the rate of return can be expected. In the long run, the optimum number of spawners for each particular spawning area was the most important

October 23, 1950. Engineering crews surveyed the areas, widths, velocities and gravel composition of all the major sockeye and pink salmon spawning areas. The lower Adams River survey crew consisted of George McNaughton, Jim Jensen, Bill Ellis and Jim Pyper (left to right).

October 5, 1950. George McNaughton and Ross Stewart measuring depths and velocities on the lower Adams River spawning area.

December 11, 1951. George McNaughton, Jim Jensen and Roy Hamilton using a suction pump for measuring egg deposition in relation to spawner density and gravel quality on the lower Adams River.

August 25, 1945. The
confluence of the Chilcotin
and Fraser Rivers. Nearly all of
the soil is fine-grained silty
material that is easily eroded
by run-off and river action.

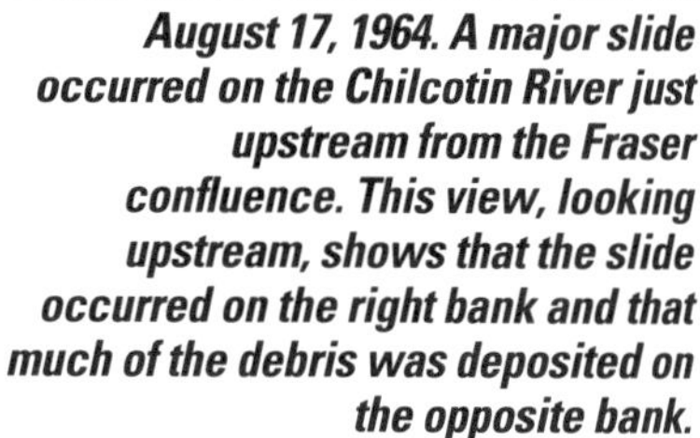

August 17, 1964. A major slide
occurred on the Chilcotin River just
upstream from the Fraser
confluence. This view, looking
upstream, shows that the slide
occurred on the right bank and that
much of the debris was deposited on
the opposite bank.

August 25, 1964. Looking downstream at the slide origin on the
steep silty bank. The slide, which
occurred at the beginning of the
Chilko sockeye run, dammed the
river for several hours and delayed
the sockeye for an estimated
period of six days.

management goal each season. For the most part this could be controlled.

As might be expected, subsequent examinations of escapement-return data for such major races as Chilko and Adams River did indeed show considerable variation. When recognized and understood these did not deter the Commission from using the optimum number of spawners as a management goal.

Under normal conditions, the arrival of fish of a particular race to their spawning grounds is spread out over at least 30 days. The 1946 Adams River run was delayed at Hells Gate and in the Thompson River rapids for about 10 days because of low water levels (IPSFC 1965a). This delay, associated with a late-timed migration up the Fraser River, is believed to be responsible for the failure of the 1950 Adams River run. The extreme lateness of the 1958 Adams River migration up the Fraser River and extended duration of migration upriver related in part to abnormally large dispersion in salt water (Gilhousen 1960, IPSFC 1965a), were probably important factors contributing to the poor return in 1962. On the other hand, the 1950 Adams escapement* was on time and though only 1.2 million spawners reached the river, this small escapement produced a large return of about 9,000,000 fish in 1954. In 1954 the escapement number (2,000,000) was near the optimum, and because of the unique escapement pattern to leave the Strait of Georgia and swarm through the lower river in only two and one-half days, the duration of arrival time at the spawning grounds was much less than normal (IPSFC 1965a). This short migration spread of a normally-timed run produced a modern-day record return of over 15,000,000 fish in 1958.

In 1964, the Chilko escapement was completely blocked when a huge slide dammed the Chilcotin River. When upriver passage was possible following a natural cut-through by the

river, the arrival duration of the escapement was very short, with nearly all arriving within a short time period. The rate of return for the 1968 run was exceptional—probably due to the compactness of the 1964 escapement and the high efficiency of egg fertilization and deposition as well as high egg to fry survival.

Prevailing management philosophy had been that escapement should be provided from all segments (timing) of each run, with the majority of escapement from the peak; however, it was found that exceptional-to-maximum production was achieved with compact, short-duration arrival.

In general, management plans took into consideration the environmental characteristics prevalent during each year and the requirements of the race (and the cycle-year of each race) that were the major components of that year's run. For example, on the Chilko River dominant cycle (1984), if the run was early, greater fishing effort was exerted on the early portion of the run to reduce pre-spawning mortalities. However, in this specific case, the abundance, timing and needs of the spawning population at the south end of Chilko Lake had to be considered to provide for maximum spawning success. If river temperatures were high during an early-timed migration, even more effort was made to reduce the early portion of the escapement. If the run time was normal and if temperatures and water levels in the river were average, the bulk of the escapement came from the peak of the run.

Another example is the dominant Quesnel cycle (Horsefly). If the runs were early and if river temperatures were high, greater harvests from the early portion of the run reduced escapements and less fish were lost on the migration route or on the spawning grounds. On the other hand, if the runs were on normal schedule, escapement was spread over all segments of the run—early migrants to McKinley Creek and the upper Horsefly River, later migrants in the lower Horsefly River, and the latest component in the Mitchell River. In

---

* Adams River denotes Adams River/Little River spawning areas.

the Quesnel system, the volume and timing of sockeye escapement was tailored to current conditions on migration routes and in spawning areas.

Previous observations of the dominant lower Adams River race indicated the following management strategy: for late-timed runs (in the outside fisheries) which occur after summers of low Fraser River flows (this produces extended durations of upstream migration, Gilhousen 1960), the early portion of the run was to be protected for escapement, and intense fishing effort applied to the latter portion of the run. With a normal-timed run, effort was made to reduce fish escaping during the tail-end of the run. Experience showed these as unproductive spawners, likely to dislodge eggs deposited by previous peak-of-the-run spawners.

In summary, current in-year management objectives included environmental factors, cycle-year requirements, and escapement needs. By the late 1970s timing of the summer runs was either normal or later than average, therefore this management concern was reduced.

Greater emphasis on compact escapements rather than on escapements spread over the entire arrival period was probably warranted. In the Fraser system, the extreme early and late sockeye escapements were far less productive spawners than those near the peak of the run.

The cyclic relationship of dominant, sub-dominant and off-year returns is important too (Ward and Larkin 1964). The maintenance of these relationships is critical to the sustained optimum production of the race. For the Adams and Horsefly River runs, the subdominant run occurs the year following the dominant run whereas in the Chilko population, the subdominant run occurs in the year preceding the dominant return. Optimum escapement of the dominant lower Adams River race is about 2,000,000 fish whereas the subdominant run should have an escapement of about 700,000 fish and considerably less in the other two off-years.

Ricker (1987) characterized the 1913-1914 blockages and slide as a "fortuitous beneficial effect" in that some of the upriver dominant year races shifted to other cycle years off the 1913 cycle. These changes resulted in more even run distribution amongst the four cycle years and eliminated conflict with odd-year pink salmon runs, however, the possible effect these changes had on total productivity has not been evaluated.

Plankton (fry and fingerling food) abundance in major nursery lakes has also been used to evaluate year-class productivity. The size of the lake-rearing area, its zooplankton production and competition and predator species were considered in determining optimum escapements and maximum productions for each race. In addition, where data are available, knowledge of the survival rates from fry to smolts and from smolts to adults is important in understanding the reproductive and survival successes of a particular race, from escapement to return.

. . . .

PREDICTIONS

Pre-season

When the Commission began to recommend sockeye fishing regulations, industry was also interested in the expectations of run-size for each season. In 1955, the Commission held its first annual meeting with the industry, and run-size predictions continued to be presented at Commission annual meetings each December thereafter.

Returns were first predicted on escapement-return relationships from previous cycles. In subsequent years, smolt counts and jacks became useful indicators of adult return abundance. Analysis of the marine growth patterns on scales in relation to the ratio of three-year-old jacks in one year to the numbers of four-year-old sockeye the following year ($4_2/3_2$) were also utilized. Environmental factors were found to be related to smolt abundance and timing, and were used to estimate marine survival rates. Since sophisticated data were not available for all stocks, the more conventional analysis of escapement-return projections were used for the less-abundant races. In other cases, air temperatures in the interior of British Columbia during egg-incubation periods were correlated with smolt production and adult returns. High flows in the Fraser River during smolt out-migration were found to be positively related to smolt survival. For pink salmon, the relationship of discharge height of the Fraser River during peak spawning, sea surface water temperatures in Georgia Strait, salinities, coastal river discharges and other factors possibly related to pink salmon production were all used to predict the abundance of returning runs. Visual surveys of relative abundance of pink salmon fingerlings in the Gulf Islands areas in Georgia Strait and northern Puget Sound were also conducted.

A summary of predictions for 1970-1985 is shown in Table 2. In most cases, returns were greater than predicted, occurring about 69 percent of the time. Overall, sockeye returns were about 27 percent greater each year than predicted and for pink salmon about 23 percent more fish returned than were predicted.

Forecasting forced Commission scientific staff to examine all relevant data available for each race. Important management options were recommended, decisions made and implemented to provide a fishery and to enhance the stocks based on these analyses.

IN-SEASON ADJUSTMENTS

While pre-season predictions were usually accurate enough for industry needs, greater in-season reliability to update estimates of total run size and the proportions of major races in the run was also necessary. The abundance of races varied substantially within years and between years, creating a complex management challenge.

Previous cycle-year events and numbers and stock catch-per-unit-effort data for comparison with the current year's data was also needed, particularly the comparison of data at or near the peak of the runs. Timing—the

TABLE 2. *Predictions and returns of Fraser River sockeye and pink salmon, 1970-1985 (millions).*

	SOCKEYE		PINKS	
	PREDICTION	RETURN	PREDICTION	RETURN
1970	6.800	6.164		
1971	4.800	7.696	1.500-5.000	9.768
1972	3.000	3.708		
1973	5.200	6.878	4.000-6.000	6.789
1974	5.000	8.616		
1975	5.500	3.684	5.000-6.000	4.894
1976	3.000	4.341		
1977	7.000	5.835	4.400	8.249
1978	6.500	9.432		
1979	3.800	6.426	15.000	14.404
1980	3.200	3.133		
1981	6.000	7.743	9.000	18.685
1982	10.000	13.989		
1983	6.500	5.242	20.000	15.346
1984	3.200	5.918		
1985	9.000	13.881	16.000	18.864

peak-abundance days at a standard site—of the current year's return was critical.

Because of changes in numbers, deployment and efficiency of the various fishing gears from year to year and the migratory habits of the fish, data analysis covering many years did not always produce consistent patterns. The most useful information obtained was total catch by area and average catch per purse seine, adjusted for competition between different numbers of vessels fishing.

The first information on relative abundance of Fraser River sockeye and pink salmon runs was taken from troll fishery catches along the west coast of Vancouver Island. In more recent years, with exceptionally high rates of diversion to the north, purse seine catches at Queen Charlotte Island helped identify migration paths and timing of Fraser stocks when used with other early-season information on catches closer to or in the Convention Water fisheries.

Troll catch information on sockeye was useful only in a general way because of the variabilities in that fishery, but it did provide early information for in-season confirmation or adjustment of pre-season predictions. In some years trollers would target on sockeye. This was important to the analysis, but so was the availability of the fish to the gear. The landfall of sockeye along Vancouver Island—the point where the schools came close to shore—was variable. This had a pronounced effect on catch. The primary effort of the troll fishery was directed toward chinook and coho salmon, but this was affected in later years by the abundance, availability and increased market value of sockeye and pink salmon.

The Canadian purse seine fishery in Area 20 at the entrance to Juan de Fuca Strait developed in the mid-1940s. Catch information gave reliable information useful for in-season analysis of the size of the Adams River sockeye and pink salmon runs (Figures 10,11). But predictions were restricted to the major races, such as the Chilko and Adams River sockeye populations and the pink salmon run.

Important but missing information, especially in later years when northern diversions of sockeye and pink salmon through Johnstone Strait were high, was the number and percentage of fish migrating through this northern approach. Catches in this fishery outside Convention Waters had to be analyzed for computing total run size. The percentage of Fraser River sockeye migrating toward the Fraser River through Johnstone Strait ranged from a low of about one percent in 1954 to a high of about eighty percent in 1983 (IPSFC 1984).

By the 1980s the Commission was using computer programs to produce run size estimates and escapements. Estimates of escapements into the lower Strait of Georgia from the Johnstone Strait fisheries were needed because those fish would be subjected to further fishing in the Strait and in the river. Further discussion of Commission management procedures can be found elsewhere (Woodey 1987).

Management of Fraser River sockeye populations by individual genetic races was developed and perfected by the Commission. This management philosophy was an important component of the rehabilitation of the runs in combination with the contribution of fishways and commercial fishing closures.

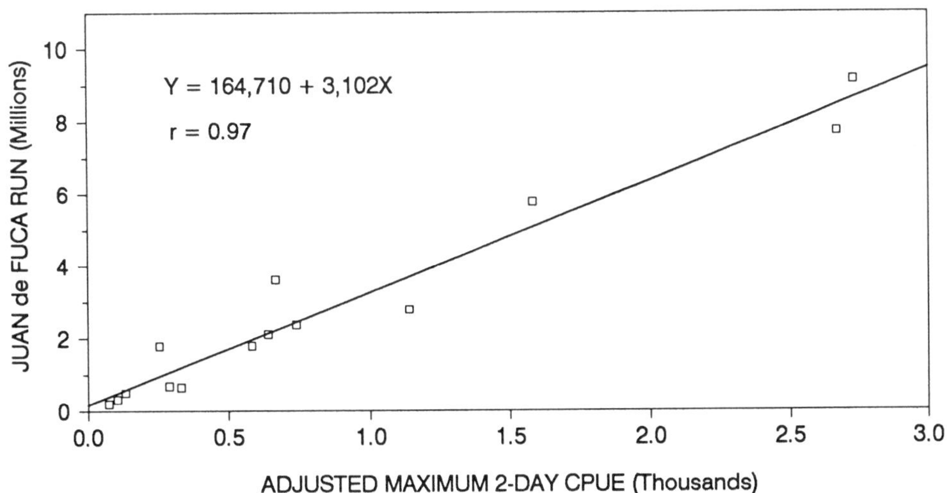

Figure 10. Abundance of Adams River sockeye salmon migrating via Juan de Fuca Strait in relation to the maximum 2-day catch per purse seine in Area 20. Updated from Woodey 1987.

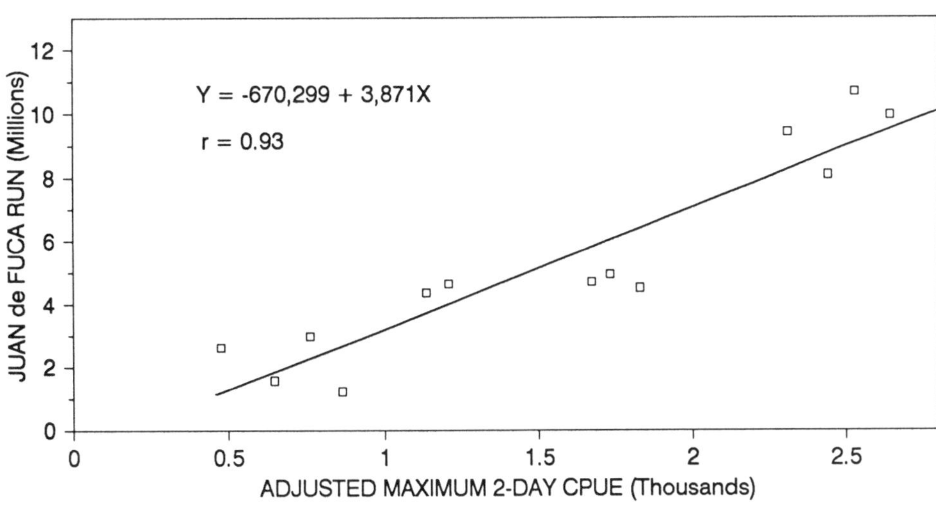

Figure 11. Abundance of pink salmon migrating via Juan de Fuca Strait in relation to the adjusted maximum 2-day catch per purse seine in Area 20.

Pink Salmon Protocol

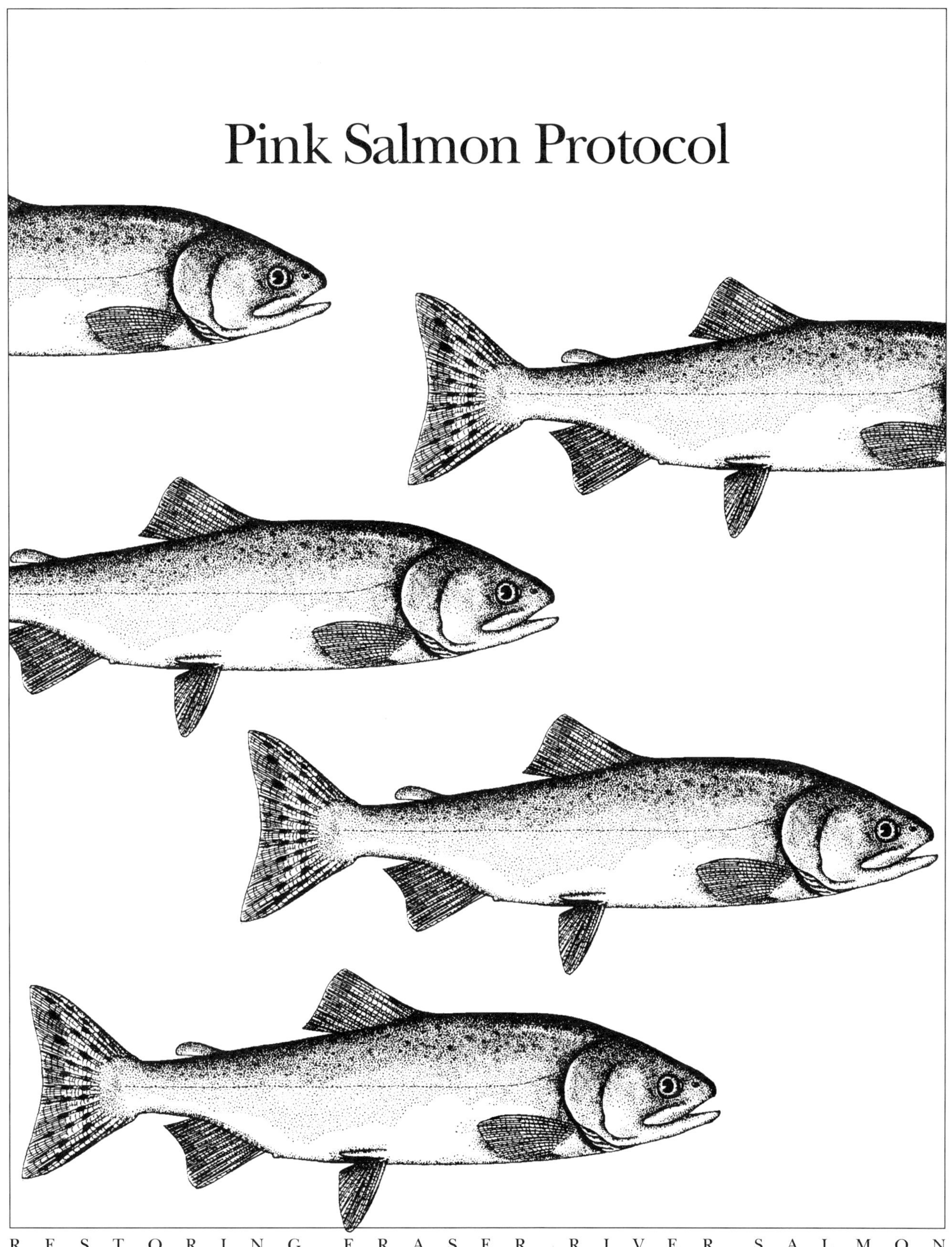

By the 1950s, the governments and fishing industries of both countries were confident that the Commission's programs and accomplishments were being successfully applied to the restoration of the Fraser River sockeye stocks. The Fraser River pink salmon runs at that time were considerably diminished compared to the abundance of the runs before the Hells Gate slide and there was a general feeling that the Commission's responsibilities should be broadened to cover this stock. Pink salmon, however, were not depleted to the extent sockeye were in the 1930s.

Preliminary discussions on the increasing competitiveness for pink salmon were held by representatives of the U.S. and Canada in the mid-1950s. The earlier informal cooperative measures between the Canadian Department of Fisheries and the Washington State Department of Fisheries were rendered inadequate by increased fishing pressure in each country.

By early 1954, the Minister of Fisheries for Canada, James Sinclair, was actively promoting Canada's interest in establishing a pink salmon treaty. He pointed out the similarity between the sockeye situation 20 years earlier, when the Americans balked at a new treaty because they were catching the bulk of the fish (65 percent). Until Canada caught the majority (in 1936), the Americans did not agree to a treaty. Sinclair (1954) stated:

> Today an exactly similar situation prevails with the pink salmon. Until the last year or two, the Americans have had first crack at them as they came through Puget Sound and the American fishermen have got from 80 to 90 percent of the fish.

The actual landings in Convention Waters for U.S. fishermen from 1945-1951 was 70 percent. Sinclair pointed out that by 1953 Canadian fishermen were correcting the situation by "going out to the western approaches to Juan de Fuca Strait and catching the fish in Canadian waters before they enter American waters." By 1953, Canada was catching 46 percent of the catch and in 1955 Canadian fishermen had taken 47 percent of the Convention pink salmon catch.

There is no doubt of Canada's intent with respect to the clout intended from the Straits fishery. Sinclair said at the same meeting:

> As Minister of Fisheries, I want now to state quite deliberately that I hope our Canadian fishing fleet goes out and catches a much greater number of pinks off the West Coast, for once we get the bulk of this run, I think we will find, as with the sockeye, that our American friends will realize the value of an international commission to conserve the fisheries and divide the catch equally between the two nations ... To correct the situation, I promise you that you will be permitted next year ... to use exactly the same gear in these waters as the Americans are using that year in Puget Sound ... I feel sure that this program will help speed an agreement on joint measures for pink salmon.

A meeting with fishermen of both countries in White Rock, B.C., in May 1955 resulted in strong endorsement of the Sockeye Commission and a desire for inclusion of pinks in an international treaty.

In May 1955, the Fisheries Association of British Columbia went on record for the first time in support of a pink salmon treaty. There had been some discussion about placing pink salmon under the International North Pacific Fisheries Commission but it was not given serious consideration. It appeared to be understood that the pink salmon stocks did not face the same crisis as the sockeye had.

Another important aspect of improving the pink runs to the Fraser, was the additional leverage this had in discouraging hydroelectric development on the river. By the mid-1950s, interest in Fraser River hydroelectric power had grown considerably but due to the importance of the sockeye fishery and the effective conservation measures put in place by the Commission, the fishing industry had consid-

The Pink Salmon Protocol was signed in Ottawa on July 3, 1957 by His Excellency Livingston T. Merchant, United States Ambassador to Canada (on the left) and the Honorable E. Davie Fulton, Acting Secretary of State for External Affairs of Canada. Standing, left to right, are George R. Clarke, Deputy Minister of Fisheries of Canada, and the Honorable J. Angus MacLean, Minister of Fisheries of Canada.
Courtesy Pacific Fisherman.

erable influence in discouraging such development. With Fraser River pink salmon added to international control and management, the fishing industry argument for continued safeguards against encroachment of the power interests was that much more powerful and effective.

Sinclair again did not attempt to hide the Canadian government's intent. At the U.S. National Institute of Fisheries meeting in New Orleans, Louisiana, he said: "Our Canadian fleet will make a maximum effort in this fishery out on the high seas" (Sinclair 1955).

This deliberate effort by Canada to bring about a pink salmon agreement was the same kind of effort that would be seen 20 years later in Canadian non-Convention Waters, again, with the purpose of forcing the Americans into a new treaty.

It took only a few years to complete the Pink Salmon Protocol. On May 23, 1957, the Committee on Foreign Relations unanimously approved the signed December 28 Protocol, without amendment, and in its report to the Senate recommended ratification. The United States and Canada agreed on December 28, 1956, to amend the Sockeye Salmon Fisheries Convention to include Fraser River pink salmon stocks. This agreement became effective with an exchange of ratifications on July 3, 1957. The agreement was commonly known as the Pink Salmon Protocol (Appendix A).

The provisions of the new agreement included that regulations to manage pink salmon fishing be adopted immediately, instead of waiting the eight-year period stipulated in the 1930 Sockeye Convention.

Article IV of the 1930 Convention required Commission orders to be uniform within any one of three designated sub-areas. Article II of the Pink Salmon Protocol amended that provision so it would permit orders to be issued for any portion of the Convention area. This eliminated a problem in 1956 when the Canadian government refused to approve the Commission's regulatory recommendations for that year because the Area 20 fishery was to be closed when the Fraser River would be open to fishing (Sinclair 1956). The Commission changed its recommendations to conform to the 1930 Convention. Article III of the Protocol amended Article VI of the Convention Provided that, except for orders adjusting the closing and opening of fishing periods and areas and for in-season emergencies, regulations made by the Commission shall be subject to the approval of the two governments. This meant that pre-season regulations had to be approved by the governments. No such approval was required by the 1930 agreement. Under Article IV of the protocol, the Commission would be required to regulate the sockeye and pink salmon fisheries to allow, as far as practicable, an equal portion of the catch to be taken each year by the fishermen of each party. The Advisory Committee to the Commission was increased to six members from five by Article V of the Protocol, which also added a fifth branch of the fishing industry (processing) to the previous four (purse seine, gillnet, troll and sport fishing), required by the 1930 Convention.

In the United States Senate report it is stated that:

> The International Pacific Salmon Fisheries Commission, which has managed the sockeye salmon fisheries in the same area since 1937 to the great benefit of both countries, has earned the confidence of both the United States and Canadian fishing industries. (Committee on Foreign Relations 1957)

The report further stated that:

> The protocol (sic) has the support of all

elements of the Northwest fishing industry of the State of Washington, and of the public affected thereby.

Two Provisions included in the Pink Salmon Protocol were as follows:

Provision I

1. The Parties shall conduct a coordinated investigation of pink salmon stocks which enter the waters described in Article I of the Convention for the purpose of determining the migratory movements of such stocks. That part of the investigation to be carried out in the waters described in Article I of the Convention shall be carried out by the Commission.

2. The sharing cost shall not apply to the above investigations except for the part performed by the Commission.

3. The Parties shall meet in the seventh year after entry into force of this Protocol to examine the results of the above investigations.

Provision II

1. The Commission has the right to record information on stocks of salmon other than sockeye or pinks incidental to its activities on sockeye and pink salmon.

These two provisions were Articles VI and VII of the Pink Salmon Protocol.

One of the interesting, and perhaps subtle, omissions of the 1957 Pink Salmon Protocol was the change in Article VII. The first sentence of Article VII of the 1930 Convention stated: "that they should share equally in the fishery." This was replaced by Article IV of the Pink Salmon Protocol which did not contain the first sentence contained in Article VII of the 1930 Convention. The second sentence of Article VII of the 1930 Convention was changed somewhat in the Pink Salmon Protocol as follows:

> The Commission shall regulate the fisheries for sockeye and for pink salmon with a view to allowing, as nearly as practicable, an equal portion of such

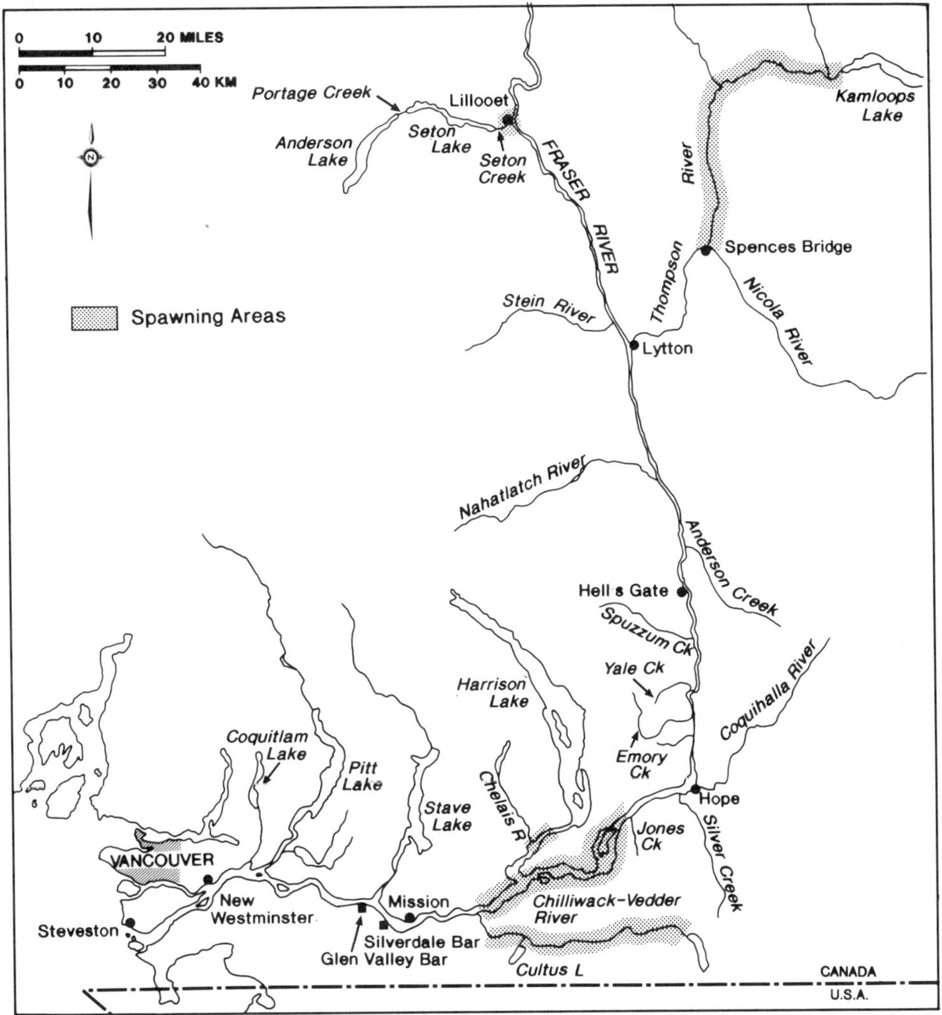

Figure 12. Principal pink salmon spawning areas of the Fraser River system. From Ward 1959.

sockeye salmon as may be caught each year and an equal portion of such pink salmon as may be caught each year to be taken by the fishermen of each party.

What, if any, effect this change in Article VII had on the Parties' views or obligations with respect to sharing equally in the fishery after 1957 and the large catches made outside the Convention area thereafter is not known.

STATUS OF THE STOCKS

The Fraser River pink salmon run consists of two major components, the early and late runs. Part of the early-run pinks spawn above Hells Gate in the Seton Creek area and in the Thompson River. Another portion spawns in the main Fraser below Hope. Late-run pink salmon spawn in the Chilliwack-Vedder and Harrison Rivers (Figure 12). In the Report of the Provincial Fisheries Department for 1913 it is stated:

> Notwithstanding the fact that this was a humpback (pink) salmon year, and that former years of their runs, millions of humpback spawned in Seton Creek and the Thompson River, not a single one of that species reached these streams or any other stream north of the Fraser River canyon this year. (Babcock 1914a)

Although it had generally been believed that no pink salmon appeared above Hells Gate in any cycle year following 1913 until the Hells Gate fishways became operative in 1945, this apparently was not the case as reported earlier. In most years, no pink salmon were able to migrate past Hells Gate until the fishways were completed. Starting in 1945, pinks began to appear on upriver spawning grounds on a regular basis and by 1955, thousands were reported in both Seton Creek and the Thompson River by Canadian federal fisheries officers.

In the case of sockeye, fish which did not get past Hells Gate or that were excessively delayed were probably not productive spawners; however, many early-run pinks blocked at Hells Gate could have moved back downriver and spawned in the main Fraser River spawning grounds below Hope. Einarsen (1930) documented pink salmon spawning in these areas on October 10, 1929. How effectively they spawned is not known. This area had the potential to accommodate large numbers and from 1981-1985 there were from 2.3 to 5.3 million spawners using these grounds.

Pink salmon are weaker swimmers than sockeye and Hells Gate conditions had a greater impact on them (Williams *et al.* 1989a). Whether or not the early-run spawned in the main Fraser River prior to the Hells Gate slide or spawned below Hells Gate as a result of the block was not definitely established.

Catch data suggested Hells Gate had a strong impact on the abundance of pink salmon (Rounsefell and Kelez 1938). Possibly some of the early-run spawning stock had not been well established in the main Fraser River during this period and that the progeny of the fish forced to spawn in the area were not initially well adapted.

Catches of Fraser pink salmon in the mid-1930s and early 1940s in the Convention area were one half as large as were the catches in the mid-1940s to 1957 (Vernon 1958). Improved migration through the fishways provided access to spawning grounds above Hells Gate. The average catch of Fraser pinks was about 8,000,000 fish each year from 1945-1955. These data do not include catches in Johnstone Strait or the escapement. Thus, the average Fraser River pink salmon run in the 12 years prior to the Commission's involvement with pinks (1957) was likely about 11,000,000 fish. This suggests that Fraser pink salmon runs had declined after 1913 to a much lower level of abundance until about the mid-1940s. The runs increased in the decade before the Commission assumed control in 1957. In 1957, the first year of escapement enumeration, almost half of the total escapement was composed of early run pink salmon spawning in the main Fraser River.

It was clear in 1957 that the spawning populations frequenting Seton Creek and Thompson River had been greatly reduced; the total returns to the Fraser River were about one-third of the estimated number prior to the Hells Gate slide. If fishing in both countries had continued without constraints, there would have been very serious stock depletion.

Fishways, more spawners above Hells Gate from 1945-1955, a shorter and less complex life span (two years), and the fact that segments of both early and late runs were still intact, all contributed to the potential for a more accelerated restoration of pinks.

RESEARCH PROGRAM

The authority to manage Fraser River pink salmon came about one month before their 1957 return to the Fraser. With twenty years of experience in scientific investigations on sockeye, the Commission was obliged to quickly undertake its new responsibilities under the Pink Salmon Protocol.

The following initial research program was approved by the Commission in 1957:

1. The establishment of a daily statistical record of pink salmon catches by areas

within Convention Waters.

2. Initiation of a coordinated program to study the migratory movements of pink salmon which enter Convention Waters. Undertaken by the Canadian Department of Fisheries, the Fisheries Research Board of Canada, the Washington Department of Fisheries and the Commission, the study determined rate, direction, and timing of movement, the daily catch by area, and escapement of individual pink salmon stocks entering Convention Waters.

3. The establishment of reliable methods to enumerate spawning pink salmon.

4. Recording of environmental factors during the freshwater phase of pink salmon.

5. Determination of the size and physical characteristics of the individual spawning areas.

6. Development of indices of freshwater survival in adult production from the major spawning areas.

7. A statistical analysis of the history of the fisheries to establish potential productivity of Fraser pink salmon stocks.

8. Determination of the factors affecting survival of young pink salmon in the estuarial areas adjacent to the mouth of the Fraser River.

The first seven of these were initiated in 1957, either in whole or in part. Enumeration of the escapement was developed in 1957 and was used for each pink salmon run (Ward 1959) and results clearly established that the timing of the upriver stocks (Seton Creek - Thompson River) was the same as for the main Fraser spawning population. It appears that the restoration of the Seton Creek and Thompson River populations that started in 1945 came from fish destined to the main Fraser spawning grounds.

The 1957 program established, for the first time, the difference in migration timing of the early and late run stocks in the Fraser system (Ward 1959).

In accordance with the Protocol, a coordinated pink salmon tagging program was conducted in 1959. It extended from Queen Charlotte Strait in the north, to Admiralty Inlet in Puget Sound (Vernon et al. 1964). Over 53,000 tags were applied to pink salmon. Recoveries totalled 31,018 tags in catches throughout the study area; an additional 1,700 were recovered on various pink salmon spawning grounds. This tagging program extended findings from previous tagging investigations in the same area (Pritchard 1932, Pritchard and DeLacy 1944, DeLacy and Neave 1948, Manzer 1958, Milne *et al.* 1959). The 1959 program was the most comprehensive, single year, marine salmon tagging program ever undertaken and involved the combined efforts of the Fisheries Research Board of Canada, the Washington Department of Fisheries and the International Pacific Salmon Fisheries Commission.

It established abundance and migration routes of pink salmon to the Fraser River system, to non-Fraser Canadian streams and to American streams (Figure 13, 14). Of the 10,264,000 fish entering the study area, 6,459,000 (63 percent) entered the Fraser, 2,533,000 (25 percent) were bound for other Canadian streams and 1,272,000 (12 percent) were returning to American streams (Vernon *et al.* 1964). Fraser fish made up the majority of the southern approach run through Juan de Fuca Strait and formed 64 percent of the total of all pink salmon runs to the study area. On the northern approach (Johnstone Strait), Fraser pinks were almost as numerous as Canadian non-Fraser fish. American pinks were more numerous than Canadian non-Fraser fish on the southern approach but contributed very little to the northern approach run. The timing of the study area pink salmon runs in 1959 via Juan de Fuca and Johnstone Straits is shown in Figures 15 and 16, respectively.

The study emphasized the need for escapement enumeration in many spawning areas. On the basis of the 1959 results, returns in

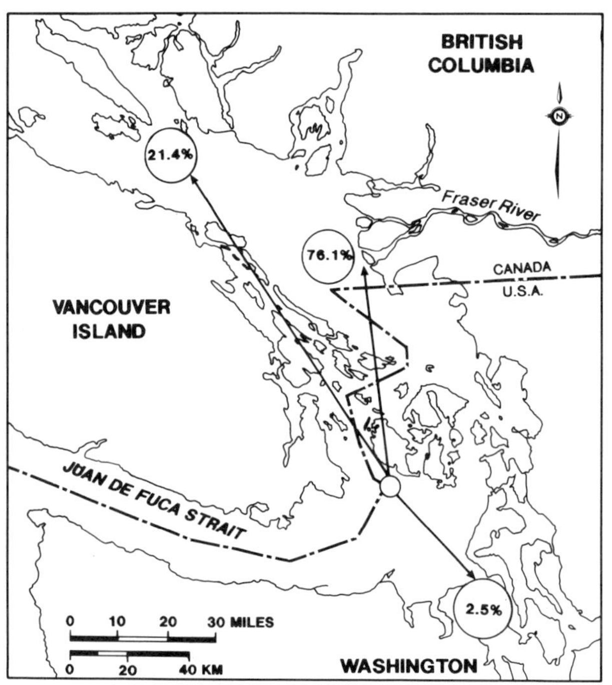

Figure 13. Stream recoveries from 1959 pink salmon tagging at Salmon Banks expressed as percentages of the total stream recoveries. From Vernon et. al., 1964.

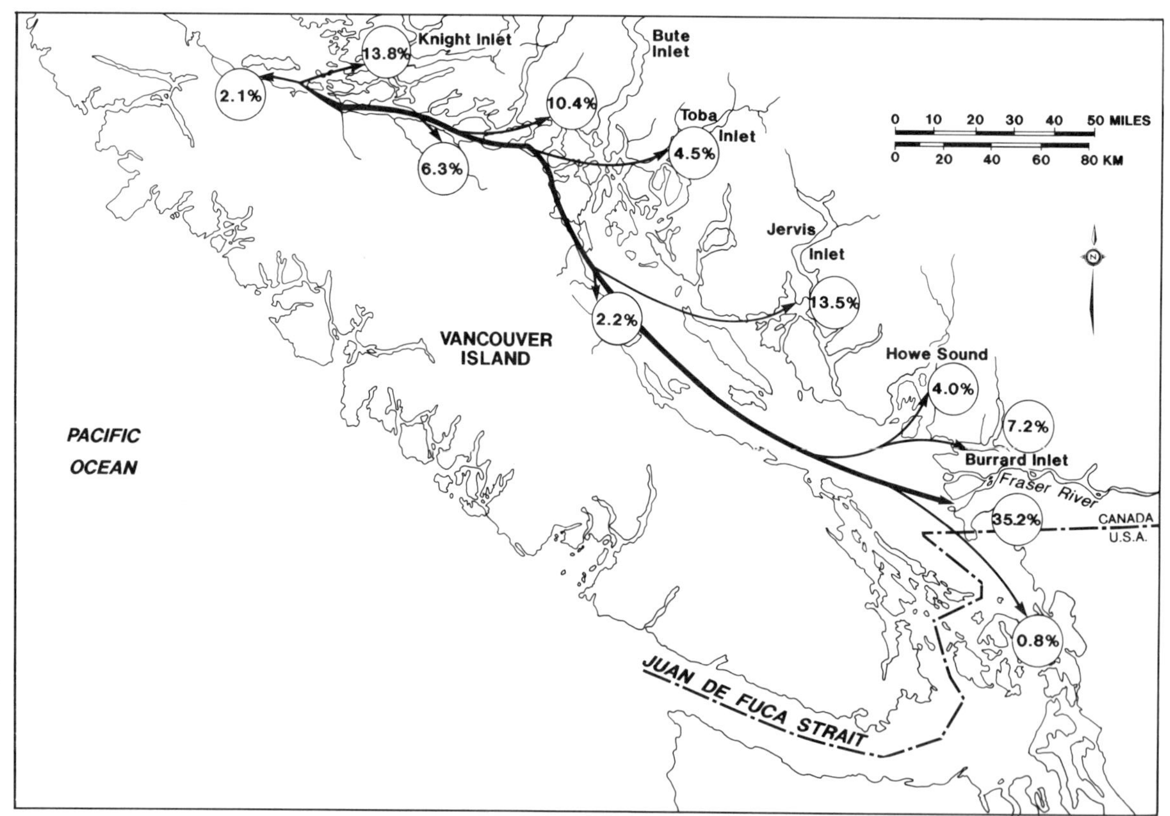

Figure 14. Stream recoveries from 1959 pink salmon tagging in Johnstone Strait expressed as percentages of the total stream recoveries. From Vernon et. al. 1964.

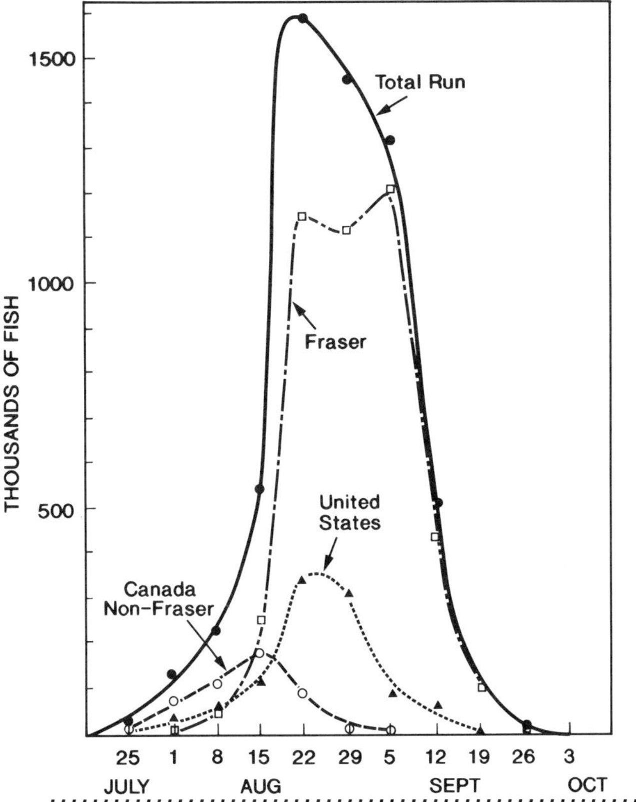

Figure 15. Passage times of southern approach pink salmon runs to Canada non-Fraser, Fraser and United States streams in 1959 as shown by their weekly abundance in the Juan de Fuca Strait fishing areas. From Vernon et. al. 1964.

subsequent years were analyzed and the various components were calculated for each year's return. This method was also extended to the 1961 run based on a limited tagging program in that year (Hourston *et al.* 1965).

ESCAPEMENT ENUMERATION

Managing salmon stocks required accurate assessment of stock size; both catch and escapement had to be included. Pink salmon produced in the Fraser River system were separated into five major spawning stocks and numerous smaller runs that spawn in small streams. One of the assessment difficulties was the lack of a method to determine the numbers of pink salmon spawning in the main Fraser River because not all early-run pink salmon remained to spawn in that area. To resolve this, a unique program was established. The entire escapement into the Fraser system was determined by a tag-recovery program performed in the lower reaches of the Fraser River near Glen Valley, about 30 miles east of Vancouver (Ward 1959). Tags were applied here throughout the migration. In other major spawning grounds such as Seton Creek, Thompson River, Chilliwack-Vedder River and Harrison River, a localized tag-recovery program similar to that established for sockeye was carried out. Several pink salmon populations in small, clear streams were enumerated by using a maximum live-count multiplied by an established index. This method had first been applied to sockeye spawning populations in the Stuart Lake system. The indices were determined from tag-recovery estimates in locations where live counts were also made. This assumed that the maximum live-count, plus total dead salmon recovered to

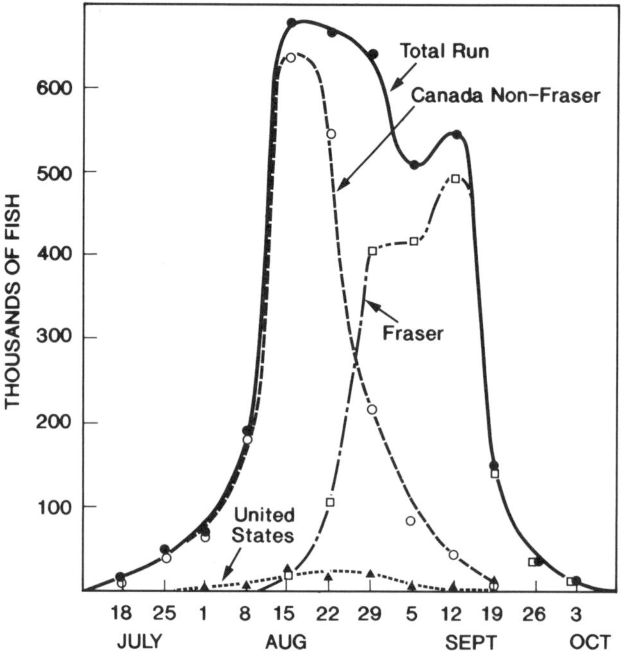

Figure 16. Passage times of northern approach pink salmon runs to Canada non-Fraser, Fraser, and United States streams in 1959 as shown by their weekly abundance in the Johnstone Strait fishing areas. From Vernon et. al. 1964.

that point, represented the same proportion of each escapement. The live-count method was generally used on populations with escapements below 10,000 fish.

Once the numbers of spawners were determined for all spawning areas (except the main Fraser) then escapement for the main Fraser area was obtained by subtracting the sum of all other populations from the total escapement estimate from Glen Valley tagging.

Total pink salmon escapement estimates into the Fraser River system were obtained for each odd-numbered year from 1957-1985 (Appendix E). The escapements varied from a low of 1,093,000 in 1961 to a high of 6,461,000 in 1985. Escapement above Hells Gate increased from a low of 103,000 in 1959 to a high of 1,828,000 in 1981.

The Commission found that pink salmon were spawning as far upriver as the lower reaches of the Quesnel River. This was the longest migration distance travelled by Fraser pinks. Escapements of 3,000 and 4,000 fish were observed there in 1965 and 1971. It is likely this was the maximum distance pinks could travel due to energy and migration time limits. It is interesting that from 1979-1985, the maximum escapement to the Quesnel River was 400 pink salmon, and in 1985, none were present. In all these years the runs were large (small fish) and in 1985 very difficult migration problems were encountered in the Fraser Canyon.

Pink salmon usually do not migrate through lakes to spawn; however, they have been observed in small numbers in Gates Creek, Birkenhead River, Adams River and Lower Shuswap River, involving migrations through at least two lakes.

The Commission carried out one pink salmon egg transplant experiment in 1983 using Harrison River donor stock planted in egg boxes by the Alouette River. The Alouette River had formerly contained a pink salmon

September 15, 1957. Commercial fishermen were employed to seine pink salmon on Glen Valley Bar in the lower Fraser River. The fish were released after being tagged and fish were examined in all parts of the watershed after spawning to recover the tags. The total Fraser River pink salmon escapement was then calculated from this information.

October 18, 1957. Although large numbers of pink salmon spawn in the Seton, Thompson, Harrison and Chilliwack-Vedder River, most spawn on gravel bars in the main stem of the Fraser River between Hope and Chilliwack. The depressions shown in the photo are false redds dug by maturing pink salmon as water levels dropped.

April 5, 1962. A commercial gillnetter was employed to operate floating traps at Mission in the lower Fraser River in order to capture pink salmon fry. The total population of pink fry was calculated from the data obtained, thus providing information for managing the commercial fishery during the next pink salmon run.

run but had been barren for many years. Over 3,000 fish returned to the Alouette in 1985.

FRY ENUMERATION

Pink salmon fry have a much simpler early life than sockeye fry. Many sockeye fry populations have complex migratory patterns, both downstream and upstream, to lakes where they feed for one or two years before migrating seaward, whereas pink fry migrate directly to sea upon emergence from the gravel. The number of migrants is important as a measure of the success or failure of a spawning population and of escapement strategies. Fry enumeration also provides useful information to evaluate environmental effects on spawning, egg incubation and early fry survival.

Establishing total numbers of downstream migrating fry was important for predicting adult returns and for evaluating those escapement levels that resulted in optimum production. The Commission needed information on the number of fry produced in the main Fraser River below Hope and estimates of fry production for each of the many spawning streams above.

Through the years, there were attempts to obtain egg-to-fry survival rates at Seton Creek, Thompson River, Vedder River and Harrison River. Because of the nature of these systems, this was a difficult and dangerous task during the spring freshets. While some information was obtained in the upriver spawning areas, more was obtained for the late run stocks. This was reported for the 1961 and 1963 brood year escapements into the Vedder and Harrison River systems (IPSFC 1965). Based on fry-per-spawner data, useful information was obtained on desired escapement levels into these rivers.

It was clear, therefore, that a method of fry enumeration—a method far different from any procedure previously used—was necessary. An estimate was obtained by sampling the migrating fry in the lower reaches of the main Fraser River. Not only did a large body of water have to be sampled, but the area was subject to changes in flow direction and velocity due to tidal fluctuations. The area chosen was near the city of Mission. To overcome fluctuations in river flow, a pontoon-mounted trap was moved upstream at 2.5 ft/sec for fifteen minute periods on set transects (Vernon 1966). The results were augmented by operating a similarly propelled gear fitted with conical nets which could be fished to depths of 13.5 ft. At greater depths, migrating fry were captured by stationary gear. From the data obtained on lateral, vertical and diurnal distribution of fry, the daily trap samples were weighted so that they represented fry numbers per unit volume of water during each 24-hour period. Daily fry migration numbers were then determined from the daily volume of seaward river flow estimated at discharge gauging stations at sites above tidal influence. Fry migration extends from early March to mid-May with most activity in April. The total numbers of fry calculated by this method had been obtained for each brood year escapement since 1961. Results are shown in Appendix O.

Fry population estimates led to the eventual determination of optimum escapement levels for individual stocks and for total escapement and were used in association with environmental indices to predict total Fraser River pink salmon returns.

ENVIRONMENT - PREDICTIONS

Factors affecting the abundance of pink salmon were examined (Vernon 1958), such as climatic effects on the freshwater environment and early marine residence, and any subsequent relationship to abundance of adult returns. In freshwater, water temperatures during winter months from December to February were inversely related to abundance of the early segment of returning runs (r = -.8027).*

---

* A perfect relationship exists when r = 1

Peak winter stream flow was inversely related to subsequent abundance of the late segment of the run (r = -.8345). Early summer seawater temperatures in Georgia Strait were inversely related to subsequent total abundance (r = -.8595). Salinity during the same period in Georgia Strait tended to be directly related to abundance. Multiple regression analysis of seawater temperature and salinity with total catch indicated a standard error in the order of 24 percent of the mean actual catch (R = 0.8920).

The preliminary analysis by E.H. Vernon set the stage for further investigations in other areas. However, as Vernon pointed out: "...use of the relationships isolated in forecasting catches must be made with caution until these relationships are tested with future data and until the underlying mechanisms of population control are better understood." The breakdown of some of these relationships occurred as soon as 1965. A poor return resulted even though environmental conditions in Georgia Strait were considered favorable in the early summer of 1964 (IPSFC 1966).

In 1971, a run of one and a half to five million was forecast. The actual return was 9.8 million. The accuracy of predictions needed to be improved. By 1973, after examination of additional environmental variables, and with the availability of six cycles of survival data, a hypothesis emerged that increased reliability in forecasting. This hypothesis was concerned primarily with the availability of food at various stages of the pink salmon life cycle from fry to adults, and involved consideration of four factors (IPSFC 1974):

1. The condition factor (length vs weight) of the fry at Mission as they migrated downriver to Strait of Georgia. The better the condition, the better the survival.
2. The timing of the fry seaward migration. Later migration in May being more favorable than April.
3. The volume of run-off from the Fraser River from April through July. Low run-off being more favorable than high.
4. The abundance of zooplankton in the area of the Pacific Ocean frequented by pink salmon.

Essentially, the larger the fry, the faster they escape predators. The later the fry migrate out of the river in the spring, the greater the abundance of food (zooplankton) in the Strait of Georgia; hence, faster growth. Also, the lower the Fraser run-off, the greater the primary production in the Strait of Georgia because of increased light penetration due to a shallower and smaller plume of turbid water (Parsons 1972). Increased food supply in the Pacific Ocean would be beneficial to pink salmon growth and survival.

Correlation of these factors for the six cycles from 1961-1971 was made (Table 3) and estimates of survival of pink salmon fry to adults corresponded closely with the survival actually recorded (IPSFC 1974).

These factors resulted in a close prediction for the 1975 pink run (5-6 million prediction versus 4.9 million actual). But by 1977 the prediction was considerably off—a 4.4 million prediction versus 8.2 million actual. Further understanding of survival factors was neces-

TABLE 3. *Comparison of predicted and actual Fraser River pink salmon survival from fry to adults.*

Brood Year	1961	1963	1965	1967	1969	1971
Predicted Survival Percent	3.6	0.9	4.5	1.3	5.1	2.9
Recorded Survival Percent	3.8	0.8	4.7	1.7	5.0	2.8

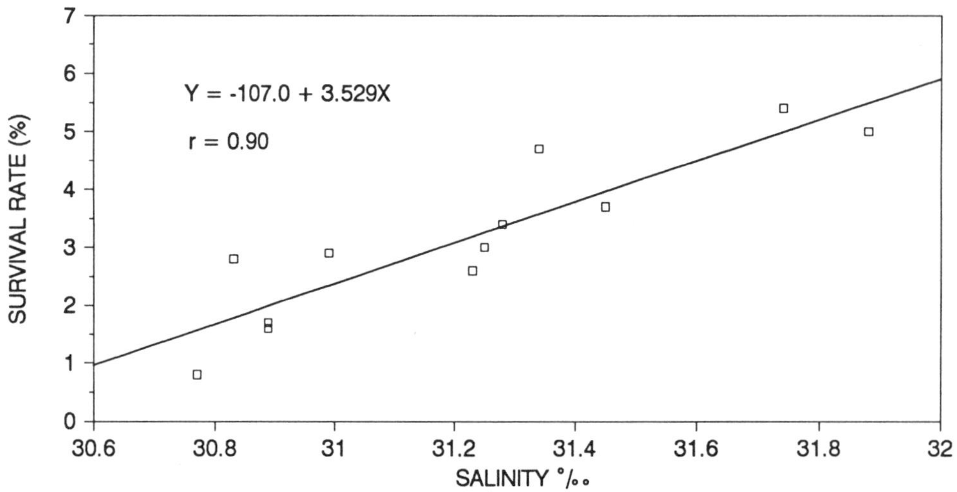

Figure 17. Fraser River pink salmon ocean survival in relation to Amphitrite Point, Race Rocks, and Neah Bay salinity during fingerling migration, 1963-1985.

sary. In 1979, the prediction was very close (15.0 million versus actual 14.4 million). In 1981 the prediction was 9.0 million and the actual run was 18.7 million fish.

The much higher-than-predicted marine survival for the 1981 pink salmon return showed that the multi-factor method used in making the forecast was not reliable. Again, there was a need for further examination of prediction methods for Fraser River pink salmon returns. Unfortunately, there were no data available on the numbers of young pink salmon entering coastal waters off Vancouver Island; therefore, environmental parameters had to be re-examined.

The average of salinities at Amphitrite Point, Race Rocks and Neah Bay from June to September were examined for the period when pink fingerlings were moving seaward through those areas. In general, the relationship (Figure 17) showed an increasing survival with higher salinity; thus the average salinity factor was used in 1983 and 1985 as the basis of the predictions. In 1983, the prediction was for a 20 million return. The actual run was 15.3 million—one-third overestimated. In 1985, a 16 million forecast was made and the actual run turned out to be 18.9 million—18 percent underestimated. The predictions of two years were realistic, however, in fitting into the long-term relationship, with a correlation of r = 0.895 for all years.

Other information, such as the late spring-early summer discharge of three Washington State coastal rivers (Hoh, Chehalis and Quinault) was also examined. A significant relationship between discharge and marine survival was found. The higher the discharge, the lower the survival (r = -0.918).

In predicting Fraser pink salmon returns the Commission learned there was not yet just one reliable method to use for consistently forecasting pink salmon returns. It appeared for a number of years (1961-1985) that the salinity-survival relationship held considerable promise but how long it will be useful for predictions only time will determine.

Management of the Harvest

When the Commission assumed responsibility for the management of the fisheries in 1946, many races were depleted. The Convention fishing areas were primarily at Salmon Banks and Point Roberts in the United States and in the Fraser River in Canadian waters. The average weekly fishing schedule in both countries had been at least five days each week. Purse seines were the primary gear used in United States waters and almost all of the sockeye harvested in Canadian areas prior to 1946 were taken by gillnet.

CHANGES IN GEAR TYPE, ABUNDANCE AND EFFECTIVENESS

Canada

Gillnets were the first type of commercial fishing gear developed in the region (Rounsefell and Kelez 1938), and for many years, from about 1873, this was the primary method of catching sockeye. By 1900, 3,683 gillnet licenses were issued in Canadian Fraser River fisheries. That number declined to 1,803 in 1934. Gillnet boats were initially powered by sails and oars. Soon after the turn of the century, gasoline engine power was available and used by the majority of gillnetters.

When the Commission assumed control of the commercial fishery in 1946, there were 1,843 gillnets fishing in the Fraser River area. Their number in Canadian Convention Waters gradually declined but with the change from linen to nylon nets in the 1950s their effectiveness increased. By that time several hundred additional gillnetters were fishing the Sooke and Port San Juan areas of Juan de Fuca Strait.

From 1946 to 1949, the Fraser River gillnet share of the catch in Convention Waters decreased from an average of 89 percent from 1938 to 1945 to an annual average of 80 percent. Further decreases in the gillnet share occurred and from 1981 to 1984, it dropped to only 66 percent (Figure 18 and Appendix F).

Although there were only about 750 Canadian gillnetters fishing from 1982-1985, individual efficiency was high because of improved net materials, mobility and fishing strategies.

Purse seines in Canadian waters were present as early as 1917 but there were few used. By 1942 there were more seines and due to the abundant Adams River run and its greater vulnerability to the deep seine nets than to the shallower gillnets, about 41 percent of Canada's Convention Waters catch was taken by

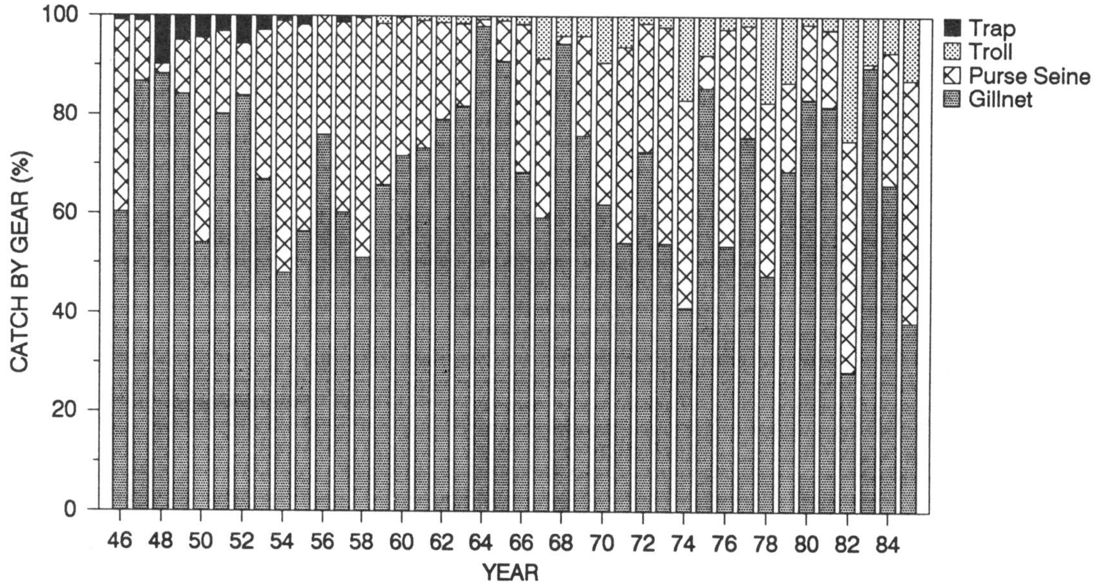

Figure 18. Sockeye salmon catch by gear in Canadian Convention Waters, 1946-1985.

September 1954. Gillnetters and other marine traffic on the Fraser River just below Pattullo Bridge.

August 6, 1961. Fraser River gillnetters waiting for the first drift at 8:00 a.m. Monday morning.

A Fraser River fisherman off Steveston in 1947 picking up his gillnet.

American purse seines at Point Roberts in 1954 at the height of the Adam Rivers run.

September, 1954. The American purse seiner "Lemen II" making a set at Point Roberts for Adams River sockeye.

September, 1954. A Lummi Island reef net. These nets, which are raised when a school of salmon swims over them, are much more effective during migration of summer runs than later in the season.

September, 1954. An American purse seiner lifting the last fish from a set at Point Roberts.

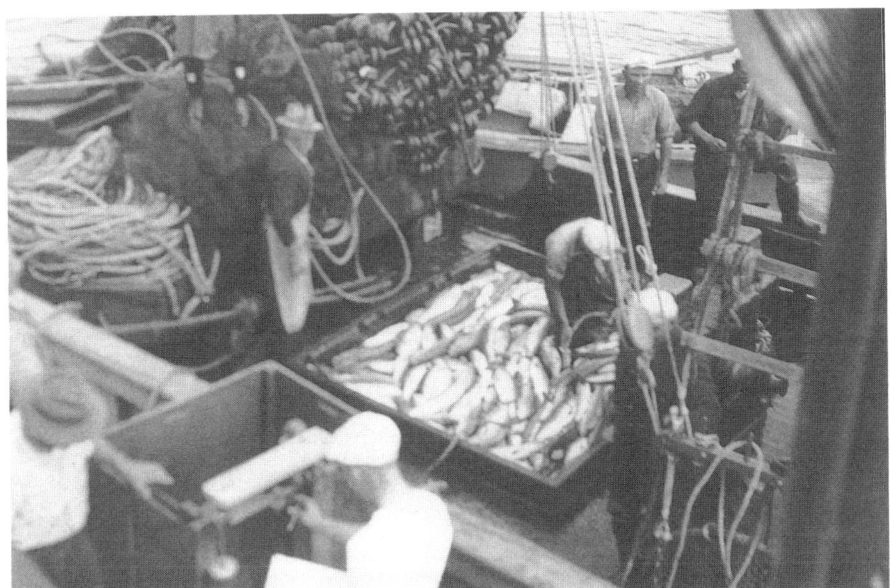

August 18, 1946. Preparing to unload a full-load catch of 8317 Fraser River sockeye from the purse seiner "Artic" to the packer "Pacific Foam".

.July 31, 1961. Canadian purse seines and packers waiting in Port San Juan, Vancouver Island, for another day of fishing in Juan de Fuca Strait.

purse seines. Most, primarily delaying sockeye, were taken in the Strait of Georgia near the mouth of the Fraser River. By the mid-1940s, Canada was encouraging a seine fishery on migratory stocks in waters near the entrance to Juan de Fuca Strait (now Area 20). This was believed to be in response to the United States purse seine fishery on Swiftsure Bank. With the development of the Area 20 fishery, Canada would be in a position to get an early or "first crack" at the fish.

In 1946, the Adams River dominant cycle year, 132 purse seines took about 39 percent of Canada's Convention Waters catch. On other cycle years, purse seines had little impact on migrating stocks. Catch percentages for purse seines in 1947, 1948 and 1949 were only 12, 2 and 11 percent, respectively.

In the 1950s large hydraulic-powered pulleys replaced muscle-power on purse seines and this significantly increased the numbers and efficiency of seines in the Juan de Fuca Strait fishery. In 1954, 300 purse seines were actively fishing in Canadian Convention Waters. No significant changes in the number of gear units took place during the next 30 years; technological improvements have made the present-day purse seiner very efficient.

At times, the Canadian government policy of encouraging the development of the Area 20 fishery clashed with Commission fishery management. It was desirable to catch late-run sockeye and pink salmon in the seaward location because of the higher quality. Summer-run sockeye could all be taken in the Fraser River area with no loss in quality. From a management perspective, no fishing was required in Area 20 before about the second week of August.

United States interests (sports, commercial and State authorities) were vocal in their discontent with the Strait fishery, especially with respect to coho catches. A high percentage of the coho were of U.S. origin. The fishery was important to the Canadian side as a visible show of Canadian net fishing power in front of the United States fishery, which was fishing on Canadian stocks. The contentious aspects of the fishery also permeated the Canadian commercial fishing industry. There were numerous conflicts over division of catch between Area 20 fishermen and Fraser River gillnetters. Although the battles usually involved seines versus gillnets, it even extended to conflict between river and "West Coast" gillnetters.

On many occasions, representatives of both countries expressed concern to the Commission about the effect of the bycatch in the Canadian Juan de Fuca Strait fishery on other species, specifically coho and chinook.

In early summer, large numbers of immature chinook were sometimes gilled in the seine webbing, an observation distressing to both commercial and sports fishing interests. Although not easily predictable, the catches were definitely harmful to chinook stocks.

At other times, large numbers of immature coho were taken with sockeye. Although marketable, these were taken before they reached maturity and many reached the fishing plants badly belly-burned because of the high food content in their stomachs. In September, there was concern about the sometimes large number of mature coho salmon being taken by both seines and gillnets in Area 20 during directed sockeye and pink salmon fisheries.

The Commission was accused from time to time of allowing certain fishery openings, particularly in Area 20, for the purpose of catching coho salmon (Haring 1981). The Commission was consistently concerned with this issue and made every effort to avoid excessive capture of non-target species. In-season catch information, historical catch data, and current test fishing results were reviewed prior to any recommended opening. As a general policy, the Commission attempted to prevent fisheries where other species catch might exceed target species catch. Only on rare instances did the vagaries of the migration behavior and abundance patterns of the various species produce larger than projected catches of the non-target species.

Traps in Canadian Convention Waters were permitted for many years but were never as effective as in the United States. In the 1940s only four or five traps fished in the Sooke area. Efficiency varied from year to year, probably due to the magnitude of the races involved and their migration patterns. From 1941 to 1952, annual trap catches as a percent of total Canadian catch varied from a high of 9.9 percent in 1948 to a low of 0.7 percent in 1954. The largest single season's catch was 130,000 sockeye in 1941. The development of traps in Canadian waters was largely in response to the great number of traps in the United States. With the abolition of traps in the United States in 1935, it was surprising that traps persisted in Canada until 1958, despite measures to remove them.

Sockeye catches by troll fishing were negligible prior to the 1960s. By 1970 the catch had increased to 145,000 fish or about nine percent of Canada's Convention catch. The catch continued to increase until 1982 when 825,000 sockeye were taken by Canadian trollers in Convention Waters. This was 25 percent of the total Canadian catch. In 1982, 2,179,000 Fraser sockeye were taken by coastal Canadian trollers. Adams River and other late-run sockeye were consistently vulnerable to troll gear. By 1985, when the large, summer-run Horsefly River race returned, trollers had improved gear and technique and were able to take 603,000 Fraser sockeye in Canadian Convention Waters (13 percent of the total landings). Troll gear caught 1,206,000 Fraser sockeye in all Canadian waters. Such catches were an important management concern.

Traps took a small fraction of the pink salmon run prior to the Commission's control of the fishery in 1957. From 1951 to 1955, the percentage catch averaged about three percent, and catch reached a maximum of 126,000 pinks. By 1957, the last year of traps, the catch and percentage had declined to 31,000 fish and one percent.

Prior to 1957 the pink salmon troll catch was small. Pinks were an incidental and non-targeted species. In 1957 the catch was only 41,000 fish (1.6 percent). In the 1960s the demand for fresh, troll-caught pinks increased and in 1963, 440,000 (10.5 percent) pink salmon were taken in Convention Waters by Canadian trollers. By 1979 a record 1,547,000 (37.4 percent) pinks were landed. In 1979 the Washington-British Columbia coastal troll fishery reached a record catch of 3,744,000 pink salmon.

Gillnets caught most of the pink salmon in Canadian Convention Waters prior to the Commission's regulatory involvement. In 1945, almost all the catch was taken by gillnets in the Fraser River. Thereafter, the gillnet catch percentage declined and by 1957 the share was 43 percent. From 1957 to 1985, the percentage harvest and total pink salmon harvest by gillnets continued to decline (Figure 19), even though total returns and catches increased significantly. For example, from 1959 through 1965, during a period of very low production (runs averaged 4.2 million each year), gillnet annual catch averaged about 454,000 or 26.5 percent of the total catch by all gear. By 1979-1985 the Fraser pink salmon run size had quadrupled; however, gillnet average catch (277,000) and percentage (7.5 percent) were even lower than in earlier years. The catch for 1983 was omitted because of unusual circumstances of that year related to "El Nino", the displacement of warm southern waters to the north.

In later years, purse seines were the dominant gear taking pink salmon in Convention Waters. About 55 percent of the catch was taken in 1957. During the early 1960s the annual percentage averaged about 61 percent each year; similarly in the late 1970s and early 1980s. Purse seines and trolling gear had replaced gillnets.

United States

With the elimination of traps in the U.S., purse seines immediately became the dominant gear type for taking sockeye. From 1935 to

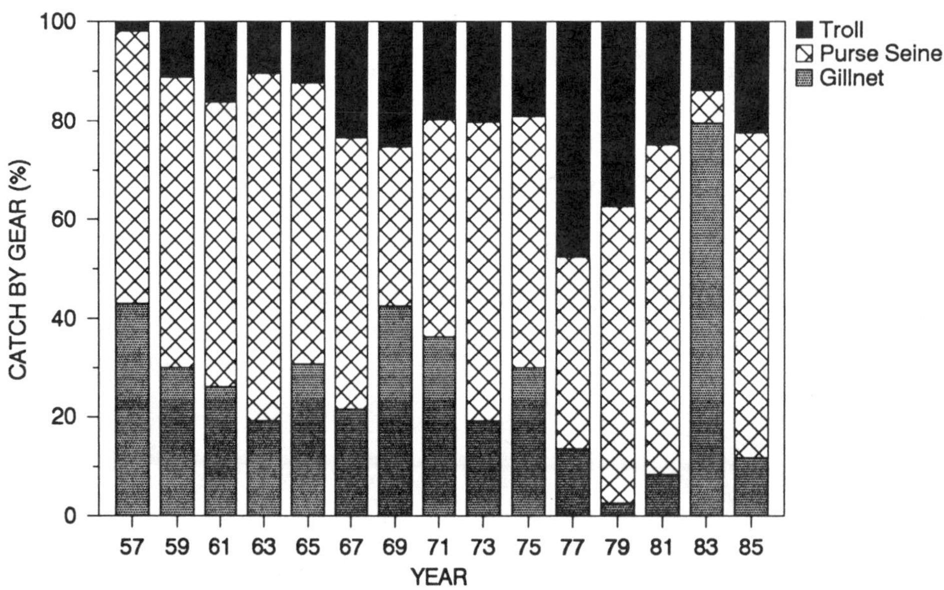

Figure 19. Pink salmon catch by gear in Canadian Convention Waters, 1957-1985.

1945 seines averaged 87.6 percent of the total sockeye catch. In the first year of Commission management for sockeye (1946), seines took almost 95 percent of the total United States sockeye catch in Convention Waters. By the 1980s (1982-1985) seines were averaging only about 48 percent of the U.S. catch (Figure 20). In addition, about one-half the earlier number of purse-seines (about 150 vessels) were fishing.

Although gillnets in the U.S. were the minority gear in the 1930s and early 1940s (1.4 percent of the catch in 1946), the numbers increased rapidly to more than 1,000 in 1977 from 58 in 1946. The percentage of catch increased to almost one-half of the U.S. sockeye catch from 1982 to 1985 from four percent in the late 1930s and early 1940s. Actual landings increased to a record catch of 1,424,000 in 1985 from only 50,000 in 1946.

Reef nets in contrast to the mobile gillnetters and purse seines are fixed to a specific shallow water site. The fish are captured by lifting the net once they have followed leads enticing the fish toward the net. Effectiveness is dependent on availability of fish and on winds and strong tides. Few adequate sites exist for this gear in Puget Sound. From 1935 to 1945 reef nets averaged eight percent of the annual catch. In the early period of Commission management, reef nets ranked higher than gillnets in relative harvests, averaging seven and one-half percent of the annual catch. Improvement in technology of other gear and increased competition from larger numbers of gillnets and purse seines decreased the reef net catch to less than half that taken in earlier years. This happened even though the Commission granted reef nets early weekly fishing openings in the 1970s and 1980s. From 1982 to 1985, annual reef net catch averaged only 2.7 percent of the United States total. Operational costs for a reef net were much less than for other gear.

Troll fishermen in the United States have never taken significant numbers of sockeye. Normally, the catch would be in the hundreds for any particular year. A maximum catch of 4,663 fish was taken in 1959. The area of coastal sockeye landfall (Vancouver Island) and non-availability to U.S. trollers were probable reasons for the insignificant catch.

Most of the pink salmon catch in United States waters prior to 1957 was by purse seine. From 1951 to 1955 this gear captured an average of 87 percent of the total catch each cycle year. Only minor changes have occurred in catch distribution through the years as the

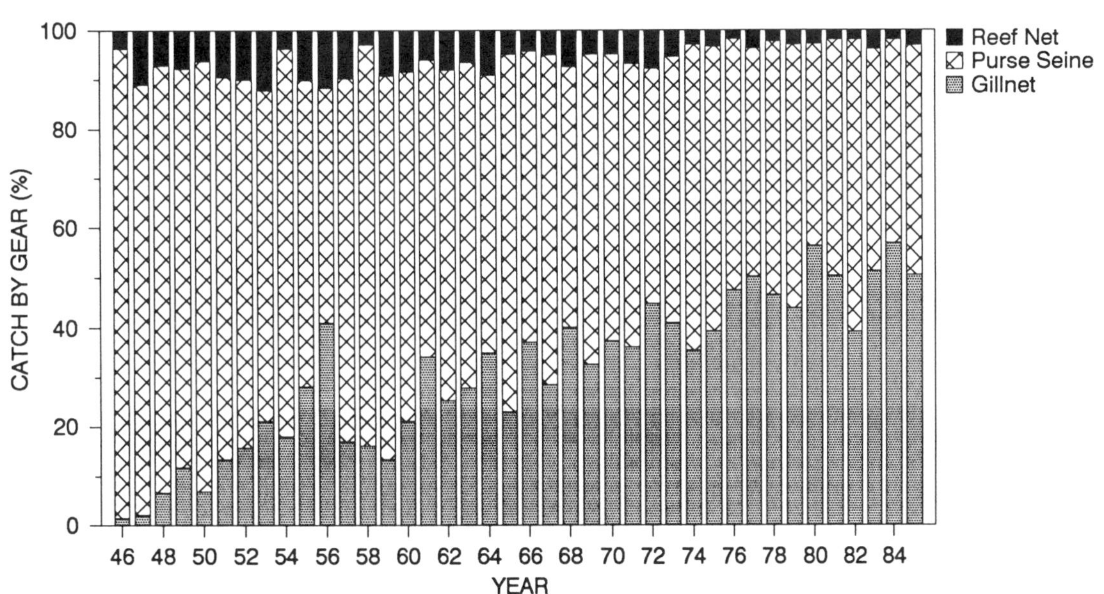

Figure 20. Sockeye salmon catch by gear in United States Convention Waters, 1946-1985.

average percentage catch from 1979-1985 was much the same at 84 percent (Figure 21).

Gillnets have always taken a low percentage of the pink salmon catch. In 1951; for example, the pink salmon gillnet catch was only two percent of the total landings by all gear. The average percentage catch from 1951 to 1957 was five and one-half percent. There was little change over almost 30 years of Commission management. The average percentage catch from 1979 to 1985 was 8.7 percent.

Reef net gear in the early 1950s caught fair numbers of pink salmon (e.g., 410,000 in 1953) and averaged about seven percent of the total annual catch from 1951 to 1957. Total catch decreased significantly and by the 1980s, maximum annual catch of about 80,000 fish was the norm and percentage taken from 1979 to 1985 decreased to an annual average of 2.4 percent.

In most years troll fishermen did not take large numbers of pink salmon. In 1951 the catch was 30,000 fish (0.6 percent). By 1957 the catch was 165,000 (6 percent) and from 1961 to 1965 during low Fraser pink runs, the catch ranged from 11 to 14 percent of the U.S. total. The largest catch occurred in 1963 (500,000); however, most of that catch was made from an unusually large Puget Sound run and availability of fish (because of U.S. origin) to the U.S. troll fleet may have been greater off the Washington coast. In the more recent years (1979-1985), troll catch has averaged only 4.8 percent each year, despite greatly increased Fraser River pink runs.

The Commission had no authority over type and quantity of fishing gear in Convention Waters. This was a national or domestic responsibility. From 1946 to 1957, when waters were opened by the Commission, it was open to all types of authorized gear to fish simultaneously. In 1957, each gear type was given daily fishing times. To a great extent, the schedule followed traditional fishing times. Purse seines and reef nets fished during the day and gillnets at night. The Convention did not specifically give the Commission the authority to set fishing times for each gear. However, harvesters in both countries requested this and the governments gave approval for the Commission to adopt such a scheduling system. In a broad sense, it could be construed that the Commission became involved in allocation of catch by gear, but this was never the intent. Setting different times for different gear reduced gear conflict caused by large numbers of competing vessels.

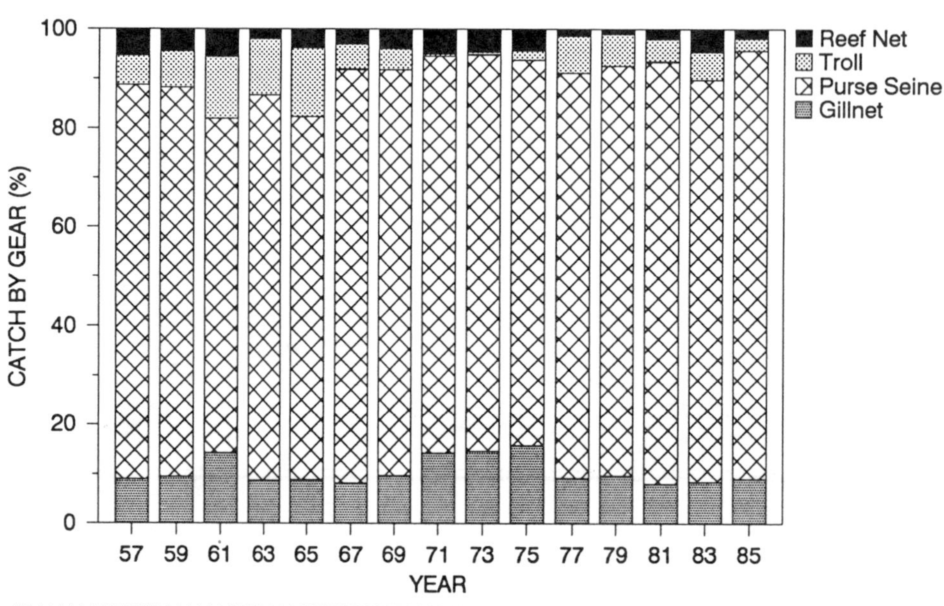

Figure 21. Pink salmon catch by gear in United Sates Convention Waters, 1957-1985.

The Commission attempted to schedule nearly equal fishing hours for each net gear in Convention Waters. However, in the early years, trollers fishing the high seas areas of Convention Waters enjoyed almost unlimited fishing hours per day because of their small incidental catch of Fraser sockeye and pink salmon. Later, they targeted these species (only pinks by U.S. fishermen); as well as their traditional target species, coho and chinook salmon (depending on in-season timing and fish availability). Inside net fishermen in both countries were upset about the Commission's lack of control over the offshore troll fisheries.

To a certain extent, regulations affected catch distribution between gears. For example, in Canada, the Area 20 sockeye fishery was closed in years of small runs. This was the reason for less than two percent catch by purse seines in those years (Figure 18). The low seine catch in 1975 was due to an industry strike and the low catch in 1983 was due to a diversion of 80 percent of the sockeye through Johnstone Strait. For both Canadian gillnets and purse seines, there were wide yearly fluctuations in percentage catch but no great difference in long-term catch percentage over a 40-year period. However, the catch in the essentially unregulated troll fishery had increased significantly since 1966 (Figure 18).

In the U.S., trends evident in catch distribution by gear-types, as shown in Figure 20, were primarily the result of changes in numbers of gear and their efficiency and not due to regulations. Domestic laws permitting increases in gillnet vessels meant that the purse seine catch monopoly from 1935 to 1945 became, by the 1980s, an almost equally shared fishery with gillnets.

On many occasions the Commission expressed strong concerns to both governments about the increasing gear units and the increased efficiency of individual boats, especially in the United States. Royce *et al.* (1963) discussed the excessive fishing effort present in U.S. waters. The capacity to harvest the resource increased in a spectacular fashion. Increases in fishing power from the 1940s to the 1980s precipitated great reductions in fishing times. Despite this, the Commission was able to fulfill its obligation to divide catches and obtain the necessary escapements. The alternative of having the authority to control the numbers of fishing gear and the social-political intensity of dealing with that responsibility would probably have been even more difficult and inappropriate for an international body. However, as a result, over-capitalized and excessive harvesting capacities developed and existed in both countries for thirty years. The difficulties caused by this did not prevent the Commission from carrying out its responsibilities and stands as an example, as will be shown later, that resource protection and stock increases can take place despite an over-capitalized industry. Lacking control of fishing effort, the Commission exerted stringent control of the fisheries primarily through area and time of fishing restrictions. Fortitude of the managers was required.

Attempts have been made by both countries to reduce the number of fishing vessels. The most concerted effort was made in British Columbia with a reduction of gillnet vessels. However, this simply transferred more fishing power to seine vessels.

In April 1946, the Commission passed a resolution to both governments expressing concern about possible Japanese interception of Fraser sockeye outside the three-mile boundary of Convention Waters, but this never became a major problem for the Commission.

The abstention issue within the International North Pacific Fisheries Commission (INPFC) was of some concern to the IPSFC in the mid-1950s. Japan had agreed to restrict its fishing expansion if it could be shown that North American salmon stocks were being fully utilized under a scientific management regime. Both Canada and the United States prohibited net fishing for their nationals outside the surf line in 1957. From a North American point of view, abstention from high seas salmon fishing was an integral part of conservation. The Commission (IPSFC) expressed concern about possible problems of an extensive off-shore high seas fishery by the Japanese. This was handled effectively by the International North Pacific Fisheries Commission; however, "abstention was without precedent in international fishery law" (Jackson and Royce 1986). The INPFC established a provisional boundary line at 175°W prohibiting salmon fishing by Japan east of that line. This line, in effect from 1952 through 1977, was moved 10° further west in 1978. Fraser River sockeye and pink salmon were protected from Japanese high seas net fishing as a result of these lines, and were being fully harvested.

SURF LINE AGREEMENT

In the early 1950s the Commission became increasingly alarmed at the expanding high seas net fishery developing within Canada and the United States. The seaward extension of the net fisheries, coupled with dramatic increases in gear efficiency and numbers of vessels, seriously challenged the Commission's ability to manage the stocks properly. This was reviewed in late 1955 and a strong statement was forwarded to both governments detailing the problems created by high seas fisheries for Fraser River salmon. Of greatest concern was the potentially effective high seas gillnet and purse seine fishery in Canadian Convention Waters. The percentage of the Canadian catch taken in Juan de Fuca Strait increased to thirty-five percent by 1956 from three percent in 1944. At times, the United States purse seine fishery off Swiftsure Bank was effective on certain stocks but on the high seas, the Canadian fleet would have had the greater catch potential because of the migratory pattern of Fraser salmon.

Two potentially serious management problems existed in 1956: a developing high seas net fishery and the inability to close the Juan de Fuca Strait net fishery in Canadian Convention Waters because of the limitations of the Convention. The treaty called for uniform fishing openings throughout each country's Convention Waters. It was not practical; for example, to have two days fishing in Canadian waters of the Juan de Fuca Strait with two concurrent open days in the Fraser River, particularly on summer-run sockeye. The Commission made two recommendations:

1. that all high seas areas and the adjacent territorial waters excepting rivers, bays and estuaries off the coasts of the United States and Canada be closed to all salmon net fishing;

2. that it was necessary for the Commission to have the right to control fishing in any or all parts of Convention Waters, except waters set aside as national salmon preserves where no commercial salmon fishery is permitted.

Through the efforts of the Commission, the Pacific Marine Fisheries Commission in the United States and other concerned parties, all high seas salmon net fishing of the Eastern Pacific Ocean and the territorial waters adjacent thereto were closed. This was the Surf Line Agreement (Anon. 1957). It was agreed that "The bilateral action of the two governments in preventing the development of such a net fishery avoided a situation which would have made the continued scientific management of the Fraser River sockeye fishery impossible" (IPSFC 1958). Whether the Commission could have coped with a regulated high seas net fishery is a matter of speculation, especially in view of later developments in the Canadian troll fishery which occurred west of Vancouver Island both in and out of Convention Waters and in the highly efficient Johnstone Strait net fishery. The latter, in many years, was a major fishery outside of Convention Waters. In effect, the high seas net fishery of Swiftsure Bank involving fishermen of both countries was replaced by the intensive and difficult-to-evaluate Canadian net fishery in Johnstone Strait. In almost all years, catches of Fraser River sockeye and pinks in Johnstone Strait exceeded those taken by net fisheries in the Swiftsure Bank areas prior to 1957. In many years exceptional catches of several million fish were taken outside Convention Waters compared with the much smaller numbers taken in earlier years at Swiftsure Bank. In addition, the Swiftsure Bank fishery was controlled by the Commission while the Johnstone Strait fishery was not.

The Protocol for Fraser River pink salmon also gave the Commission new and necessary authority to regulate fishing in specific Convention areas by separate regulations. This authority greatly facilitated rational regulation of the fisheries and management of the resource. However, the Surf Line Agreement was clearly deficient in that it did not provide the Commission the authority to control the troll fisheries on Fraser River sockeye and pink salmon in the high seas approaches to Convention Waters.

EXPANSION OF FISHERIES

While the actions taken in 1957 by the two governments created some stability with respect to net fishing gear in Convention Waters, other types of fishing gear began to target on Fraser River sockeye and pink salmon. This seaward expansion had its initial impact primarily in Canadian waters, but catches in later years had a significant effect on fishermen of both countries, particularly in the United States.

CANADIAN TROLL FISHERIES

The pursuit of Fraser River sockeye by the offshore troll fishery did not begin immediately after control of net fishing on the high seas. For example, only 1,700 sockeye (0.1 percent of the Convention Waters catch) were taken in 1957. Even in 1958, with the great abundance of that year's run, only 4,900 sockeye were captured by Canadian trollers in Convention Waters. However, by the late 1960s, increased gear, effort, strategies and interest were directed towards sockeye, primarily in the years of large Adams River returns. In 1967 the catch was 162,000 sockeye (8.6 percent of Convention Waters catch). By 1982, the total troll catch reached a record 825,000 fish (25.2 percent of the Canadian catch) in Convention Waters. Market demand for troll fish was increasing. By 1984 and 1985, trollers had developed gear and techniques that made them highly effective on non-Adams River races, such as Chilko River and Horsefly River sockeye populations. In 1984 the Convention catch was 114,000 fish

156

(7.2 percent of the Canadian catch) and for 1985, the catch increased dramatically to 603,000 fish (12.9 percent of the Canadian catch), primarily of Horsefly River origin. Large numbers of sockeye were also taken by Canadian troll fisheries north of 49°N latitude outside Convention Waters. The Canadian troll fleet had developed into a major fishing force.

The sockeye harvested by the high seas troll fishery in the early 1980s were far greater than those taken by the high seas net fisheries in the mid-1950s. Although the Commission expressed concern about the unregulated troll fishery, the effectiveness and harvests of the off-shore fishery continued to increase. Ironically, this was contrary to the intent of the 1957 Surf Line Agreement. The loophole in the agreement enabled Canada to increase its catches outside Convention Waters and to benefit at the expense of the United States.

The story of the rapidly expanding, almost exploding, troll fishery did not end at the Convention Waters northern boundary at 49°N latitude near Tofino, B.C. For example, in 1982 a record 2,179,000 Fraser sockeye were taken by the coastal troll fishery, with as many as 1,100 trollers (day and trip (ice) boats) participating in the fisheries. Only 727,000 of these sockeye were taken in Convention Waters. This catch comprised almost 23 percent of the total commercial catch of Fraser River sockeye by both countries in 1982. It was apparent that a major effort was being mounted by Canada to intercept Fraser River sockeye before the fish migrated into Convention Waters. Fish taken outside the Convention boundary did not have to be shared with the United States. It was probably not coincidental that the increase in the non-Convention troll fishery took place from 1970 to 1985 during treaty negotiations on salmon interceptions.

The Commission made several unsuccessful attempts to regulate the troll fishery in Convention Waters, particularly in the Adams River cycle years. However, because of the international situation and concerns for other salmon species taken by trollers, these efforts were ineffective.

Following the Surf Line Agreement, troll fishermen developed methods and lures designed to target on pink salmon stocks. At the same time, there was a dramatic increase in market demand for fresh, troll-caught pink salmon. By 1963, 939,000 pink salmon were taken by trollers in Convention Waters. Substantial catches were also made outside the Convention area north of 49°N and south of 48°N. The impact of this fishery was overshadowed to a great extent in that year by the unexpected large return of over 10,000,000 pink salmon to Washington State streams (more than twice the size of the Fraser run).

While most scientific experts were trying to explain the unprecedented return of the Washington State pink salmon runs, net fishermen in both countries were expressing deep concern and agitation over the increased troll catch of pink salmon by trollers. Increasing numbers of "their" fish were being caught before they reached the inside net fisheries. The Commission was urged by the majority of its Advisory Committee members in 1963 to write the governments for assistance. The governments were non-committal, possibly because:

- the coastal troll fishery was targeted primarily on other salmon species;
- the catch by troll fisheries was not a conservation problem;
- the Canadian Government probably did not want to lose the "first crack at the fish" advantage and the accompanying negotiating power that position provided inside and outside the Convention area;
- sockeye and pink salmon caught outside Convention Waters (almost entirely by Canadian fisheries) were not part of the catch that had to be shared with the U.S.;
- restricting the trollers' Fraser sockeye and pink catch would have redirected

their efforts back to coho and chinook salmon—a high proportion of which were stocks originating in U.S. rivers and Washington State officials were not anxious to encourage such action.

The pink salmon coastwide troll catch continued to intensify and by 1967 Canada took 975,000 fish within the Convention boundaries and an equal number outside the area. In addition to 194,000 pinks taken by U.S. trollers in Convention Waters, 102,000 fish were taken off the Oregon coast. In 1963 some pinks were caught as far south as California.

In 1969, in response to industry concern over increasing pink salmon troll catch and the Commission's increasing management problems, the Commission submitted another report describing the troll catch situation for both species. As in the past, the two governments did not take definitive action, apparently because of the complexity of the problems involving other salmonids and fisheries both inside and outside Convention Waters.

Thus, the outside troll catch intensified in the late 1970s and 1980s, but there was little done to control or reduce it. Inside net fishermen were increasingly frustrated over the seven-day-per-week troll fishery compared to their one or two day net fishery per week. The 1963 record catch by United States trollers of about one-half million pink salmon never recurred, even though the Fraser River pink salmon run increased dramatically in later years. In 1971 only 13,000 pink salmon were taken by United States trollers in the Convention area. The largest catch by those trollers after 1963 was 261,000 fish in 1979. It appears that the size of the U.S. troll catch was dependent upon the numbers of U.S. pink salmon returning, and not on the Fraser run.

The pink salmon troll catch in Canada, however, continued to increase. By 1977 almost one-half (47.6 percent) of the Convention Waters catch was taken by trollers. In the following cycle year (1979), a record catch of 1,547,000 pinks was taken in Convention Waters by troll gear. The catch taken outside the Convention area was even greater. The total coastal troll catch by both countries, extending from Grays Harbor in Washington State to the northwestern tip of Vancouver Island, was a record 3,744,000 pinks. Most of these were taken by Canadians—a great increase when compared to the 1957 Washington and British Columbia catch of 278,000 pinks. The troll fisheries had become a major factor affecting the numbers of sockeye and pink salmon which eventually reached Convention Waters.

The coastal troll fisheries in Canada had great impact on the runs, since the Fraser River sockeye and pink salmon runs were more available to Canadian troll fishermen. The intent of the Surf Line Agreement was clearly eroded and the exploitation of sockeye and pink salmon in the high seas continued to increase from 1957 to 1985. Sound management was threatened with uncontrolled exploitation of stocks of undetermined origin and abundance. Moreover, the rationale for expanding coastal high seas troll fisheries on mixed stocks was inconsistent in some respects, with the U.S. position on high seas fishing by Japan on Bristol Bay sockeye stocks. A significant high seas uncontrolled interception fishery by any gear is not compatible with rational fishery management. Although the large troll catch created user-group conflicts and difficult management problems for the Commission, the Commission adjusted to the situation, and controlled fishing in Convention Waters to ensure adequate escapement and serious conservation problems were avoided.

CANADIAN NET FISHERIES

Prior to the Sockeye Convention in 1930, most of the Fraser River sockeye runs were caught by both Canadian and American fishermen in areas close to the mouth of the Fraser River, within waters later defined as Convention Waters. In 1946, almost 97 percent

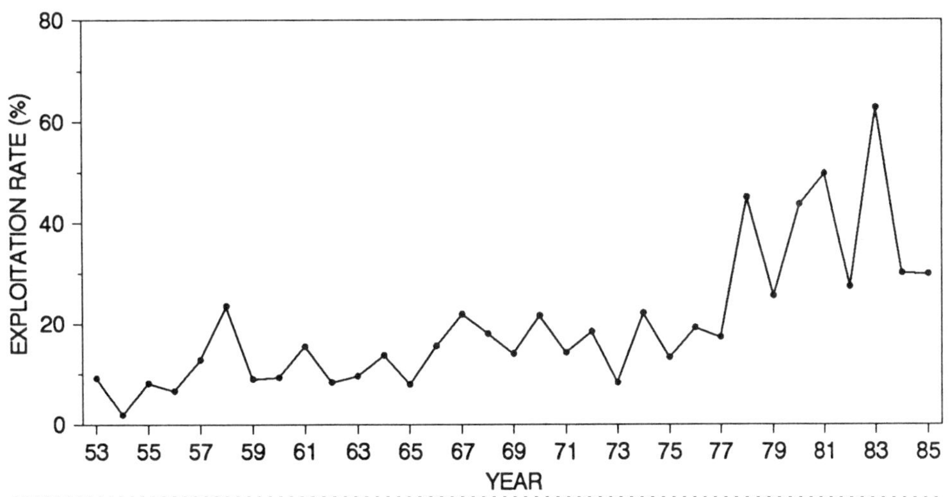

Figure 22. Combined exploitation of Fraser River sockeye salmon in commercial fisheries outside of Convention Waters and in the Fraser River Indian food fishery, 1953-1985.

of the total run reached the Convention areas with only a small percentage caught outside. It was the intention of those who put together the treaty that the annual Fraser sockeye catch would be divided equally between the fishermen of the two countries. The two governments originally envisioned the Commission having almost complete management control over harvests (with the exception of a minor fishery known at that time to occur in Johnstone Strait north of Convention Waters). This was substantiated by the treaty itself, where the monies for the Commission's programs were to be funded on a fifty-fifty sharing arrangement. The Indian food fishery in the Fraser River was not discussed in the Convention. Sharing also extended to enforcement. Regulations were to be enforced in each country by its own fishery officers.

In 1958 there was a strong migration from the north through Johnstone Strait. More than four million sockeye were taken outside Convention Waters (IPSFC 1959). The percentage of the run harvested outside the Convention area increased to almost 25 percent. This did not appear to cause consternation, probably because of the large run and because both countries had had an excellent fishing season. There was a gradual but steady increase

in effort outside Convention Waters both by troll and net fishermen. From 1981 to 1984, 42 percent of the total annual Fraser sockeye run was caught either before the fish reached the Convention area or after the fish passed commercial fishing areas, when they were caught in the Fraser River Indian fishery (Figure 22).

From 1978 to 1984, opportunities for increased catches outside Convention Waters were enhanced by oceanographic events which caused sockeye to migrate toward the Fraser River from very northerly areas. Canadian fishermen capitalized on this with delight and enthusiasm. A catch of 17.6 million sockeye was taken during the years 1978-1984. This catch was excluded from the divisible obligations of the treaty because they were taken before the fish entered Convention Waters. The Canadian fishing industry urged the Canadian government to permit maximum utilization of Fraser River sockeye stocks even to the point of admitted jeopardizing (through overfishing) local, non-Fraser populations (Department of Fisheries and Oceans, personal communication). Serious overfishing of Chilko River sockeye which almost occurred in 1980 because of the pressure to continue fishing in Johnstone Strait (IPSFC 1981) was averted by the Commission.

The escalation of non-Convention fisheries on Fraser River stocks peaked in the late 1970s and early 1980s. Although the troll fishery contributed significantly to the increase, the major impact was caused by the net fisheries, almost exclusively in Johnstone Strait. In 1983, because of El Nino, less than 25 percent of the total catch was made in the Convention area (Table 4). From 1978 through 1984, an average of 49 percent of the total commercial catch was made outside the Convention Waters boundaries (Table 4).

TABLE 4. _Percent of total Fraser River sockeye salmon catch made in non-Convention area fisheries._

Year	Percent
1978	60.3
1979	28.4
1980	55.8
1981	58.0
1982	35.4
1983	75.5
1984	30.0
1985	_32.7_

In 1983, an estimated 96,000 Fraser-origin sockeye were captured in such distant fisheries as the United States fishery at Noyes Island, near Dixon Entrance, Southeastern Alaska. In the same year, an estimated 230,000 Fraser-origin sockeye were taken by Canadian net fishermen inside the surf line off the west coast of the Queen Charlotte Islands. In 1984 and 1985 an estimated 127,000 and 83,000 Fraser sockeye were taken in the respective years in Areas 1 and 2W (Queen Charlotte Islands).

The Commission was severely tested in its ability to cope with major changes in catch distribution and fishing effort. In addition, Indian fisheries in both countries were causing concern. (See Chapter 15).

Strikes by fishermen or cessation of fishing caused by shore workers' disputes were not addressed by the 1930 Sockeye Convention nor the amending 1957 Pink Salmon Protocol. As a result, the Commission was obligated to establish its own policies to respond to situations created by strikes or shut-downs in fishing operations. The many policies adopted were a reflection of the make-up of the Commission at the time. Since membership on the Commission and the Advisory Committee changed from time to time, policies were periodically reviewed and changed. This was one of the few areas where politics was exerted by various Commissioners and Advisors. In some instances it was agreed that fish caught by the non-striking country during a strike or shut-down would not be counted in the division of catch. The striking fishermen were very vocal about the Commission's influence (indirect as it was) on the labor relationship between companies and the union.

Strikes caused problems; for example, when it appeared that a strike might end (most strikes were in Canada), Canadian interests pressured the Commission to open Canadian waters earlier and longer than United States waters to improve the balance of catch. This caused antagonism between Commissioners, Advisory Committee members, the fishermen's union and fishermen of both countries. In later years, the Commission's position was that all sockeye or pink salmon taken during the strike or shut-downs were to be counted in the division of catch. Once the strike was over, fishermen of both countries would resume normal fishing operations under the pre-season schedule. If there were surplus fish to be harvested, the country that was behind in division of catch would be given extra fishing time. However, at the same time, the country that did not strike would not be penalized by losing their scheduled fishing days. The Commission believed this policy was consistent with Article VII of the treaty dealing with the equal portion of the catch. "As nearly as practicable" was interpreted to mean that it was neither practical not to count the fish taken by the non-striking country during a strike, nor was it practical to shut the industry down in the non-striking country to make up the deficit.

On several occasions, a union-sponsored "strike relief" fishery was requested in the Fraser River area. The union and the Commission had several disagreements over this. The Commission's position was that it could not involve itself in the bargaining process by approving fishing times in excess of the pre-season schedule. In 1963, a serious confrontation took place between the Commission and the United Fishermen and Allied Workers' Union (UFAWU). In its Annual Report to the two governments, the Commission stated:

> Since the 1963 strike of Canadian fishermen was the second in succession on this cycle year, the Commission considered it obligatory to inform the industry that it was interfering with the natural interrelationship of the four cycle year sockeye populations by upsetting the principle of equal or consistent annual fishing pressure established in 1951. (IPSFC 1964)

The strike was partially responsible for allowing 1,002,000 sockeye, about three times the maximum desired escapement for the cycle, to reach Chilko River and Chilko Lake. As a result of the early migration, high water temperatures and excessive numbers an estimated 900,000 Chilko fish died without spawning. The Commission blamed the union. The union countered that Canadian fishermen were not given additional fishing time in the Fraser River during strike relief fishing and that if such additional fishing time had been granted the over-escapement would have been reduced. Escapement would have been less; however, over-escapement would not have been prevented.

Strikes also caused administrative problems, including the sometimes unwarranted union interference insisting that they approve test-fishing operations and the handling and distribution of the salmon taken in test fisheries.

Despite the clash between Commission and union, one beneficial aspect of the over-escapement during the 1963 strike was the establishment or advancement of an early-run sockeye spawning stock at the south end of Chilko Lake. By the third generation of this cycle in 1975, the spawning population had increased to 55,000. In 1987, an escapement of 181,000 sockeye was reported (PSC 1988b). Whether or not this race existed in earlier years is not clear. Also, the 1967 return of Chilko sockeye was the largest on the cycle since 1899; this occurred despite the large loss of spawners in the brood year 1963. One can only speculate as to what the 1967 return might have been if the proper escapement had been obtained in 1963.

During Commission management from 1946 to 1985, a total of 4,280,685 Fraser River sockeye was taken by United States fishermen during Canadian strikes. Despite all the hassling in strike years, the Commission made attempts to equalize the cumulative catch between the American and Canadian fishermen. For the eleven strike years in the 27 seasons between 1952 and 1978, division of catch favored United States fishermen by 1,998,555 sockeye. Since the total catch during strike periods by U.S. fishermen was 4,280,685 sockeye, opportunities were given to Canada to catch an additional 2,282,130 sockeye at a later time during the strike years. Strikes had a minor effect on the pink salmon catch division since they were usually early in the season and had been settled before the pink runs arrived.

Although disruptive in the management of the fisheries, the immediate and long-term effects of the strikes were not all negative. It would have been helpful if provision had been made in the treaty to cover this aspect of the agreement, thereby avoiding the considerable time and effort necessary to resolve each strike issue. This would have also eliminated the necessary review of previous policies, the setting of a current policy, the posturing by affected individuals, and the antagonism between the industries of both countries each time there was a strike.

. . . .
DIVISION OF CATCH

Article VII of the Convention stated:
The Commission shall regulate the fisheries for sockeye and for pink salmon with a view to allowing, as nearly as practicable, an equal portion of such sockeye salmon as may be caught each year and an equal portion of such pink salmon as may be caught each year to be taken by the fishermen of each Party.

The fishing industries of both countries scrutinized catch figures closely—both during and at the end of the season. At times, the issue of equal division of catch was pre-eminent with everyone in the industry. For some, a mediocre season with a close division of catch was more acceptable than a good season with substantial difference in catch.

Although both were critical of unequal division, Canadian industry was usually the more so. Complaints were accompanied by comments that "the fish are bred and reared in Canada and belong to Canadians." Industry representatives would point out that Americans do not participate in the costs of protecting the watershed, i.e. fighting pollution and opposing power dams, etc. It did not matter that in the 1980s, when large numbers of Fraser sockeye and pink salmon were taken by the Canadian industry outside the Convention area, the Commission was still obligated to achieve division of catch in Convention Waters. This unconcern held even though outside fishing activities (not under the control of the Commission) were very intensive and disruptive to the Commission's management task.

Each year the Commission had to obtain the necessary escapements and divide the catch equally. In order of priority, the Commission always considered resource conservation and enhancement (escapement), as the more important.

The significance of this was evident in the early history of Commission management. The Adams River run in 1950 was a failure, even though it followed a large return and spawning escapement potential in 1946. On September 6, 1950, Canada's catch in Convention Waters was less than 900,000 fish, compared to a total season's catch of 4.2 million in 1946. Canada was more than 300,000 fish behind the United States catch. There was pressure for a continuation of fishing in Canada while closures were placed on United States fishermen. In addition, no one could be sure whether or not there were "millions" of sockeye delaying in the Strait of Georgia. Despite strong urging, Commission Director, L.A. Royal, refused to recommend more fishing in Canadian areas. Even though there was an unprecedented closure from September 7 to 25, a minimum escapement of about 1.2 million sockeye reached the lower Adams and Little River spawning grounds. This emphasis on conservation paid big dividends; in 1954 about 9,000,000 fish returned to the Adams River. The escapement of over two million spawners from that run produced the largest single race return on record of about 15 million sockeye in 1958. This record still stands.

For 40 years of Commission management between 1946 and 1985, the total cumulative catch was within one-half of one percent of the fifty-fifty target. United States fishermen took 66,605,899 sockeye (49.64 percent) and Canada took 67,585,133 (50.36 percent) (Appendix G_1). From 1957 to 1985, on odd-numbered years only, the United States pink salmon catch totaled 37,145,683 (51.19 percent) and Canada's was 35,421,449 (48.81 percent) (Appendix G_2).

Since the division of catch in 1985 was under a new sharing* agreement (not fifty-fifty) as set by the Pacific Salmon Treaty, the catches

---

* Division of catch for sockeye by each country by cycle year is summarized in Appendix H_1 and total Fraser sockeye catch in all waters from 1946-1989 is listed in Appendix H_2. Total catch of Fraser pink salmon from 1959-1987 is listed in Appendix H_3 with division of catch in Appendix H_4.

for that year should not be included. For the years 1946-1984, the U.S. sockeye catch was 63,682,556 fish (50.31 percent); 62,908,988 fish (49.69 percent) were caught by Canadian fishermen in the Convention area. The pink salmon catch from 1957 to 1983 was 33,281,028 pinks (50.83 percent) for United States fishermen and 32,195,388 pinks (49.17 percent) for Canada.

During the odd-numbered years when pink salmon were involved, there were in-season proposals to bargain pink salmon for sockeye salmon if division of sockeye catch could not be made. One proposal was that if Canada was ahead on sockeye, the United States fishermen should be given more pink salmon than Canada to equalize the total catch. While the Commission could not openly support such a proposition, it did not want one country to be behind on catch division of both species (no matter how small the difference in catch) at the end of the same fishing season.

It was realized by most that yearly variations in the fifty-fifty sharing were rapidly forgotten and that any reasons to explain the discrepancies would be immaterial. The bottom-line, as viewed by industry and others, was the cumulative yearly catch totals for each country. As mentioned earlier, during the strike years the United States was 1,998,555 sockeye ahead of Canada in total catch, even though the United States took over four million sockeye while the Canadian industry was shut down. From 1946 to 1984 the Canadian industry catch was behind the U.S. by only 773,568 sockeye. Without jeopardizing the rights or benefits of either country's industry, catch equalization was nearly achieved. There was an unspoken understanding within the Commission that they were dealing with a Canadian resource and recognition of that fact (although not specifically required by the treaty) was an underlying principle in the Commission's actions. In later years, U.S. fishermen were upset if they were short on catch division because of the large catches taken by Canadian fisher-

men outside the Convention area. On occasion the Commission debated whether the Convention required division of catch each year or on a cumulative yearly basis. The general opinion was that it was intended to be on a yearly regime.

In discussing the complexities of the IPSFC management system and specifically on obtaining adequate escapements and division of catch, Crutchfield *et al.* (1969) stated: "The requirement virtually dooms the Commission to a series of second-best solutions, year by year, since only the most miraculous of coincidences could assure that closures to balance the catch would result in optimum escapement." The Commission was, however, able to obtain division of catch and reach escapement objectives.

The inclusion of sport caught sockeye and pink salmon into the divisional catch for each country was occasionally mentioned by some Commissioners, but usually in jest. Relatively few sockeye were taken in Convention Waters although there was increasing effort in recent years targeting sockeye at the mouth of the Fraser River. Pink salmon were susceptible to sports gear and a large catch was reportedly made in U.S. waters in 1963, the year of the largest U.S. run on record. Total sports catch by species in Washington State did not become available until 1972 and odd-numbered year pink catch from 1973 - 1985 ranged from 21,000 to 99,000 fish (State of Washington Department of Fisheries 1989). Sports catch data for sockeye and pink salmon for Canada were not available prior to 1985. The Commission never formally considered regulatory control of the sports fishery in Convention Waters or inclusion of the catch in division of catch. Sports catches were not included in computation of total run analysis except in 1985 for pink salmon.

The International Pacific Halibut Commission (IPHC) solicited legal interpretation from the two federal governments which resulted in adoption of sports fishing regulations starting in 1973 under the authority of the

Halibut Convention. The IPHC has implemented sports fishing regulations for each year since 1973 (The International Pacific Halibut Commission 1987).

VARIABLE MIGRATION PATTERNS

A prevailing concept, prior to the Commission's existence, was that the vast majority of the Fraser River sockeye runs migrated towards the Fraser River through Juan de Fuca Strait. There was, however, evidence of sockeye migrating into the Strait of Georgia from the north in the 1910s, 1920s and 1930s (Babcock 1914a, Gilhousen 1960). In those years the fishery in Johnstone Strait was small, migration was low, and there were only small numbers of Fraser sockeye. Thus the impact on the annual runs was considered inconsequential.

Still later, from 1946 through 1956, even though fishing intensity in Johnstone Strait was increasing, there was little effect on Fraser stocks. In 1957 however, significant numbers of Fraser River sockeye were captured and in the next year (1958) more than 4,000,000 Fraser sockeye were taken at the northern end of Georgia Strait before the fish reached Convention Waters. The total return of almost 19,000,000 fish in 1958 was exceptionally large. It was later calculated that about 35 percent of the total run migrated through the northerly route. This was certainly the largest recorded number of fish and the highest percentage up to that time. In contrast, for the large 1954 run, the parents of the 1958 returning fish, only about one percent came through Johnstone Strait. In 1958 the high diversion rate, the large numbers of sockeye, the unexpected reduction in fishing efficiency due presumably to aberrant migration behavior, and increased regulatory restrictions resulted in an increased escapement from the Johnstone Strait fishery. This contributed, in part, to at least 1,500,000 surplus sockeye (mainly Adams River race) arriving at the mouth of the Fraser River (IPSFC

1959). The late timing, the large number of sockeye and their poor quality resulted in a Canadian industry price dispute. A surplus number of sockeye of poor spawning quality migrated up the Fraser River to the Lower Adams River spawning grounds. An electric fence was installed at the mouth of Adams River to prevent approximately 1,000,000 late-arriving poor quality spawners from reaching the Adams River. Most of these fish spawned along the lake shore beaches where Mitchell (1925) had previously observed large numbers of sockeye spawning. The fence was a "necessary function of proper management" (IPSFC 1959) and did not contribute to the poor return in 1962.

In the twenty years following the large 1958 migration through Johnstone Strait, the migration proportion using the two routes was fairly normal (75-80 percent via Juan de Fuca Strait) from 1959-1977,* but in 1978 a large diversion of sockeye through the northerly strait took place. This continued for many years. The proportion of the run approaching the Fraser River through Johnstone Strait in 1978 was unprecedented—an estimated 58 percent of the run used that route. The diversion rate in 1979 was 30 percent, the largest on record for that cycle. In 1980, a record 70 percent of the Fraser sockeye migration came via Johnstone Strait. Following the pattern of the three previous years, an estimated 67 percent of the 1981 Fraser sockeye run returned through Johnstone Strait. The migration pattern in 1982 was normal with an estimated 22 percent of the total run migrating via Johnstone Strait. Oceanographic conditions such as El Nino prevailed in 1983 and it was estimated that a record 80 percent of the Fraser sockeye run returned through Johnstone Strait. In 1984 migration from the north was estimated at 31 percent. The same percentage occurred in 1985.

The percentages of Fraser River sockeye

---

* The percentage migration in 1972 reached 34 percent but the total return was much lower than in 1958.

164

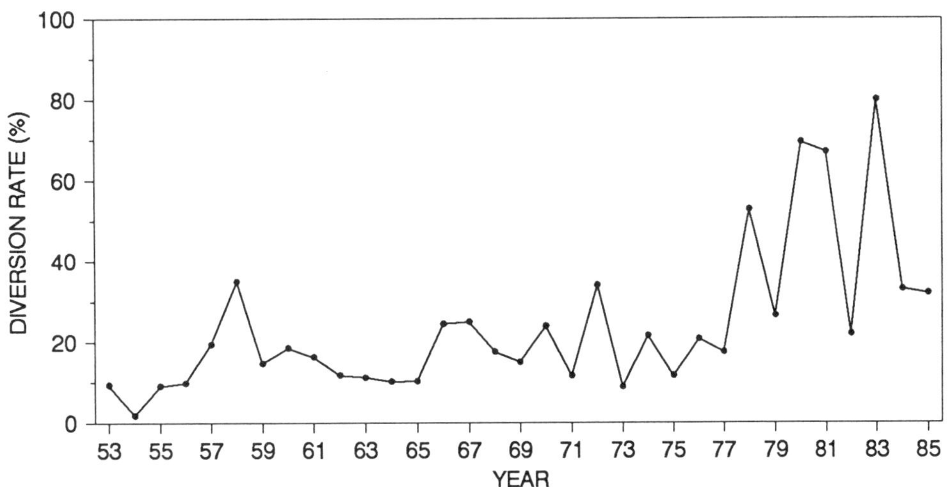

Figure 23. Proportion of Fraser River sockeye salmon migrating via Johnstone Strait, 1953-1985.

migrating via Johnstone Strait from 1953 to 1985 are shown in Figure 23.

The high diversion rates of these later years, along with an intense and highly efficient fishing fleet, dramatically influenced the numbers of Fraser River sockeye that actually reached Convention Waters, causing serious conservation and management problems.

In 1980 when 70 percent of the total Fraser sockeye run migrated through Johnstone Strait, Commission staff expressed concern about heavy fishing on the major component of the run, the dominant cycle-year Chilko stock. It was fortunate (following Commission staff urging) that Canada Department of Fisheries and Oceans reduced previously scheduled fishing time in the Strait. This permitted adequate sockeye to reach the mouth of the Fraser River, where they were protected for escapement. Because of the large harvest north of the Convention boundary, the gillnet catch in Canadian Convention Waters (primarily in the Fraser River area) was the lowest on the cycle since 1928 (IPSFC 1981). There was a strong outcry from Fraser River gillnetters, but the problem was outside the Commission's area of control or influence.

During the 1983 fishing season—a season of a record 80 percent diversion of the sockeye run through Johnstone Strait—excessive fishing in non-Convention Waters resulted in inadequate escapement. Three quarters (75.5 percent) of the total commercial catch in 1983 was taken in non-Convention fisheries outside Commission control. The total spawning ground escapement of 976,000 sockeye was more than 500,000 fish below the pre-season goal and 432,000 lower than the brood year escapement. Only 44.1 percent of the total run reached Convention Waters, the lowest percentage on record. Lack of Commission control and appropriate restriction of the fishery in Johnstone Strait resulted in significant overfishing of Fraser sockeye. The Commission informed both governments in 1983 of the serious management problems. The governments offered no assistance.

Variations in the percentage of the Fraser River pink salmon run returning through Johnstone Strait have been much less compared to sockeye (Figure 24); however, in 1983, a record 66 percent of the pink salmon run approached the Fraser via the north route. Earlier, from 1959 to 1981, it varied only slightly from the 30 percent figure.

Scientists speculated as to why the proportion of sockeye approaching the Fraser River via Johnstone Strait varies. The temperature of the North Pacific Ocean was the first environmental factor considered (IPSFC 1959). Previous examination of this phenomenon suggested a natural periodicity of 10 or 11

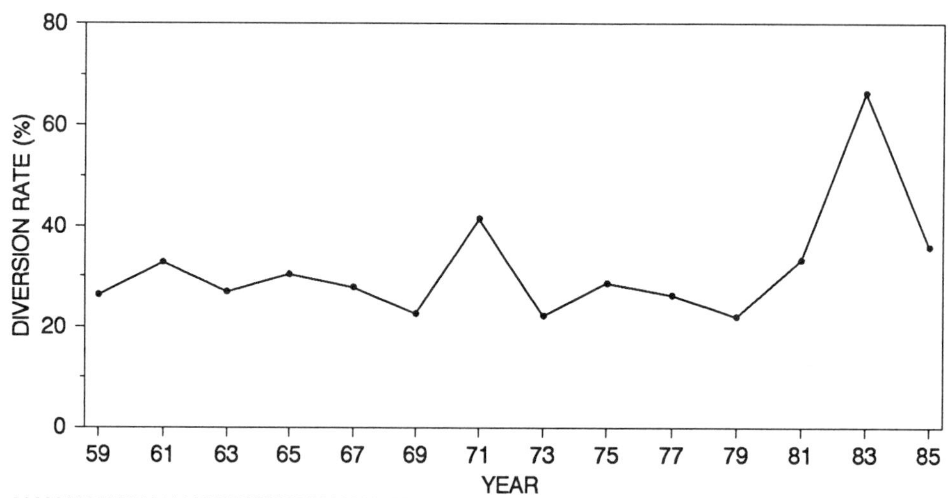

Figure 24. Proportion of Fraser River pink salmon migrating via Johnstone Strait, 1959-1985.

years, similar to the periodicity of sunspot maxima (Gilhousen 1960).

By the late 1970s, influences such as water transport along the British Columbia coast, salinities, sea levels, a downwelling index, river discharge, atmospheric pressure, and coastal rainfall were looked at. With the high diversion rates in 1978, 1980 and again in 1981, it was evident that such multiple factors were involved. Most analyses of temperature and pressure showed that the highest correlations (r = .72 to .74) for sea temperatures at Neah Bay (March - April) and sea level barometric pressure at 39°N - 146°W (January-March) occurred in late winter and early spring (Blackbourn 1984). This suggested sockeye probably make their migratory route decisions before the summer season. Although no single factor was involved, ocean temperatures produced as good a correlation as any other indicator.

Oceanographic conditions in 1983 were highly unusual. There were record high sea levels and water temperatures, as evidenced by El Nino. The percentage diversion in 1983 (Figure 23) of about 80 percent was the highest of any year on record. Sea surface water temperatures were about 3°F (1.7°C) above average. This warming factor seemed to be present whenever high diversion rates occurred. The relationship between these temperatures on the west coast of Vancouver Island and diver-

sion rates for the years 1953-1983 (Figure 25) shows a significant statistical relationship (r = 0.66). Temperature alone, however, still accounted (statistically) for less than half of the relationship.

Although the high diversion rate in 1983 was definitely associated with the warm water brought in by El Nino, the 1980 and 1981 diversions were not. Large northerly diversions also occurred in 1958 and 1978. The spacing between these migration patterns seems to occur no less frequently than 10 to 12 years. It was found, however, that the unusual migrations of 1980 and 1981 could apparently occur in any year. They were not predictable until the spring of the return year.

Further analysis of the Commission's data on sockeye migration through Johnstone Strait was made by Hamilton (1985). He found "a very significant . . . correlation between the diversion rate and the temperature change in the coastal waters off Vancouver Island over the last 18 months of the salmon's ocean residence." He identified 1915, 1926 and 1936 as years of probable large sockeye diversions. The best evidence to date suggests ocean temperatures as an important factor in affecting migration routes. It is important to continue studies to identify and understand the reasons for the diversions as well as to establish management plans to harvest and protect the stocks. Addi-

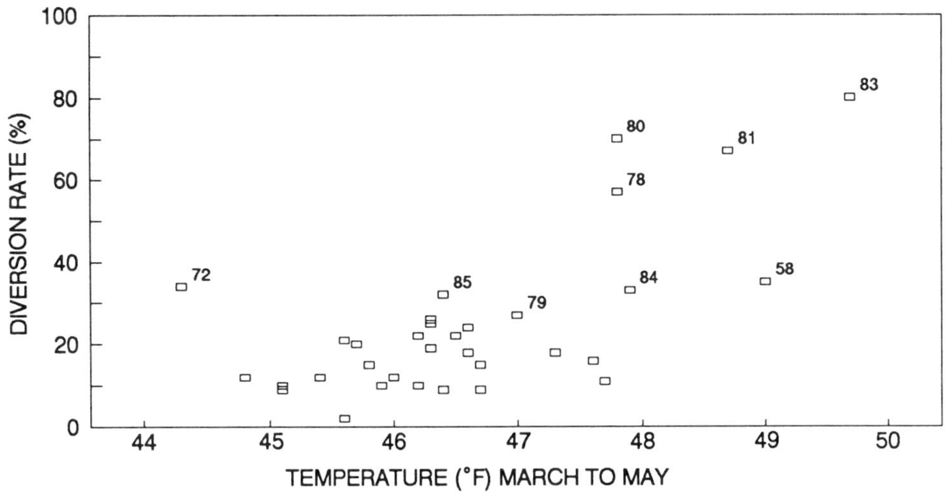

Figure 25. Proportion of Fraser River sockeye salmon migrating via Johnstone Strait in relation to sea surface temperature at Kains Island and Langara Island from March to May of the return year (1954-1985).

tional discussion of these migratory variations can be found in Wickett (1979).

VARIATION IN TIMING OF THE RUNS

The pre-season fishing regulations for harvesting Fraser River sockeye were set prior to the season and were based on expected or normal timing of the various races. The Commission realized in the 1950s and 1960s that there could be, in any particular year, wide fluctuations in appearance of the individual races in the fishery. While the timing could definitely be established during the season, it was also important to be able to predict timing well before the fishing season starts. Timing and relative abundance for one of the four cycles (1984 dominant Chilko run) are shown in Figure 26.

A major aberrant timing occurred in 1958 with the early Stuart runs about ten days later than normal, and the Chilko and Adams River populations at least three weeks later than normal. Another extreme variation took place in 1963 when the Chilko run peaked in Puget Sound on July 23, almost two weeks earlier than normal (August 2). Thus, with the 1958 Chilko run peaking (late) in Puget Sound

on August 21-22 (Henry 1961), there was one month's fluctuation in the arrival time of a particular stock. Since all races generally follow each other in chronological order and spacing within a season—with few exceptions—it can easily be seen that determination of expected timing well in advance of the fishing season is very desirable for the management agency.

While many factors were examined to explain variations in timing of the runs, the most useful environmental factor identified was the surface water temperature in the mid-Gulf of Alaska early (January-May) in the year of return. When these temperatures were above the long-range average, the tendency was for the runs to return later than normal.

Surface temperatures of the Pacific Ocean at Station P for the months January to May were above average in 1958 and 1981. It was hypothesized that warm temperatures during these months displace (Figure 27) maturing sockeye further to the northwest (Blackbourn 1987). If this was the case, then it followed that in their return migration sockeye would have further to migrate and would reach the Fraser River later. There was also a tendency toward higher diversion via Johnstone Strait when the runs returned late.

Oceanographic data that may be associ-

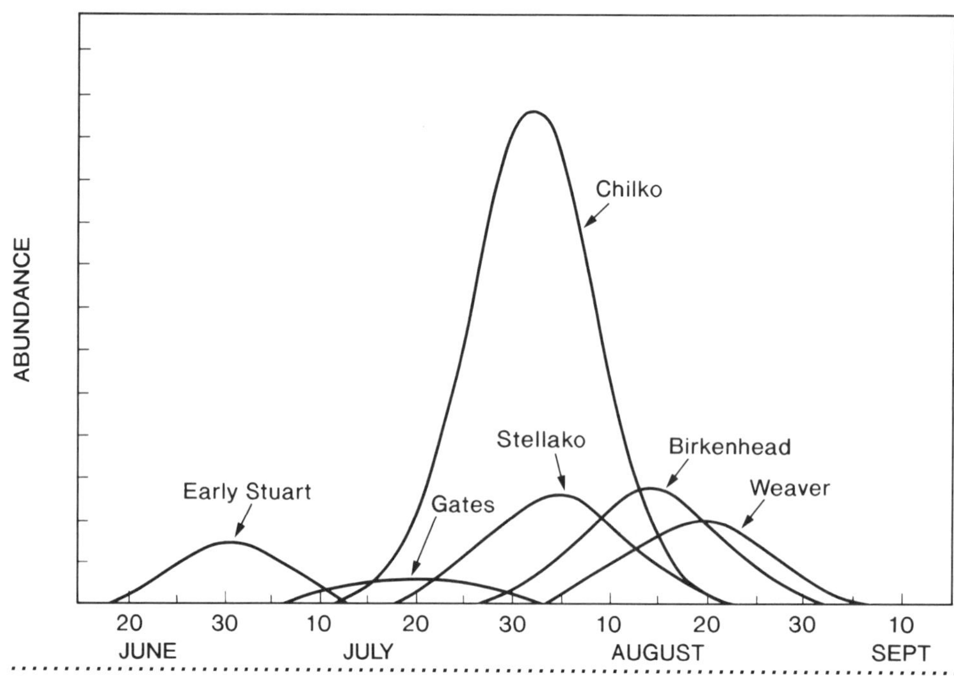

Figure 26. Timing and relative abundance in Juan de Fuca Strait of some of the sockeye salmon runs on the 1984 cycle.

ated with timing and direction of migration of the stocks have been studied only in recent years. The findings have provided useful insights into possible factors affecting migration behavior. The emphasis on oceanographic research must continue. The need was recognized by the Commission in 1959, following the anomalies of the 1958 season. In its report to the governments in 1959 the Commission stated:

> . . .on the basis of existing evidence, fish size, distribution, timing and path of adult migration appear related to oceanic environment and further knowledge of the cause of anomalies in those characteristics is essential to scientific management of the salmon resources. . . .The required physical data on estuarial and oceanic environment is not now available for the proper management of our fishery resources. There is an immediate need, of considerable economic and biological significance, for a 'Western Bureau of the Sea'. (IPSFC 1960)

The significant fluctuations in timing and migration routes, especially in the late 1970s and early 1980s will likely occur again, and prior knowledge to predict these events is essential.

Unusual migration patterns and behavior also occur in the inside waters; for example, the infrequent, but sometimes significant migration of Adams River sockeye through Haro Strait instead of through United States waters (Salmon Banks). In addition, the three major races, Stellako, Chilko and Adams River, had peak catches on the same day in 1979 (August 6) in United States waters. This was the first time on record that these three races yielded peak seasonal catches on the same day in the same area. This took place as a result of delaying Stellako River sockeye, which also contributed to a large catch in Canadian waters on the same day. This clearly illustrates the necessity of having accurate racial catch data.

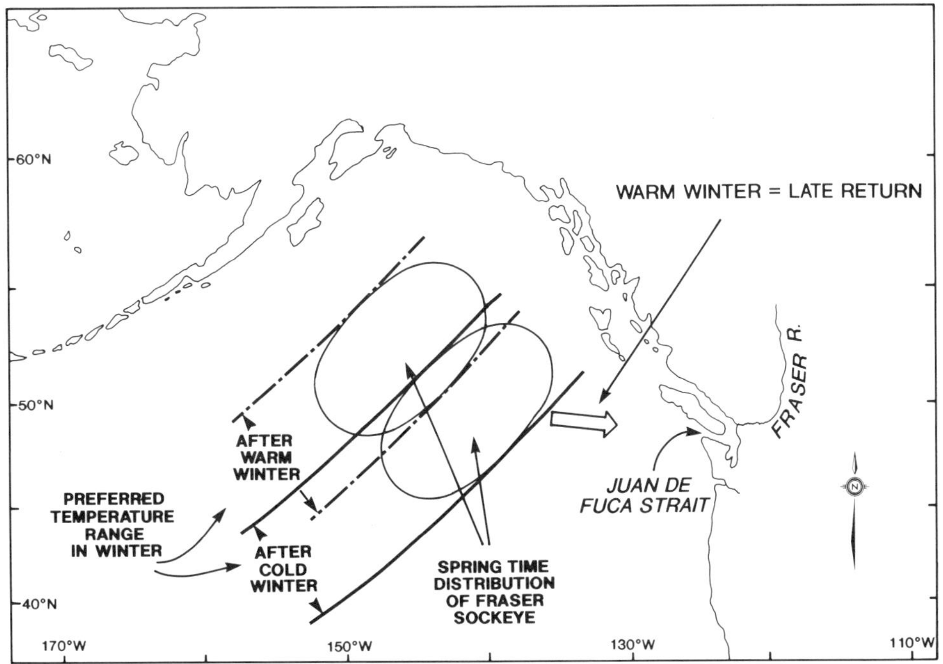

Figure 27. A simple temperature-displacement model of the return timing of Fraser River sockeye salmon. From Blackbourn 1987.

VARIATION IN SIZE OF SOCKEYE AND PINK SALMON

The size of individual sockeye returning to the Fraser River varied considerably from year to year. From 1946 to 1985, the average weight of four-year-old sockeye ranged from a low of 5.12 lbs. in 1959 and 1983 to a high of 7.21 lbs. in 1951. One of the cycle years always had large sockeye and that was the dominant lower Adams River runs (1982-1986 cycle). The average weight of four-year-old fish was 6.44 lbs. (range of 5.9-7.0 lbs.). The other three cycle years had similar average weights: 1983-1987, 5.73 lbs.; 1980-1984, 5.76 lbs.; and 1981-1985, 5.71 lbs. Killick and Clemens (1963) examined oceanographic data in relation to size of sockeye. They found an inverse relationship between sea surface temperatures and size of sockeye (cold sea temperatures were related to large size of fish).

While environmental factors appeared to affect the size of sockeye there were also genetic differences between the races. Adams River four-year-olds were always large fish. Pitt River four-year-olds and five-year-olds were, however, probably the largest of all races for their ages. Nadina River sockeye were one of the smallest of all the races of sockeye. Although there were yearly variations in the average weights of each race, there was a consistent small-size or large-size that was characteristic of each race. In general, the pattern of size carried through from the beginning to the end of each season.

Changes in average annual weight of pink salmon were greater than observed in sockeye salmon. From 1957 to 1985 average weight of pink salmon varied from 6.72 lbs. in 1961 to 4.31 lbs. in 1983. Fraser River pink salmon average weight was negatively related to the numbers of returning fish—the greater the number, the smaller the fish (IPSFC 1984). Although the effect of numbers on size of individual fish was consistently evident, environmental factors also affected the size of fish. In 1983, when El Nino occurred, the smallest-sized pink salmon returned.

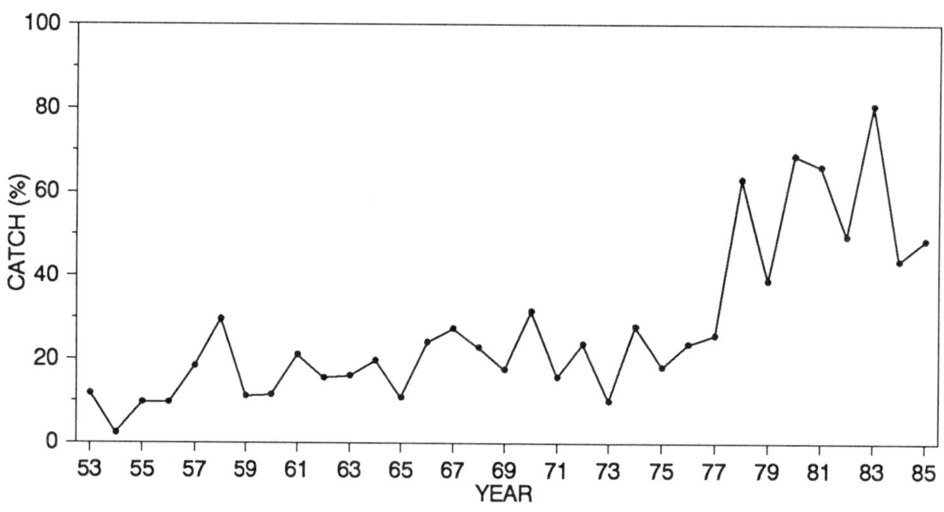

Figure 28. Proportion of Fraser River sockeye salmon catch taken outside of IPSFC regulations, 1953-1985.

CONTROL OF FISHERIES

The problem of diminishing control of the fisheries confronting the Commission (for reasons stated earlier) was further exacerbated in 1977 when the United States government began to unilaterally regulate the treaty Indian fishery in United States waters (discussed in Chapter 15).

In the late 1950s, the Commission asked both governments for clarification of the Canadian government's responsibility in providing escapement of Fraser sockeye and pink salmon that were migrating through Johnstone Strait. No clear policy statement was received. The Commission also asked for definition of its responsibility to provide escapement of pink salmon originating in Washington State that approached their spawning grounds via the Convention Waters in Juan de Fuca Strait. Again no direction was given to the Commission.

From the inception of Commission regulations in 1946 to the mid-1950s, it was common for 90 percent or more of the total harvest of Fraser sockeye to be taken in Convention Waters under Commission regulations. By 1983, control of the Fraser catch was reduced to the extent that a record 81 percent of the total sockeye catch (Figure 28) and 79 percent of the

pink salmon catch was taken under regulations promulgated by agencies other than the Commission. Multi-agency management was contrary to the initial treaty discussions and both governments sought to eliminate conflicts arising from multinational control of the fisheries.

The Commission's responsibilities, as defined in the Convention, did not require it to respond to the biological needs of other species. In addition, it could not officially acknowledge that it was considering the requests of the national authorities with respect to species outside the jurisdiction of the Commission. Where and when possible, the Commission responded to the input provided by the national technical authorities but the overriding management responsibility was always toward Fraser River sockeye and pink salmon. The Commission did whatever it considered possible and appropriate to do to manage its stocks while at the same time assist and be sensitive to the protection of other species. For many years in the 1960s, the Canadian government consulted with the Commission for approval of large mesh net fisheries specifically for the harvest of Fraser chinook in the Fraser River. These openings were additional to the normal fishery periods given by the Commission for sockeye salmon fishing.

Environmental Problems and Research

After most of the fishway construction was completed, the Commission's attention was directed toward defining the environment's role in stock productivity. Industrial development and increased human population within the Fraser watershed in the 1950s and 1960s were potential threats. The Convention called for the protection and preservation of sockeye and pink salmon of the Fraser River, and the Commission had to know the impact of each industrial or domestic development proposal. Although federal and provincial governments had specific control over land and water uses in the watershed, it was vital that the Commission advise the federal governments on technical aspects and recommend controls.

A decision of significant importance took place on September 22, 1938. A request (by resolution) was made by the Commission to the Canadian government that the Commission be notified of any proposed activity in the Fraser River basin that might affect sockeye salmon. It established an authority and a responsibility for the Commission to evaluate the impacts of proposed developments on the environment and to advise the Canadian government on these matters. At that time there was no facility for detailed studies of salmon responses to environmental changes.

THE SWELTZER CREEK FIELD STATION

For several years the Commission planned to establish its own research station with equipment and facilities necessary for environmental research. Sweltzer Creek, a short distance downstream from the outlet of Cultus Lake, was chosen as the site and the Sweltzer Creek Field Station was completed in 1961. It provided research facilities for studies on the tolerance limits of sockeye and pink salmon to environmental changes due to natural causes, enhancement, flood control projects, and power development. The station's Environ-

ment Conservation Division also investigated forest spraying, mosquito control, log driving, marine disposal of sediments, effects of decaying bark on incubating salmon eggs, responses of adult sockeye to thermal shock, effect of heavy metals on juvenile sockeye and pink salmon and many other biological and environmental problems. The research laboratory had a small, highly specialized staff capable of investigating and defining complex environmental situations and problems. Also housed in the facility was the staff of the Commission's Biology Division.

PULP MILL EFFLUENT STUDIES

By 1961, proposals to operate pulp mills in the British Columbia interior were being considered by a number of companies. The Department of Fisheries of Canada received numerous enquiries on requirements for waste disposal under terms of Canada's Fisheries Act. The Department and the Commission had already carried out intensive studies on the possible impact of pulp mill effluent on the salmon resource. A report outlining recommended procedures and acceptable practices and levels of effluent compounds was available. It formed the basis for effluent-control facilities for pulp mills established in the mid-1960s at Prince George and Kamloops. The Commission published many reports on its investigations; one showed that the toxicity to salmon of pulp mill effluents could be substantially reduced by biological treatment (Servizi *et al.* 1966).

The policy adopted and sustained was that all wastes from industrial plants should be treated by whatever procedures necessary to reduce toxicity to a defined minimum. Initially, one of the companies was reluctant to accept these strict standards and planned to utilize the Fraser River's capacity for dilution. As a result, the government adopted a policy that receiving waters at the point of effluence

February 29, 1968. Sweltzer Creek
discharges into the Chilliwack-Vedder
River, where severe floods accentuated by
logging and other watershed
developments threaten the pink salmon
resource.

should be maintained in a high-quality state (as a margin of safety) and that effluent entering the river should be non-toxic to fish immediately encountering the effluent. There was no preferential treatment given to any industry based on location of plant or time of construction. The immediate impact of effluent on the receiving waters and the fish in them was the only criterion of meeting environmental standards.

It is easy to overlook the importance of policies adopted in the early 1960s that maintained and preserved water qualities at a high level. The facts of record-size sockeye runs and large escapements are more easily identifiable and recognized. Nevertheless, assisting in the development of these policies was an important contribution by the Commission.

By 1968 four pulp mills were operating in the Fraser River basin. Monitoring effluent outfalls was required and at times samples showed that specifications for biochemical oxygen demand (BOD) and toxicity were not being met. Commission staff counseled and assisted mill personnel by recommending modification of in-plant operations to correct treatment procedures and reduce outfall toxicity levels.

Not all applications for mill construction were approved, such as a proposed mill at Ashcroft, B.C. on the banks of the Thompson River which was rejected because of serious threat to salmon populations in the Thompson River.

As studies of pulp mill effluents progressed, new compounds were discovered and understandings developed on the effects of different wood varieties on the composition of the effluent. Studies indicated that approximately four to five days treatment was sufficient to detoxify mill effluent when it was held in aerated lagoons (Gordon and Servizi 1974).

A federal-provincial task force, assisted by Commission staff, was formed in the early 1970s to study the odor, foaming and excessive algae growth in the Thompson River and Kamloops Lake. It was seen by many as a serious biological threat, and the public wanted the river clear and free of algae. A 1976 report concluded that an increase in phosphorous-loading caused the increase in algae. As a result, the Weyerhaeuser Canada Limited pulp mill, upriver from Kamloops Lake, stopped adding phosphorous and nitrogen to the aerated lagoon in 1976. This reduced nutrient levels which discolored the water from being added to the Thompson River. Detoxification of effluent was unaffected since nutrient needs were satisfied from other sources in the mill. Phosphorous was also removed from Kamloops municipal sewage effluent by precipitation with alum.

Studies continued to evaluate conditions in the Thompson River. Egg-to-fry survival of pink salmon was high; algal growth had not, however, disappeared and the mill was directed by the Pollution Control Branch to undertake studies to reduce the amount of brown-colored water discharged into the Thompson River.

During the approximately 20 years that Kraft-process pulp mills were operating in the Fraser River basin, there were numerous breakdowns in plant operations or procedures, resulting in effluents that did not meet pollution control standards. In spite of these, there has been no known instance of Fraser River sockeye and pink salmon losses because of effluent toxicity.

Research conducted by the Environment Conservation Division played a vital role in the development of pollution control criteria, not only for the Fraser River area, but also for all of British Columbia and Canada. Dr. Loyd A. Royal, as Director of the Commission with knowledge of pulp mill problems in Washington State, was instrumental in establishing these protective measures in British Columbia. Studies led to a definition of toxicity of pulp mill effluent under the Fisheries Act for Federal Pulp and Paper Regulations. Research led by Commission staff, in cooperation with industry, federal and provincial agencies, was a major factor in the adoption of the required minimum four and one-half day treatment of Kraft-process pulp mill effluents in aerated lagoons. Research, which finished in 1984, also resulted in criteria for detoxification of effluent from thermomechanical and chemical thermomechanical pulp mills.

Study results are contained in Commission reports (Appendix I) published from 1965 through 1985 and also in other publications as referenced in the Commission's annual reports. Other Commission publications and reports are also listed in Appendix I.

PRE-SPAWNING MORTALITIES

One of the more difficult and frustrating problems the Commission encountered was the loss of sockeye that had escaped all fisheries, in certain years and on certain spawning grounds, but which died before depositing their eggs. It was particularly disheartening to see the fish survive all of life's vigors, escape the fisheries, and reach their ultimate destination, yet fall short of their goal to reproduce. Substantial numbers of unspawned and dead sockeye were observed in 1949 and again in 1953, mainly in the Stuart Lake and Quesnel Districts, and the problem exploded in 1961 when several major races suffered high pre-spawning mortalities. Extensive pre-spawning losses occurred: early Nadina River, 86 percent; Bowron River, 60 percent; Chilko River, 31 percent; Horsefly River, 62 percent; and Stellako River, 31 percent. Water temperatures were high and the runs had arrived earlier than normal. The rehabilitation potential for future cycles was greatly diminished by these losses. The Commission intensified its investigation into the problem following a loss of 90 percent of the Chilko run in 1963. Both male and female sockeye died unspawned.

A team of experts in the medical, veterinarian and fish-pathology fields was mo-

bilized. One of their first findings was that columnaris disease was prevalent in several locations of the Fraser system (Wood 1965). This disease, caused by the myxobacterium *Chondrococcus columnaris,* was believed responsible for most of the large loss of Chilko sockeye in 1963 and for the large mortality at Horsefly River in 1961. This was the first time the disease had been identified in Fraser River sockeye. Columnaris had been found on Columbia River sockeye (bluebacks) in the 1940s and 1950s contributing to pre-spawning mortality (Fish 1948 and Craddock 1958).

Initial studies were centered on the Horsefly River population because of its great potential and because the pre-spawning losses of this race had been 30 percent in 1953, 62 percent in 1961 and 47 percent in 1965. In every dominant year, a high mortality rate was associated with warm water. For example, in 1961 the temperature in the lower Horsefly River reached 72°F. Normal acceptable water temperatures for Fraser sockeye during spawning is approximately 55°F. Examination of the data suggested that if water temperature could be maintained below a daily maximum of 57°F, the mortality of unspawned sockeye could be greatly reduced. Therefore, in 1966 the Commission recommended that cold water be siphoned from McKinley Lake (tributary to the Horsefly River) to cool McKinley Creek. This was a prototype temperature-control study to test the effectiveness of cooler water in controlling the disease before proceeding to a much larger and more expensive program involving drawing cold water from Crooked Lake to cool down a major portion of the Horsefly River (IPSFC 1966a).

Construction for the McKinley Creek temperature-control project was completed in June 1969, prior to the August arrival of sockeye. The project maintained water temperatures below 57°F for at least the upper half of McKinley Creek. The project successfully controlled columnaris disease. Despite the physical success of the project, however, a pre-spawning

October 22, 1968. A temperature-control structure was built on McKinley Creek, tributary to Horsefly River, to test the feasibility of providing more favorable spawning ground temperatures.

June, 1969. A 5 ft.-diameter pipeline was placed in McKinley Lake to draw water from below the thermocline to provide temperature control for a test area in McKinley Creek.

Commissioner Richard Nelson (on the left) and Advisory Committee member Ken Fraser (third from left) inspected the newly completed McKinley Creek temperature-control facilities in 1969 along with staff members Jim Pyper and Loyd Royal.

loss of 65 percent of the sockeye occurred in McKinley Creek. It appeared that a virulent bacterial gill disease was responsible for the mortality. Further investigations were necessary in light of the 1969 findings (Williams 1973).

During the course of the investigations it was determined that suckers (*C. catostomus*) residing in the lower Fraser River carried the bacteria causing columnaris (Colgrove and Wood 1966). It was also revealed that the causes of pre-spawning mortality of Horsefly River sockeye were complex. Perhaps a major cause was that the small number of sockeye of this race (1,000 fish) which existed in 1941 did not provide adequate genetic diversity to adapt to the environment. Other factors were also implicated, such as temperature in the lower Fraser River during upriver migration, timing of the run, (early timing being unfavorable), other infections, degree of sexual maturation and spawning ground water temperatures (Cooper 1973). The most critical factors appeared to be the timing of the runs and the cumulative temperature regime fish encountered during upstream migration associated with spawning ground temperatures.

The Commission considered chemical therapy to overcome the disease problem. Any possible drug use would have to have large-scale applicability to be practical. In 1970, Ni-Furpirenol (P7131), a nitrofuran compound effective against some fish diseases, was tested for its effectiveness in controlling pre-spawning mortalities of Birkenhead River and Chilko River sockeye. It was not considered a practical therapeutic agent for Fraser River sockeye.

Since the earliest fish arriving on the spawning grounds always suffered the highest pre-spawning mortality, it was logical that a histological examination of the early and peak segments of the run be performed. Studies were conducted on Chilko fish in 1971. Cortisol-cortisone measurements, physical parameters, body lipids and various tissues were examined. The early fish tended to have higher hematocrits, high plasma cortisol levels, and evidence of gill irritation. Also, a combiotic was administered to sockeye by injection. While there were some beneficial results, broad-scale treatment was not considered practical.

Pre-spawning studies continued in 1973 on the Horsefly River sockeye run. The pre-spawning mortality of 27 percent was considerably lower than in previous cycle years and there were no differences in mortality rates between the first, central, and later segments of the run (Williams *et al. 1977*). Water temperatures en route to the spawning grounds were above average, but below average at the grounds themselves. Timing of the run; however, was later than average. For the various parameters measured, there were very few differences between the first and central fish sampled in salt water and at the spawning grounds. It was concluded that cool water temperatures at the spawning grounds and later timing of the run were favorable to reducing pre-spawning losses.

In 1977 there was another pre-spawning mortality in the Horsefly River (as well as other races) of about 40 percent. The entire Fraser River escapement suffered about a 26 percent loss. With the Horsefly River runs significantly increasing in numbers, and with later timing, losses decreased to nineteen percent in 1981 and five percent in 1985. Whether or not the Horsefly race has reached the end of these major mortalities remains to be seen. It is believed that historical timing and distribution of spawners have been achieved. That should continue and should be beneficial to the race.

Pre-spawning losses in the Fraser system have decreased significantly since the mid 1970s. During this period there were significant increases in run abundance for several races, below-average size of individual fish—especially in the 1980s, and later arrival on the spawning grounds. It has not been possible to single out one key factor affecting pre-spawning losses. It appears that later timing is very important. High pre-spawning mortality and late-timed runs seldom occur simultaneously.

The problem of pre-spawning loss of Fraser sockeye was studied by IPSFC staff for many years and Gilhousen (1990) summarized the problem and completed the analysis. It appears that pre-spawning mortality is the cumulative effect of stress imposed on the fish during the transition from saltwater to freshwater and subsequently during the upriver migration to the spawning areas. Pre-spawning mortality is significantly related to earlier-than-normal migration timing and to higher-than-normal river temperatures. Other adverse environmental factors such as high Fraser River sediment load and pollution levels may also contribute to pre-spawning mortality.

One can only speculate as to what biological phenomena have taken place in the salmon populations of the Fraser system. The periodic and perhaps selective effects of Hells Gate blockages for over thirty years undoubtedly had an effect on the genetic make-up and adaptability of many populations. The original genetic composition of these populations is not known, but it is known that all fish returning to the Horsefly River in 1941 and 1945 and for several cycles later, tended to be early and spawned in the upper reaches of the river. Escapements in the 1980s were much later in timing and an increasing proportion used the lower Horsefly area. Perhaps the fishways themselves, although providing access upriver for salmon, initially accelerated arrival time into a spawning ground environment unacceptable for some depleted races. Greater stability in timing and survival has now been achieved and this should lead to higher production. On the other hand, if the beneficial effects of the recent low mortalities on the spawning ground are due primarily to oceanographic conditions in recent years (as they affect run-timing), then fluctuations in survival can again be expected. While it does not seem possible to reduce losses by treatment of fish or water supply, prior knowledge of events which are likely to take place would be valuable in each year's management program.

Pre-spawning losses for Fraser River pink salmon have, for all practical purposes, been non-existent. In 1963, although spawning ground temperatures were very high and pink salmon were infected with columnaris, especially in the main Fraser, mortality was low. It was believed that the early-run pink population spawning in the lower Fraser River was genetically similar to the population spawning above Hells Gate. If this was the case, there would have been little adaptation required for stocks gaining access to upriver areas following fishway construction. Hence, there was no significant pre-spawning mortality for either the early or late run segments of the pink salmon stock.

TOXICITY STUDIES

In addition to the Kraft-process pulp mills, industrial development in the Fraser River basin continued at a rapid pace in the 1960s. Related fisheries problems required considerable attention. Discussions were held with a chemical company regarding the establishment of a chlorine and caustic-soda manufacturing plant on Vancouver Island. Other problems involved a proposed steel mill on the North Arm of the Fraser River and its proposed disposal of flue dust in the river, studying the effects of test spraying of forest with the fungicide Phytoactin, and evaluating impacts of a sodium chlorite plant proposed for Prince George. This latter study established that sodium chlorite was far more toxic to sockeye than to other species of fish. The Commission and the Department of Fisheries of Canada studied the effects of peat disposal from Burnaby Lake to the Fraser River, and both agencies participated in technical discussions concerning an oil refinery and sulfuric acid plant proposal near Prince George. These results are documented in IPSFC annual reports.

In the 1970s the Commission investigated the toxicity of heavy metals on salmon. Mining operations and ore-processing are controlled by laws to prevent contamination of nearby and downstream waters, but some heavy metals do reach the environment. Research indicated that exposure of sockeye and pink salmon to copper during the egg-to-fry stage caused mortality and abnormally slow yolk absorption. This occurred at concentrations lower than levels found to be lethal to fry and fingerlings. Studies on the effect of mercury and cadmium showed mercury to be more toxic to developing sockeye embryos than to fingerlings, but the reverse was the case for cadmium. Initial experiments indicated that chronic exposure of eggs to zinc caused delay in hatching. Further studies on mercury showed that eyed eggs, alevins, fry and smolts could withstand concentrations that were lethal to pre-eyed developing eggs. Sockeye eggs and fry were found to accumulate copper and mercury in proportion to the level of exposure. Thus, eggs and fry taken in the wild could be used to measure their exposure to these metals in the natural environment. Tests showed that the tolerance of sockeye to copper, mercury and cadmium remained relatively constant from fry through the fingerling stages (Servizi and Martens 1978).

Important studies on municipal raw sewage effluents were undertaken by the Commission in the early 1970s. At that time much raw sewage was discharged into the Fraser River. A biological monitoring program was conducted at Mission and Steveston, B.C. where juvenile salmon were held for a week and longer in water pumped from the Fraser River. The tests also involved developing salmon eggs. No acute harmful effects were observed.

The construction of municipal sewage treatment facilities became of greater importance as the public increased its demands for a cleaner environment. Since chlorine is usually used to disinfect treated sewage, the effects of chlorinated sewage on salmon became the focus of the Commission's investigations. Tests performed using alevins, fry, smolts and adults demonstrated that residual chlorine was extremely toxic to salmon. Further studies showed that dechlorination could be achieved on a small scale by lagooning chlorinated sewage. This was an effective method of removing acute toxicity. It was not, however, considered practical for large sewage flows. A full-scale chemical dechlorination study was conducted at the Lulu Island primary treatment plant on the lower Fraser River in cooperation with the Greater Vancouver Sewerage and Drainage District and the Department of Environment. Toxicity studies showed that primary treated municipal sewage was lethal to fingerling sockeye salmon (Martens and Servizi 1974). Following chlorination of primary sewage to an acceptable level of disinfection, the toxicity to salmon increased substantially. Further studies determined that when domestic sewage was dechlorinated by lagooning following secondary treatment, it was not acutely toxic to juvenile salmon (Servizi and Martens 1974). A sulfur dioxide system to dechlorinate the entire flow of the treatment plant was installed. Performance of dechlorination was monitored chemically by testing for residual chlorine and biologically by continuous-flow bioassays of young sockeye (Martens and Servizi 1975). The tests showed that chlorine-induced toxicity could be removed by using sulfur dioxide. However, chlorination caused the formation of organic constituents in sewage that are not removed by this procedure. The residual constituents were toxic to fish at very low concentrations. By the mid 1970s, dechlorination was adopted at the Annacis Island and Lulu Island sewage treatment plants on the lower Fraser River and was recommended for other plants.

In 1977, the B.C. Pollution Control Branch issued a discharge permit for the Annacis Island sewage treatment plant that specified primary treatment and set limits on a wide range of substances, including toxic materials. Analyses of effluents in 1979 indicated that the criteria established in the 1977 permit were not being met. The B.C. Pollution Control Branch held an inquiry in 1980 to examine whether there had been compliance with per-

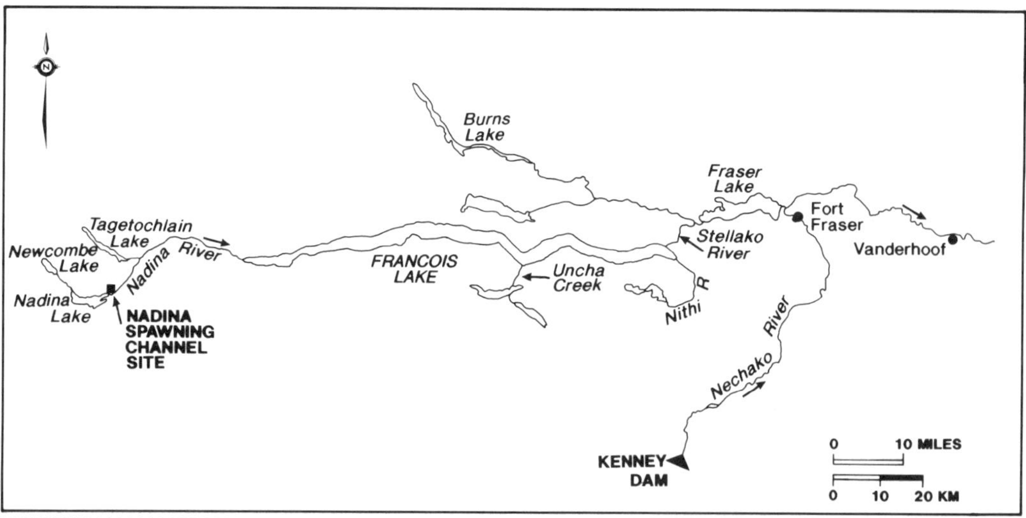

Figure 29. Location of Nadina and Stellako sockeye salmon spawning streams.

mit provisions for the Annacis Island plant. The hearings concluded that the effluent from Annacis Island and Lulu Island plants would meet acceptable objectives/criteria by 1990 if the recommendations of the Branch were followed.

The 1980 hearings resulted in charges being filed by the Province against the Greater Vancouver Sewerage and Drainage District on four counts relating to the discharge from Iona Island sewage treatment plant. A study group later produced a report containing several recommendations. By 1985 the Greater Vancouver Sewerage and Drainage District received a permit authorizing construction and operation of a deep-water outfall for effluent from the Iona sewage treatment plant. The permit specified that a pre- and post-operational monitoring program be established. The Commission, represented on the program, would assess effects of deep-water sewage discharge on the environment. Emphasis was placed on detecting subtle changes in the environment with selected species of biota, including salmon, being examined for bioaccumulation and its effects.

OTHER STUDIES

In the early 1960s log-driving occurred each spring on the Nadina and Tachie Rivers. The extent of damage to the spawning grounds was not known, but it was recommended that the drives be delayed until after the salmon fry had emerged. Expansion of log driving to the Stellako River (Figure 29) was requested in 1964 and 1965 and although this was strongly objected to by the Commission and the Department of Fisheries, effective lobbying by the company resulted in the log drive proceeding. The drive provided opportunities for extensive impact studies and for film documentation of the event. Studies at the Sweltzer Creek Field Station determined the effects of physical impact, bark, wood fibers and chemicals leached from wood on the survival of salmon and trout eggs and fry (Servizi *et al.* 1970).

A cooperative study by the Commission and the Department of Fisheries in collaboration with the Fish and Wildlife Branch of the British Columbia Department of Recreation and Conservation was carried out on the Stellako River in 1965. The log drive deposited large amounts of bark and wood fiber on the river bottom and within the spawning ground. The spawning grounds and river banks were eroded

May 22, 1965. Logs piled on the bank of Nadina River 2 miles below Popple (Tagetochlain) Creek, ready to be bulldozed into the river to start the log drive to Francois Lake and then down Stellako River.

June, 1965. The adverse effects of log driving were studied in Stellako River. Severe bed and bank erosion occurred as a result of log jams and grounding of logs in shallow areas. Close liaison was maintained with Federal and Provincial fisheries authorities in all IPSFC efforts to minimize detrimental effects on salmon production.

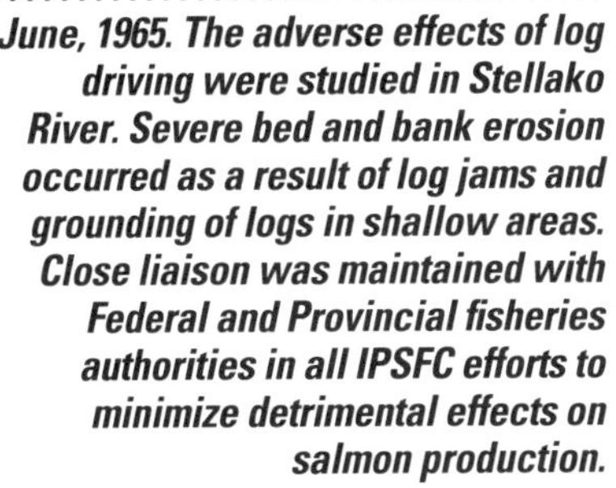

June, 1965. A log jam beginning to form in a particularly valuable sockeye spawning area in Stellako River. High velocities continually forced logs under the jam, causing deep erosion of the river bed.

due to scouring, log jams and impingement of logs. It was recommended that all log driving in the Stellako River be banned. In spite of this, political forces prevailed and log driving continued for two more years. Log driving was discontinued in 1968 because of a new log source and a change in policy by the new owners of the Fraser Lake Sawmills. Log driving continued in the Tachie and Quesnel Rivers, but because of different river physical characteristics, it was believed there was no serious damage to salmon.

Each year the Commission considered large numbers of applications for water licences and placer mining leases. In 1967 alone, 249 water licences and 143 mining applications were reviewed.

In another study of the restricted water flow created by beaver dams, a program was established to identify streams where problems existed and to suggest action required to reduce the adverse effects of the beaver populations. Many beaver dams were removed over the years.

The Commission spent several years studying a proposed ski development at Hemlock Valley upstream from the Weaver Creek spawning channel. The studies started in 1974 and centered around a sewage disposal system and the effects of road construction on erosion and water-runoff patterns. An appeal for a hearing was granted on the sewage disposal plan before the Pollution Control Board and the plan was revised so that treated sewage would be directed to a less sensitive watercourse outside the Weaver Creek drainage. A water-quality program was established and results showed that turbidity of Sakwi Creek following rainfall was much higher below the ski development than above. The Commission made many recommendations to reduce silt content in the creek by re-seeding disturbed areas and also developed guidelines for improved road construction practices.

Various mine and smelter proposals were reviewed. Considerable effort was expended to review and comment on a proposed thermal power plant at the Hat Creek coal deposits 50 miles west of Kamloops. The particles and gasses from burning coal would be discharged to the atmosphere. B.C. Hydro and Power Authority and its consultants studied environmental aspects of the project for several years. While there was to be no aqueous discharge, the gaseous emissions would have been strongly acidic. The proposal was eventually set aside.

Commission staff participated for many years in studies of "swimmers itch" at Cultus Lake and the applications of copper sulfate treatment procedures.

By 1977, Commission staff were involved in two long-term studies: acidification of the watershed and the problems caused by the prolific aquatic weed, Eurasian milfoil. Considerable research was expended on the latter as milfoil was found in Cultus and Shuswap Lakes. Both mechanical removal and herbicides were used to control milfoil growth in Cultus Lake. A diver-operated suction dredge seemed to be the most acceptable method of removing milfoil, along with bottom-covering barriers to implantation.

For many years water sampling was carried out in the Nadina River and Tachie River basins to monitor the effects of logging and road construction on suspended sediment loads and to establish baseline data.

A study was also done of early Stuart adult sockeye in relation to suspended solids, turbidity and dissolved gas saturation during their migration in 1983. This tried to identify reasons for the unexplained losses during migration. No definitive answers were obtained.

Two significantly important accomplishments which ensured a safe water environment for the future were the pulp mill effluent studies and the municipal sewage dechlorination standards. Those standards have been widely adopted. The requirement that all industrial effluent be non-toxic to fish life before discharge into the Fraser River or its tributaries was critical.

183

The emphasis of Commission investigations during its first ten years was centered on Hells Gate and other locations where fish passage problems needed evaluation and correction. There were, however, other serious problems in the watershed that required attention. To understand the problems that would be involved in protecting sockeye and pink salmon if dams were constructed in the Fraser system, the Commission conducted many experiments at various locations. Based on experiences at other locations along the Pacific coast, impacts from dams could have far-reaching effects on the stocks and, ultimately, on the fishing industry.

Methods of rearing sockeye artificially needed to be investigated to offset mortalities likely caused by dams. The effect of reservoirs needed to be evaluated as well as the more obvious problems created by impeding upstream migration and injury of fry and fingerlings during downstream migration as they passed over spillways and through turbines. Changes in flow patterns and temperature changes also required investigation.

Studies at Baker Dam (250 feet high) in Washington State using sockeye yearlings showed that 63 percent of the fish were killed after passing over the spillway and 37 percent died when migrating through the turbines (Hamilton and Andrew 1954a). These investigations were conducted from 1950 to 1952 and revealed that the dam was responsible for a decline of 55 percent of the local sockeye run. Injury and mortality were caused by abrasion on the spillway and from pressure changes and cavitation in the turbines. Also participating in the investigations were the Washington State Department of Fisheries and Canada's Department of Fisheries.

Additional studies were performed at the 130-foot high Ruskin Dam on the Stave River in the lower Fraser River watershed where 10 percent of the sockeye fingerlings were killed as a result of passing through the turbines (Hamilton and Andrew 1954b).

From 1953 to 1957, The Commission studied whether or not fish could be protected at dams by guiding them away from spillways and turbine intakes. Small-scale studies were done under laboratory conditions in Sweltzer Creek at Cultus Lake and prototype installations were tested at Baker Dam. Electrical stimuli were more effective in directing the path of downstream migrants than sound, light, shade, air bubbles or other visual stimuli. An electric screen was developed that was over 90 percent effective in guiding downstream migrants in Sweltzer Creek but it was not effective in the 250-foot-deep forebay of Baker Dam (Andrew *et al.* 1955).

Experiments conducted in 1956 at Baker Dam showed that a fine-mesh barrier only 15 feet deep by 200 feet long stretched across the dam forebay was more effective in preventing downstream-migrant sockeye from entering the turbine intake than a 50-foot-deep electrical barrier of the same length (Andrew et al. 1956a). Other studies performed at Cultus Lake and Baker Dam showed that an electrical barrier was effective for stopping the upstream migration of adult sockeye and pink salmon but also showed that the electrical stimuli were not effective for leading the fish to an alternate, safe migration route (Andrew *et al.* 1956b).

All of the foregoing studies were important because the data provided information on the critical impact of dams on the resource.

In addition to these, a comprehensive review of the available information pertaining to all previously published investigations relating to the effect of dams was required. Bulletin XI of the Commission was an epic review of all facts known to affect the fishery resources of the Fraser if dams were constructed (Andrew and Geen 1960). For 25 years this report served in forestalling dam construction.

One of the Commission's first protective actions in a spawning ground area was to alleviate a potentially serious problem in 1947 in the Birkenhead River. Engineers of the Prairie Farm Rehabilitation Administration wanted to

June 19, 1950. The threat of flood control and power dams on the Fraser River necessitated investigations of potential effects of dams. From 1949 to 1957, Baker Dam on a tributary of Skagit River in Washington State was used as a laboratory for measuring mortality rates suffered by downstream migrants in spillways and turbines and for studying possible methods of guiding fish away from these hazardous areas.

Tests were conducted in Sweltzer Creek at the outlet of Cultus Lake to determine the effectiveness of electrical, light, sound and other stimuli in guiding upstream-migrating adults and seaward-migrating smolts. At the time this photo was taken, an air bubble barrier was being tested.

Many studies of the effects of fishways, spillways, turbines, diversion flows, and other characteristics of dams were conducted at the Seton Creek hydroelectric project near Lillooet, commencing in 1956.

dry up or reduce the flow in the first half mile of the river channel to reduce flooding of the river's lower reaches. As a result of the Commission's involvement, a solution was reached which did not jeopardize the spawning area.

The Commission's input into the evaluation of various watershed proposals increased each year. A major effort was directed at protecting the watershed from the adverse effects of many well-intentioned but salmon-harming projects.

THE PROPOSED CHILKO LAKE PROJECT

In the late 1940s, the Water Rights Branch, Department of Lands, Province of British Columbia, issued a report recommending a major diversion of water from the Chilcotin district through the Coast Range to Bute Inlet on the Strait of Georgia. The Commission launched an intensive investigation and issued an interim report in 1949 detailing the serious problems created for salmon if such a project proceeded.

The proposed dam at the outlet of Chilko Lake would block upstream migration of sockeye fry, and prevent them from reaching Chilko Lake, where a one-year rearing and growing period is essential. In addition, the attractive aspects of water flow for smolts going seaward would no longer exist since water would be diverted away from the Fraser River watershed to Bute Inlet on the coast. The major spawning grounds below Chilko Lake would have been dried up and the Chilko sockeye stock, as well as a productive chinook salmon population, would have been exterminated within a short time. The Commission took a strong stance against the proposal and stated that: "...the continuance of the existing Chilko sockeye fishery requires that there be no interruption, addition, diversion, or obstruction to either the natural inflow or the natural outflow of Chilko Lake or Chilko River" (IPSFC 1949a).

Thorough examination proved that every phase of the Chilko sockeye life cycle would be devastated. Although the project did not proceed at that time, variations of the scheme continued to surface for many years. To this date, the Chilko system remains unscathed and continues to be a major producer of sockeye salmon in the Fraser River system.

THE NECHAKO-KEMANO DEVELOPMENT

Almost concurrent with the Chilko proposal, the Aluminum Company of Canada (Alcan) requested a water licence from the provincial government in 1948 to divert the entire flow of the upper Nechako watershed to the Kemano River on the coast (Figure 30). The project was designed to produce electrical power for an aluminum smelter at Kitimat using lake systems not containing sockeye (IPSFC 1953a). In 1948 the Commission investigated the possibility of opening the lakes to sockeye but had not yet taken any action to establish sockeye populations in the large lakes that were to be used for the power reservoir. The development of this system for producing sockeye was lost. The proposed diversion dam did not directly block migration to any existing sockeye spawning grounds but it appeared that favorable conditions for upstream migration of sockeye to the Fraser-Francois-Stuart Lake systems would be severely threatened by the project.

Recommendations were made with regard to flows and water temperatures that would be required to protect these stocks. An extensive report was prepared and issued by the Department of Fisheries of Canada, the Fisheries Research Board of Canada, and the IPSFC in 1951 (Anonymous 1951). It concluded that a regulated release of cooling water from the main dam was required.

The Commission wrote to the Minister of

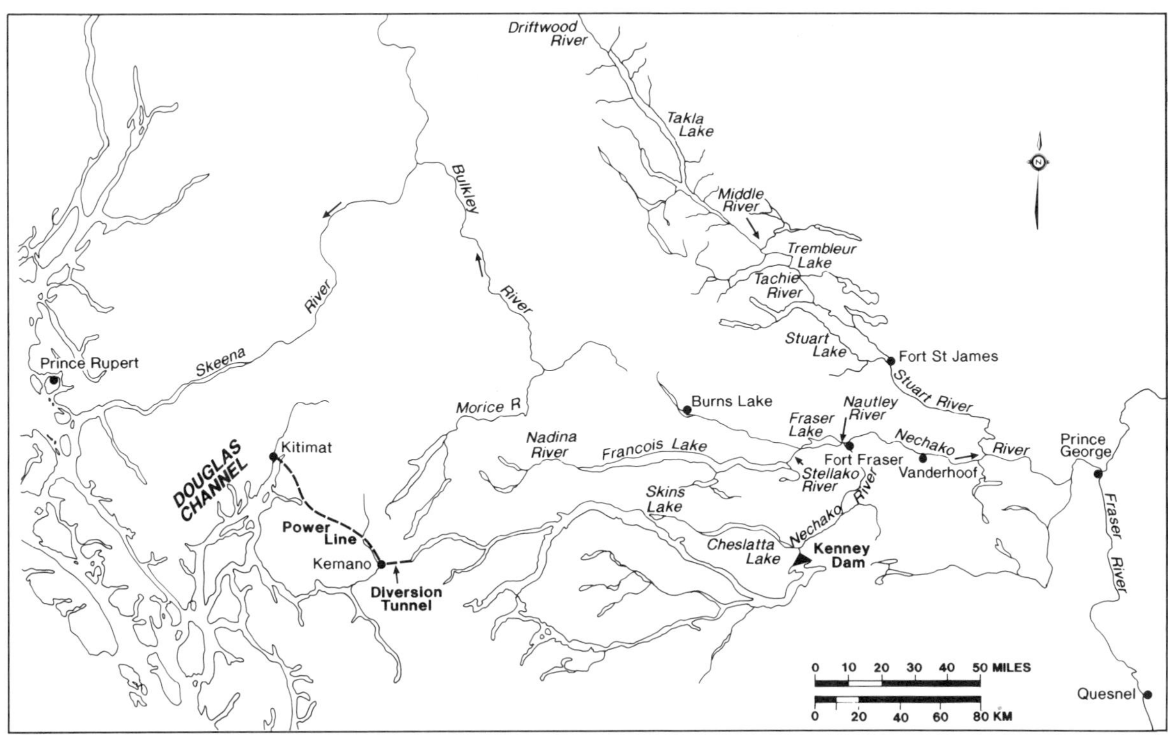

Figure 30. Nechako River system, showing the diversion to Kemano River to produce power for aluminum smelting at Kitimat and for export.

Fisheries on April 12, 1951 concerning plans for the Alcan project. An excerpt of that letter follows:

> It is the opinion of this Commission that the plans as presented are not adequate as they make no provision whatever for protecting the sockeye salmon migrating through the Nechako River to their spawning grounds in the Fraser-Francois system. We believe that in reply to the Water Comptroller's request for comment on the plans, his attention should be directed to the failure on the part of the company to make the necessary provisions for the protection of the sockeye salmon and that he should be requested not to approve the plans until provisions for adequate protection are included.

The water licence was granted on December 29, 1950 and construction was soon underway. The main dam was closed in 1953 and flows in the Nechako River below the dam were drastically reduced. During the three-year filling period of the reservoir, the company was required to provide some water storage for release during the spawning and incubation of chinook salmon in the river below the dam. When the reservoir had filled, the flow regime in the Nechako River increased because the first phase of the project diverted only one-half of the stored water. Unfortunately, there was no official policy established as to Alcan's responsibility for protecting the fisheries resource.

Because strict environmental guidelines, including minimum flows, had not been accepted by the developer, accusations of fishery damage made by the fishery agencies and denials by the company were commonplace for the next 35 years. The company engaged its own consultants for long-term environmental studies and conducted spawning ground surveys. The fisheries agencies concluded that pre-spawning mortalities of sockeye were partly due to excessively high temperatures below the dam site. Chinook salmon were also adversely affected by low flows in the upper Nechako River.

Nechako River prior to and after closure of Kenney Dam to begin water storage for Alcan's power generation project at Kemano. The Nautley-Nechako confluence is shown in the upper left photo on October 7, 1952 with 4,680 cfs in the upper Nechako and 175 cfs in Nautley River. The lower left photo shows the Nechako River about 1 mile above the Nautley confluence on October 12, 1952. By October 31, 1952, the flow at this point had dropped to 71 cfs. The above photos were taken at Vanderhoof from identical viewpoints on September 25 and October 10, 1952.

August 12, 1949. Unspawned sockeye found in Forfar Creek, a major spawning area of the early Stuart population. Even under pristine natural conditions, production of the Nadina and early Stuart sockeye populations was affected by pre-spawning mortality. The first stage of the Alcan power project diverted only about one-half of the Nechako River discharge to tidewater and much of the surplus flow was discharged down Nechako River in the summer and fall months, which limited the river temperature rise that is an inevitable result of flow reduction. The final phase is now under construction and since it will divert a much higher proportion of the flow, river temperature will be further increased and a significant increase in pre-spawning mortality is expected.

In the 1970s the Commission's investigations into the Nechako watershed problems continued and intensified because of the growing interest in a contemplated second phase (Kemano II) of development of the diversion. The B.C. Hydro and Power Authority sponsored field and office studies by the fisheries agencies into several aspects of the potential development and published seven volumes on the results of these investigations. The requirements for protection of Fraser River sockeye and pink salmon were re-examined by Salmon Commission staff and published in volume two of this series (IPSFC 1979a).

In late 1979 the Commission learned that the Aluminum Company of Canada intended to proceed with the second-stage development of the existing diversion of water from the Nechako River, including diversion from Nanika Lake in the Skeena River system into the Nechako reservoir and Fraser system.

The large scope of this proposal, termed the Kemano Completion Project, raised serious concerns for the Commission about the flow and temperature regime of the Nechako River. These concerns extended further downriver into the Fraser and included insufficient river volume for migration of sockeye, increased river temperatures, sharp temperature gradient changes, increased waste effluent concentrations, possible dredging of lower Fraser River pink salmon spawning grounds, lower discharge of the Fraser River into the estuary and the possible transfer of fish parasites because of water diversion from one watershed to another. In 1983, the IPSFC issued its final comments on the proposed project in a summary report: "Potential effects of the

Kemano Completion Project on Fraser River sockeye and pink salmon" (IPSFC 1983a). By the 1980s the courts had been enlisted to force the company to provide additional flow to protect all the fish downstream of the Kenney Dam. Confrontation between Alcan and the fisheries agencies reached a peak in the early 1980s.

By the early 1980s, the world market demand for aluminum declined and Alcan's plans to press on for the completion phase of the Kemano project were put on hold. In 1985, however, the issue was again headed for the courts, a major aspect of the trial being whether Alcan was responsible for meeting environmental standards necessary for protection of the fishery resource. However, the court case was cancelled when Canada Department of Fisheries and Oceans signed an agreement with the provincial government allowing Alcan to proceed with the second and final phase of the development.

From 1950 to 1985, a great amount of research time and expense was put forth by both sides in defining, evaluating and disputing the impacts created by this project. Much of this could have been more productively used elsewhere if strict environmental guidelines had been imposed when the water licence was first issued in the late 1940s.

PROPOSAL TO DIVERT COLUMBIA RIVER WATER TO FRASER SYSTEM

In the mid-1950s a major development was proposed involving the Columbia, Thompson and Fraser Rivers to divert 10,000,000 acre-feet of water annually from the Columbia River into the Fraser. The water was to be used for power development at a series of six low-head dams on the Thompson River and four low-head dams on the Fraser River. Additional storage of 6,500,000 acre-feet of water

was planned on six major sockeye-producing lakes in the Fraser system.

Understandably, this proposal received considerable attention by the Commission and staff. There was much public debate between some Commissioners and notable figures, such as General G. McNaughton. It also involved the Canadian Section of the International Joint Commission. The IPSFC position was staunchly given and stated without reservation that: "… the construction of ten power dams on the Fraser and Thompson Rivers would preclude the preservation and extension of all the salmon runs to these rivers above the dams" (Anonymous 1955). Senator Tom Reid, in his dual capacity as a Senator and Commissioner, played a strong role in assuring that this proposed project never materialized.

SETON CREEK POWER PROJECT

Hydro-electric projects on tributary streams to the Fraser River have been in place for many years. The Lake Buntzen facility in the Coquitlam District commenced in 1903. Another power plant was installed at Stave Lake in 1912. Both of these were located in the lower Fraser River area. Attention soon shifted to the Bridge River area about 130 miles north of Vancouver, adjacent to the Seton-Anderson Lake system. In 1927 work began on a 13,200 foot-long tunnel through Mission Mountain. Bridge River water was diverted with a fall of 1,200 feet into Seton Lake. By 1948 the No. 1 powerhouse was completed and by 1954 the facility was the largest single source of hydro power in British Columbia. A second tunnel from Bridge River to Seton Lake was put in place and by 1960 the No. 2 generating station along Seton Lake was in operation and a 180-foot rock-filled dam was constructed across the Bridge River. From 1934 to 1948, only about 30 c.f.s. of diverted water entered Seton

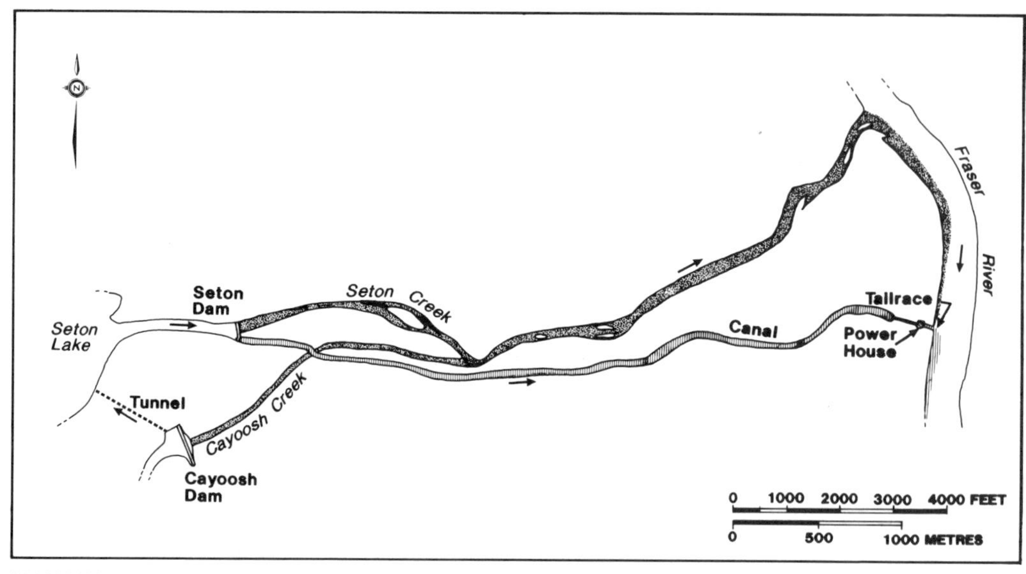

Figure 31. Layout of the Seton Creek hydroelectric installation.

Lake, but by 1954 it was increased to 2,000 c.f.s. and by 1960, 8,000 c.f.s. About 500 c.f.s. of Cayoosh Creek water was also diverted into Seton Lake (Figure 31).

The transfer of Bridge River water into the Seton-Anderson system was not considered beneficial to the system because of changes to the basic productivity of Seton Lake. Geen and Andrew (1961), in studying the effect of the diverted Bridge River water on Seton Lake productivity, concluded that the introduction of Bridge River water caused pronounced limnological changes in Seton Lake. Turbidity increased while average temperature and dissolved mineral levels were reduced. Plankton production was also greatly reduced. The cold turbid water from the Bridge River watershed had diminished the productivity of Seton Lake.

In 1953 the B.C. Electric Company requested permission to build a low dam (25 feet) at the outlet of Seton Lake near Seton Creek's confluence with the Fraser River. This project would harness water from Bridge River, Anderson Lake and Seton Lake. From the beginning, and continuing beyond completion of the dam in 1956, the Seton project has been used as an example of the cooperative approach that sometimes takes place between developers and the various fisheries agencies.

It was understood and agreed by all parties that adequate measures to protect fish were required. This agreement followed alterations to the system as noted earlier. Over the past 30 years, hundreds of meetings have taken place to resolve problems associated with protecting salmonids in the system. Considerable expense was involved in research and fish protection facilities through the years. Yet in spite of the difficulties still being experienced in providing completely safe fish passage facilities, the pink salmon and sockeye runs of the Seton Creek area are at higher levels of abundance now than they were 30 years ago. It must be pointed out though, that the spawning channel for sockeye at Gates Creek and the two pink salmon channels downstream from the Seton dam are major reasons why these runs are in such good condition. The dam and powerhouse had an adverse impact —an impact overcome to a certain extent with resource enhancement and management.

By the 1970s it was evident the Seton Creek power project could serve as an example of how the subtle effects on salmon of such a project can go undetected. Observations in the tailrace in the early 1970s showed bruised and badly cut salmon, some decapitated carcasses, and sockeye smolts impinged on the trash racks at the end of the power canal.

Studies were conducted from the mid-1970s into the 1980s in an effort to determine the cause of adult sockeye delay in the tailrace. Flow regimes in Seton Creek and plant operations were altered. Adult fish migration patterns were studied using telemetry (tagged fish with radio transmitters attached) methods. Delays were generally reduced if outflow from the plant was considerably reduced and discharge in the creek was substantially increased. This was not a satisfactory long-term operation that B.C. Hydro could accept.

Further studies of adult salmon behavior revealed that Gates Creek sockeye would move out of the tailrace area into Seton Creek without serious delay if the concentration of Cayoosh Creek water in Seton Creek for Gates Creek sockeye was at 20 percent or lower. Portage Creek sockeye did not move into Seton Creek until the percentage of Cayoosh Creek water was less than ten percent. The delay problem was solved by diverting most of the Cayoosh Creek flow by tunnel into Seton Lake (Fretwell* 1989).

Sockeye smolts that migrated down the power canal suffered a ten percent mortality going through the turbines (Andrew and Geen 1958). A possible solution was to get more smolts to migrate downstream through the dam and via Seton Creek rather than through the power canal. This was ineffective because when plant operation was shut down during the day and flow in the canal ceased, the smolt migration the following night was considerably reduced. This could result in residualism and increased smolt losses due to other factors.

This complex problem remained to be resolved.

MORAN DAM PROPOSAL

Interests in damming the Fraser River or its tributaries for electrical power development have always lurked in the background. By the mid 1950s a major proposal was put forth by Moran Power Development Ltd. The proposed structure involved a 720-foot-high dam on the Fraser at Moran, located about 20 miles upriver from Lillooet, B.C. This would have obstructed the migration route of all but one of the major sockeye runs in the Fraser system.

The overwhelming devastation to fisheries that such a structure would create necessitated extensive research to document the impact on the stocks. The enormous problems of adult salmon passage, protection of downstream migrating fry and smolts, creation of reservoirs, inundation of spawning areas, and alteration of lake rearing areas would be significantly detrimental to production. In addition, the environmental changes that would occur, though much more difficult to evaluate, would also contribute in a negative way to the well-being of all the stocks. The long-term impacts of the Hells Gate slide, Quesnel dam, Adams River splash dam and railroad construction in the Fraser Canyon should have been sufficient demonstration of the delicate balance between fish and their environment. Alternative methods of enhancing the stocks were considered as a means of ameliorating the expected losses caused by the obstruction. It was concluded that all known methods of artificial propagation could not compensate for the loss of natural upriver production of Fraser River sockeye and pink salmon if the dam was constructed (Andrew and Geen 1960). The validity of that conclusion, made about 30 years ago, remains true today.

Revised interest in the Moran dam project appeared again in the early 1970s when the British Columbia Energy Board studied the power resources of British Columbia. Another review of the problems and impacts associated with such a dam was prepared by Canada's

Department of the Environment, the Canadian Fisheries Service and the IPSFC (Anon. 1971). The document, "Fisheries Problems Related to Moran Dam on the Fraser River" concluded much the same as the earlier report by Andrew and Geen. The 1971 report stated: "The minimum effect of Moran Dam would be the destruction of all salmon and steelhead trout populations that spawn upstream from the dam." It was also estimated that there would be at least a 50 percent loss of all salmonid production below Moran dam due to environmental changes in the river and estuary.

In writing about the probable effects of the proposed Moran Dam, Dr. Haig-Brown (1972) stated:

> The Fraser salmon runs have served mankind for ten thousand years. If we give them a chance they can last as long as mankind, perhaps longer. It is their triumph that they will not fit conveniently into man's shallow technological concepts, a triumph of life and individuality that deserves to endure for its inspirational and emotional as well as its intrinsic values. To destroy them would be an act of vandalism that British Columbia cannot afford, Canada cannot afford and the world cannot afford. It would leave a burden of guilt that the collective conscience of the nation cannot sustain. To preserve them is an act of faith in the future.

Threats of hydropower development on the Fraser continue. The vigorous efforts of the fishery agencies, the fishing industries and the general public have thus far kept the mainstem Fraser River free from the devastating impact of dams. The Commission played a vital role in advising the Canadian government and keeping fishing industries aware of the inherent dangers. The fishing industries of both countries were always united in their efforts to protect the Fraser resources.

FLOOD CONTROL

Following the second worst flood on record in the Fraser River in 1948, the federal and provincial governments set up a joint Dominion/Provincial board to collect basic water discharge data and compile information previously recorded. This agency existed from 1949 to 1954, but no comprehensive report was published.

The Fraser River Board was established by the Government of Canada and the Province of British Columbia in 1955. It investigated the water resources, uses and requirements in the Fraser River basin and reported on possible flood control and hydropower production, and the effects of water control on navigation, fisheries, silting, erosion and irrigation.

The Board made several recommendations (Fraser River Board 1958), one of which was that the present system of dykes in the lower Fraser Valley be re-examined, restored and maintained. It also recommended that consideration be given to construction of several storage dams in the headwaters of the Fraser River system, none of which would be on the main Fraser River. This flood control system would consist of storage reservoirs and associated hydroelectric generation plants, which could make the system economically feasible.

One recommendation included a storage dam at the outlet of Stuart Lake, the migration route used for all early Stuart and late Stuart sockeye runs. The Commission and Department of Fisheries conducted a joint investigation of the effect of this dam and in 1962 issued a report detailing the serious consequences of such a project. It was recommended that it would be more economical to dyke the lower Fraser River Valley and as a result of these recommendations, the Board dropped the proposal for the Stuart Lake dam.

The Board did, however, at one time consider both a high and low dam at Moran in preliminary evaluation of various systems, but in the final report released in 1963 it did not recommend a dam at this site. The report emphasized that dykes should be used as the primary protection from floods in the lower Fraser, with any flood-storage and hydroelectric projects at locations that would avoid major interference with salmon. This proposal, called System E, would provide adequate flood protection without any dams on the mainstem of the Fraser River. Although it appeared that Fraser River sockeye and pink salmon would not be directly affected by the proposed System E development program, they would suffer indirect but serious losses in production. Deleterious alterations in water temperatures, water quality and flow regimes would have an impact on salmon during migration, spawning and incubation periods if the System E dams were in place.

In 1968, the Fraser River Joint Advisory Board replaced the Fraser River Board and was instructed to proceed with construction of the dykes recommended by the latter agency. In addition, the newly formed Advisory Board was to investigate further the feasibility of System E for flood control benefits alone, with no involvement of power development.

In 1969, the Province of British Columbia and the Government of Canada entered into an agreement for a ten-year program of dyke and drainage improvements in the lower Fraser Valley. The dykes would provide control of water levels two feet higher than those recorded in the 1894 and 1948 floods.

Environment Canada Fisheries and Marine Service reviewed the System E flood control proposal thoroughly and concluded that without mitigation, the proposal would result in significant reduction of salmonid production in the Fraser system. These losses would not be offset completely by mitigative measures. To date, the system of dams proposed on the Clearwater, Cariboo, upper Fraser and McGregor Rivers have not been constructed and the improvements in dyking have provided adequate flood control.

MCGREGOR RIVER DIVERSION PROPOSAL

The McGregor River enters the Fraser above Prince George, about 500 miles upstream from the estuary. In the mid-1970s, B.C. Hydro began investigating the McGregor River as a possible source of power production to be on line as early as 1984. All of the McGregor River flow would be diverted over the Fraser/Peace divide into the Parsnip River to increase power generation at W.A.C. Bennett Dam and other dams on the Peace River. The discharge of the McGregor is about eight percent of the total Fraser River flow at Hope. Although no pink or sockeye salmon spawn in the McGregor system, reducing the flow of the Fraser River system could affect both sockeye and pink salmon stocks further downriver.

There are numerous ways in which the McGregor project could have adversely affected the Fraser runs. There was a statistical relationship between the discharge of the Fraser and sockeye smolt survival—the higher the discharge, the higher the percentage return. Therefore, any lowering of the springtime flow would reduce production. Because of reduced discharge, water temperatures would increase by almost 1°F. This increase would be expected to increase pre-spawning mortality of adult sockeye. Lower discharge would extend the duration of the Adams River sockeye migration from the Strait of Georgia to the spawning ground and reduce production. Reduced flows below Hope would drop water levels on large spawning grounds of pink salmon and reduce the area available for spawning. There would be less dilution of effluents discharged to the river, thus reducing the safety factor against possible toxic effects on salmon. The joining of the McGregor River system to the Peace River system may allow the transfer of parasites and fish diseases to the Fraser system, which could have serious implications for Fraser stocks.

Analysis of the proposed project continued. However, by the end of 1977, B.C. Hydro announced the project had been shelved.

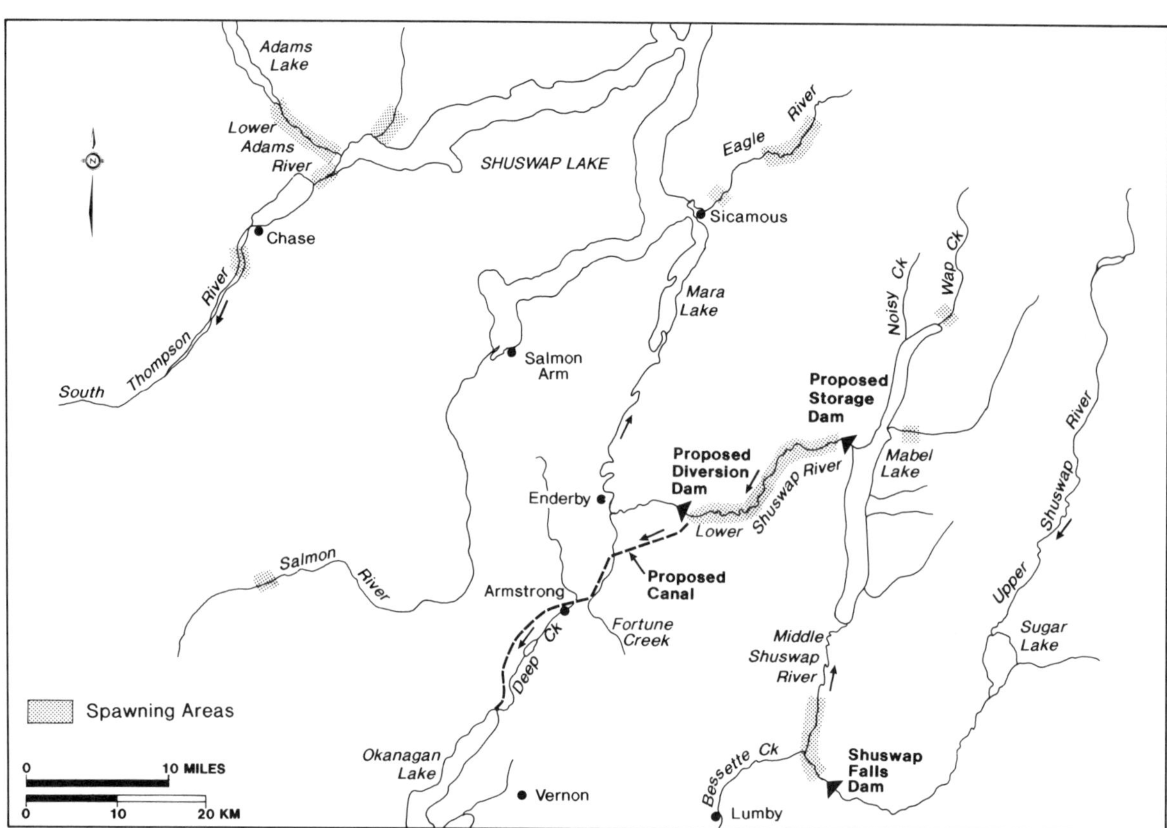

Figure 32. Sockeye salmon spawning areas in the southern portion of the Shuswap system in relation to potential diversion of flow to the Okanagan system.

LOWER SHUSWAP RIVER DIVERSION PROPOSAL

In the mid-1960s investigations were conducted by the Water Resources Service of the British Columbia Department of Lands, Forests and Water Resources on a proposal to divert water from the lower Shuswap River to Okanagan Lake (Figure 32). The purpose of the transfer would be to augment the agricultural irrigation in Okanagan River Valley. Although the drainage is tributary to the Columbia River and not a part of the Fraser watershed, a diversion would have seriously interfered with the salmon and trout stocks using the Shuswap and Okanagan rivers. In 1969 a joint report on the fisheries problems associated with the proposed diversion was submitted to the Minister of Fisheries. The report, prepared by Department of Fisheries and Forestry and the IPSFC, recommended that other sources of water be found (Anon. 1969). A federal-provincial agreement then called for a four-year study of the water resources of the Okanagan Basin. The proposed diversion never materialized.

OTHER WATERSHED WORKS

In 1950, a proposal emerged for a hydropower project on the Quesnel River. Interest was redirected to the North Fork of the Quesnel River or to other areas where no sockeye were involved. It was pointed out that a dam on the Quesnel could destroy the Horsefly River race, formerly the largest sockeye salmon run in the Fraser watershed. Today the Quesnel River provides unimpeded access for sockeye salmon as they move upstream to Quesnel Lake, Horsefly River and Mitchell River.

The Commission completed the construction of an outlet-control works at Weaver Lake on the Harrison system in 1953. The project was necessary because sockeye were prevented from entering Weaver Creek because of low-flow conditions, with resultant mortalities. The control works would store excess water in the lake for release during low-flow periods. In later years, the lower portion of Weaver Creek usually required annual maintenance to open the creek for sockeye migration.

Considerable interest was shown in placer gold mining in the Horsefly River Valley in the early 1950s. Most was to be in the general location of major sockeye spawning grounds. Cooper (1956) concluded that there would be at least three injurious impacts of placer mining operations on sockeye salmon:

1. physical destruction of the spawning grounds because of washing the gravel for gold;
2. increased turbidity of the river; and
3. increased sediment would lower survival of eggs and alevins in the gravel.

It was recommended that no operations be permitted in the river itself or any adjacent tributaries. If any work on tributaries was permitted, settling basins or outflow filtration would be required. Placer mining had occurred briefly in the Horsefly River in the early 1900s and extensively in the Quesnel River below Quesnel Lake. The report was submitted to the Chief Gold Commissioner of British Columbia and he took action to see that mining did not take place and to protect the Horsefly River sockeye run.

The production of Adams River sockeye in the three cycle-years following the record run of 1958 stayed at a low level. Investigations revealed that the flow in the river's right channel, where the major share of the run had been spawning, was considerably reduced during the winter by natural events in the river to only 10 percent of the river's flow during the incubation period. Gravel was exposed and a significant number of eggs and alevins were frozen to death. The Commission received the necessary permits to restore the volume of water in the right bank channel to 48 percent of the flow. The runs responded by increasing in 1974 and in succeeding cycles. Further adjustments in gravel (bars and channels) at the upriver section of the channel were required to maintain the proper flow distribution. Even nature could use some occasional assistance. Frequent monitoring of the natural environment is important.

The pink salmon tagging program started in 1957 revealed the extent and relative importance of the various spawning areas and it was found that the populations spawning in the Harrison River and in the mainstem of the Fraser River between Hope and Chilliwack were much larger than previously recognized. It was also discovered that extensive spawning occurred in these rivers in areas that had been regularly dredged for many years to permit towing of log booms during low-water periods. Financed by the Canada Department of Public Works, this dredging was an annual program dating back to the beginning of the century. Most of the dredging was done with a cable-drawn scraper bucket that moved river bed material from the log towing channel to the edges of this channel or to downstream areas, thus concentrating the flow into a central, high-velocity channel and tending to reduce flows in marginal areas and side channels.

In cooperation with the Canada Department of Public Works and the Canada Department of Fisheries, engineering studies were conducted for several years commencing in 1957 to determine the impact of the dredging on pink salmon production. Depths and velocities were measured and river bed materials were sampled. Pink salmon eggs and alevins were recovered from the main river channel in areas that were frequently dredged. Velocities in this log towing channel were so high that it had previously been considered unsuitable for pink salmon spawning. The turbid Fraser River water prevented visual determination of spawn-

August 31, 1954. Engineers W. Boresky and A.C. Cooper (later to become Director) explored the full length of the Horsefly River on foot and by boat to investigate effects of placer mining activities on spawning ground gravel quality.

July 31, 1946. A placer dredge operating on the North Fork of Quesnel River. This branch of Quesnel River is not a sockeye-producing stream.

ing locations but the recovery of live eggs and alevins proved that pink salmon spawned in the log towing channel in velocities of at least 6 feet per second. It was also found that towing of booms in shallow water scoured the gravel bed and caused fish to move off the spawning bed. Also, in the Harrison River, dredging of two control sections had lowered the low-water elevations of Harrison Lake. The deposition of dredge spoils on productive spawning areas at the margins of the towing channel was considered to be very detrimental to pink and sockeye salmon production in Harrison River.

Studies of river flows and depths in both the Harrison and Fraser rivers suggested that log towing would not be severely curtailed if dredging was discontinued. As a result of these studies, it was agreed that log towing would be discontinued during low-water periods, dredging of Harrison River would be restricted to June, July and August, no dredging would be done in the control sections of Harrison River and no dredging would be done in the Fraser River in years of pink salmon spawning. These restrictions were progressively applied beginning in 1957 but the major restriction on dredging of the Fraser River pink salmon spawning area was not agreed to until 1977. The major increase in the population since that date probably resulted from the restriction of dredging activities.

The Commission also attempted to protect the Chilliwack-Vedder River pink salmon population. Flooding caused by extensive logging of the Chilliwack River watershed frequently necessitated removal of river bed material and flood debris from prime pink salmon spawning area in Vedder River. Bank armoring of Chilliwack River and removal of instream timber debris was studied as a means of reducing the amount of bedload and debris deposition, thus reducing damage to the fisheries resource by frequent instream work. No action was taken, however, because the British Columbia Water Resources Branch recommended that set-back dykes be built on Vedder River to protect adjacent farm and residential properties. In cooperation with the Canada Department of Fisheries and the British Columbia Fish and Wildlife Branch, the Commission studied this proposal and it was agreed that set-back dykes would provide some protection for the fisheries resources by allowing the river to meander over a wide flood plain and by providing a permanent, large floodway, which would reduce the frequency of emergency flood repair work. With the larger floodway, debris and bedload can be removed after the salmon fry emerge from the gravel and the size of the flood channel can be maintained by periodic scalping of accumulated bedload from bars that are well above water level during low water periods. However, the frequent flooding, which appears to be more severe in recent years, still has a severe adverse effect on pink salmon production because of scouring of eggs and alevins from some parts of the stream bed and deposition of sediment and burial of eggs and alevins in other areas. This damage became so severe that the Commission began constructing a spawning channel for pink salmon on Chilliwack River in 1973. Regrettably, as will be discussed later in this book, this vital effort was stopped as a result of a Canadian government policy decision.

Logging activities in the Nadina River, Stuart Lake, Horsefly River and upper Adams River watersheds were a major concern. It was extremely important that suitable practices and protection be required in road building, river transportation and forest harvesting.

In collaboration with the Canada Department of Fisheries, meetings were held on a regular basis with B.C. Forest Service, other resource agencies and the forest companies to promote and encourage wise planning and timber-harvest procedures that would give maximum protection for the sockeye. Water samples were taken in many locations to establish base-line values to detect any future changes in water qualities.

In the early 1980s Canadian National

Railways announced that it wished to add a parallel track to its existing railroad line along the Thompson River and down through the Fraser River Canyon. The project could have had serious effects on sockeye and pink salmon migrations and spawning as a result of changes in the river shorelines and flow patterns. The Commission was involved in a federal task force responsible for coordinating the assessment of environmental impacts. This involved detailed studies of the movements and behavior of migrating Adams River sockeye, as well as of pink salmon. Any river bank materials falling into the river would increase flow velocities, fill in fish resting areas and eliminate existing and potential pink salmon spawning areas in the Thompson River. The task force stated that all adverse effects of the project were to be minimized. Where damage to the resource was unavoidable, mitigation was to be planned.

A proposed highway bridge over the Fraser River at Annacis Island, just downstream from New Westminster, posed potential problems to early migrating sockeye and pink runs. The Commission staff, along with representatives from other resource agencies, held discussions in the early 1980s with the Provincial Ministry of Transportation and Highways and its consultants concerning potential adverse impacts. Of greatest concern was the location and influence of the bridge abutments on flow patterns. Extensive investigations using radio-tagged sockeye and pink salmon provided the data necessary to protect the stocks from poor placement of bridge supports.

Many other projects and proposed projects were studied and evaluated, and the Commission gave its recommendations to the Canadian government as to the impact of all these proposals. It is believed that the commission's presence was an important factor in protecting the resource from the deleterious effects of various proposals.

One of the major achievements of the Commission (in cooperation with the Canadian government) was to prevent mainstem building of dams on the Fraser River. The proponents for dams were persistent, whether the dams were to produce power or control floods. The Seton Creek hydroelectric facility with a low-head dam (with fishway) was built in 1953 by B.C. Electric. This dam, requiring fish passage and protection capabilities, was the only such facility that was put in place during the Commission's tenure from 1937-1985.

Enhancement

April 20, 1948. The first IPSFC enhancement effort to rebuild depleted sockeye runs and barren spawning areas involved hatchery propagation. Eggs were taken on Horsefly River, the progeny were reared in Washington hatcheries and then trucked to Quesnel Lake for release. Before being released, they were fin clipped at the University of Washington School of Fisheries. The four men who did the fin clipping became long-term employees of IPSFC: Pete Peterson, Jim Mason, Roy Hamilton and Bill Tomkinson (from the left).

Article III of the treaty gave the Commission the authority "to improve spawning grounds, construct, and maintain hatcheries, rearing ponds and other such facilities as it may deem necessary for the propagation of sockeye (and, as amended, of pink) salmon." Over the years, the Commission was involved in studies and experiments involving egg incubation in gravel; transplantation of eggs, fry and smolts; hatcheries; stream improvement and spawning channels as possible direct methods of increasing salmon production. In addition, one experiment attempted to produce an even-year pink salmon run to the Fraser. In a broad sense, fisheries management and fishway construction were also enhancement techniques.

GRAVEL-INCUBATION STUDIES

The Quesnel Field Station was constructed in 1949 adjacent to Horsefly Lake to provide a facility where sockeye could be reared and transferred to locations that, at one time, had been productive. The project was primarily intended to rehabilitate the Horsefly River race. One experiment used brood stock from the Horsefly River, with the juveniles reared, marked by fin-clipping and released back into the Quesnel system. In November 1950, fingerling sockeye of this group were released at the mouth of Horsefly River. One purpose of the test was to determine if sockeye return to their native spawning area after artificial propagation. In 1953, adult sockeye from this group showed up at Horsefly Lake, and not to their parents' spawning grounds in the nearby Horsefly River (IPSFC 1954). This study proved that run-timing was inherent but that destination of the race could be changed. This provided the impetus for further transplantation experiments.

The Commission also studied the effects of gravel size, dissolved oxygen, percolation flow, water temperature and other in-gravel conditions on survival of incubating eggs. Results from initial laboratory studies conducted at New Westminster in 1953 showed conclusively that high egg survival could be obtained in a gravel substrate with high intra-gravel velocities and high oxygen concentrations (IPSFC 1978a). Similar investigations continued at the Quesnel Field Station from 1953 to 1958. Cooper (1965) showed "the importance of preventing deposition of sediments on or within a salmon spawning bed".

In addition to the conventional hatchery facilities at the station, two spawning facilities were constructed, similar to natural, lakeshore spawning grounds where there is an upwelling flow through the gravel. This facility was the first prototype sockeye channel using upwelling water flow. Survival of naturally deposited sockeye eggs was measured under various controlled flow rates and with different gravel sizes. Egg survivals were measured from spawning by both native and transplanted sockeye stocks and also from newly fertilized and eyed sockeye eggs hand-planted in the gravel. The prototype tests showed that high egg survivals could be obtained on a large scale by providing adequate percolation velocities through the gravel.

The effects of light on incubating eggs were also studied. Eggs incubated under darkness in gravel showed that the size of fry from the test areas was comparable to that of natural fry *(ibid.)*. These results were encouraging and the experiments at Quesnel led to the hatchery program at Pitt River in 1960 and eventual construction of the upwelling incubation area at Pitt River in 1962.

TRANSPLANTATION PROGRAM

The Canadian government performed more than 200 sockeye transplantations of Fraser sockeye in many areas of the Fraser River watershed (Aro 1979). Transplanting of sockeye eggs, fry and fingerlings began in 1884 and continued to the mid-1930s. About 513,000,000, including small numbers of coho and chinook, were transplanted to many locations within the Fraser, 43,000,000 to non-Fraser B.C. rivers, and 80,000,000 were imported from the Skeena River system and planted into the Nadina and Stuart Lake systems *(ibid.)*. The near-term and long-term effect these transfers had on Fraser sockeye production is not known. Smaller numbers of Fraser pink salmon eggs were also taken for transfer *(ibid.)*. The Commission gave considerable thought before establishing a similar transplantation program and concluded that two important requirements should be satisfied: (1) there should be a similar distance from the sea to the spawning grounds for both donor and recipient stocks; and (2) similar environmental conditions should exist in the two areas—the most important being water temperature.

The Canadian government's earlier experiences with hatcheries ended with the closing of all such facilities in 1937. The Commission believed that only egg or fingerling (two- to three-inch fish) transplants should be attempted. From 1947 until the early 1950s, the Commission attempted 11 fingerling transplants (Appendix J), involving several races and locations. Over one million fingerlings were released. In most instances, no adults returned. The best return was at Horsefly River when 218 sockeye returned in 1953 from a release of 94,000 fingerlings. The transplant program was soon abandoned even though success had previously been achieved in Wash-

September 2, 1953. A 16-pond hatchery was built on the shore of Horsefly Lake in 1949 to expedite the Quesnel Lake rehabilitation effort. Hatchery operations were discontinued when the natural population was established and the facilities were then used for other research projects, including the first prototype test of an upwelling-flow artificial spawning ground. Upon completion of this work, the land was returned in 1972 to the British Columbia government for park use.

September 2, 1953. The hatchery on Horsefly Lake, called the Quesnel Field Station, was conveniently located for flying hatchery-reared sockeye to barren areas. On the day this picture was taken, sockeye fingerlings of Seymour River stock were being flown to Adams Lake for release at the mouth of upper Adams River.

September 2, 1953. Hatchery superintendent Doug Hembrough holding a fin-clipped sockeye that returned to the Quesnel Field Station, where it had been reared. Efforts to use hatchery culture to accelerate rehabilitation of the Quesnel population were abandoned when the progeny returned to the hatchery outlet pipeline rather than to the Horsefly River spawning area, where the brood stock had been obtained.

One of the egg-taking crews on Seymour
River consisted of Ross Stewart (third from
left) and local workers.

ington State using sockeye fry and fingerling releases (Kemerick 1945). A possible explanation for the initial success of the Washington State program might be related to size of fish, time and area of release of transplanted fish.

Newly fertilized and eyed-egg planting experiments were begun during the same time that the fingerling transplants were being conducted. Eyed-egg plants continued over the next 35 years and involved eight donor races, involving the early Stuart, lower Adams, Raft, Taseko, Horsefly, Stellako, Seymour and Momich sockeye. Eggs and fry were planted in the Stuart Lake area, Portage Creek, middle Shuswap, Horsefly Lake, Salmon River, Barriere River, Fennell Creek, upper Adams River, a Nadina Lake tributary, Eagle River and Scotch Creek (Appendix J). During this period, almost 46 million eggs were transplanted. The best estimate of first returns to the spawning grounds from these 37 experiments, excluding estimated contributions from known natural spawners that were present in the brood year, was about 5,000 adult sockeye. Since 3,500 of those returned in 1954 to Portage Creek, it is clear that the program was not very successful.

Areas in which eggs had been buried ("planted") were examined just before the ice break-up to assess the incubation success. In the above photograph, taken in early 1959, Forrest Scott removed four feet of snow and eighteen inches of ice to examine part of the bed of upper Adams River where Seymour River eyed eggs had been planted.

The initial returns of 205 and 3,500 sockeye in 1954 to the upper Adams River and Portage Creek, respectively, were nevertheless encouraging to the Commission. In the following years, however, the transplants never produced a return of more than 500 fish to a spawning ground and in most instances no returning fish were found. There were many failures that were as unexplainable as were the occasional few returns. In most of the experiments, the planted eggs were incubated to the eyed stage using donor-area water. As knowledge was gained on the importance of imprinting on homing, the Commission subsequently used the recipient-area water to incubate the eggs. Better results might have been attained if this procedure had been followed from the start of the program. Most of these more advanced techniques though, still produced zero returns.

Studies by Brannon (1972) and Groot (1965) of fry and fingerling behavior showed that the direction of fry and fingerling migration would be an important factor in choosing donor stocks. On the basis of these studies, the wisdom of planting Taseko Lake sockeye eggs into the upper Adams River system could be questioned. Taseko sockeye are lake-shore spawners, not river spawners. Fry and smolt from Taseko Lake would have innate migration patterns opposite to those required of fry in the upper Adams system. An experiment conducted in 1959 revealed that 600,000 Taseko eggs planted in Harbour Creek, tributary to the upper Adams River, produced no return to that creek in 1963, whereas a plant of 900,000 Seymour River eggs into upper Adams River in the same year resulted in a return of about 100 sockeye to the upper Adams (Figure 33). In contrast, from a plant of 702,000 Taseko eggs into the upper Adams River in 1960, a total escapement of 162 sockeye was obtained in 1964. These results refer only to the numbers that returned to the spawning grounds— more than that were likely taken in the commercial fisheries.

In the mid 1960s, the Commission evaluated the program's status and was disappointed with the overall results of transplantation efforts. The egg transplant program was discontinued. Nevertheless, examination of the present-day status of runs where transplantation occurred, reveals some interesting and exciting results. The most notable and unexplainable returns took place in the Adams Lake region, particularly at upper Adams River and in the close-by Momich/Cayenne Creek watershed.

J. Babcock's observations in 1905 and 1909, showed that Adams Lake had previously supported a large summer-run sockeye population, originating primarily in the upper Adams River. It is a reasonable conclusion that there were no early-run sockeye indigenous to Adams Lake when the Commission came into being. Two hundred five sockeye were first seen in the upper Adams River in 1954 following an egg plant in 1950. Over the years there were 12 separate experiments involving more than 10 million eggs, using primarily Seymour River

donor sockeye. In recent cycle-years (1980 and 1984) the stock received an important genetic impetus when eggs from the nearby Momich River race were fertilized by males from upper Adams River and used to increase egg transplants (IPSFC 1985). In 1984, escapement to the upper Adams River increased to 3,502 from only 31 in cycle year 1972. In addition, an escapement of 83 sockeye was counted in 1985. A self-sustaining population has been established in the upper Adams River following many years of rehabilitation efforts.

Rehabilitation efforts on the 1984 cycle-year for the upper Adams sockeye were minimal. The 1956 egg plant from the Seymour River produced an observed return of only one sockeye in 1960. As mentioned earlier, the Taseko egg plant in 1960 produced an escapement of 162 in 1964. Since no further egg plants were made in the upper Adams River on this cycle until 1980, was the substantial return of 560 fish to the upper Adams in 1980 the result of the 1960 Taseko plant and its few fish returning over several generations? It is generally accepted that the greatly increased return to upper Adams in 1984 was partly due to natural production from the 560 spawners in 1980, and also from 334,000 Momich fry that were held and fed and then released into Adams Lake in June 1981 (IPSFC 1982). These fish were reared in floating pens near the mouth of the upper Adams River, thus imprinted with that water source.

Natural spawning of 3,502 sockeye in 1984 was given a significant boost with the inclusion of a plant of fertilized eggs in the upper Adams River of 448,000 Momich eggs and 48,000 pure upper Adams eggs (both groups fertilized with upper Adams milt)—all planted in a prepared gravel area. In addition, 372,000 upper Adams/Momich/Cayenne eggs and 44,000 pure upper Adams eggs were planted in an incubation box along the river bank. The third group of approximately 50,000 eyed eggs was transported to the Sweltzer Creek Field Station at Cultus Lake for a short-term

207

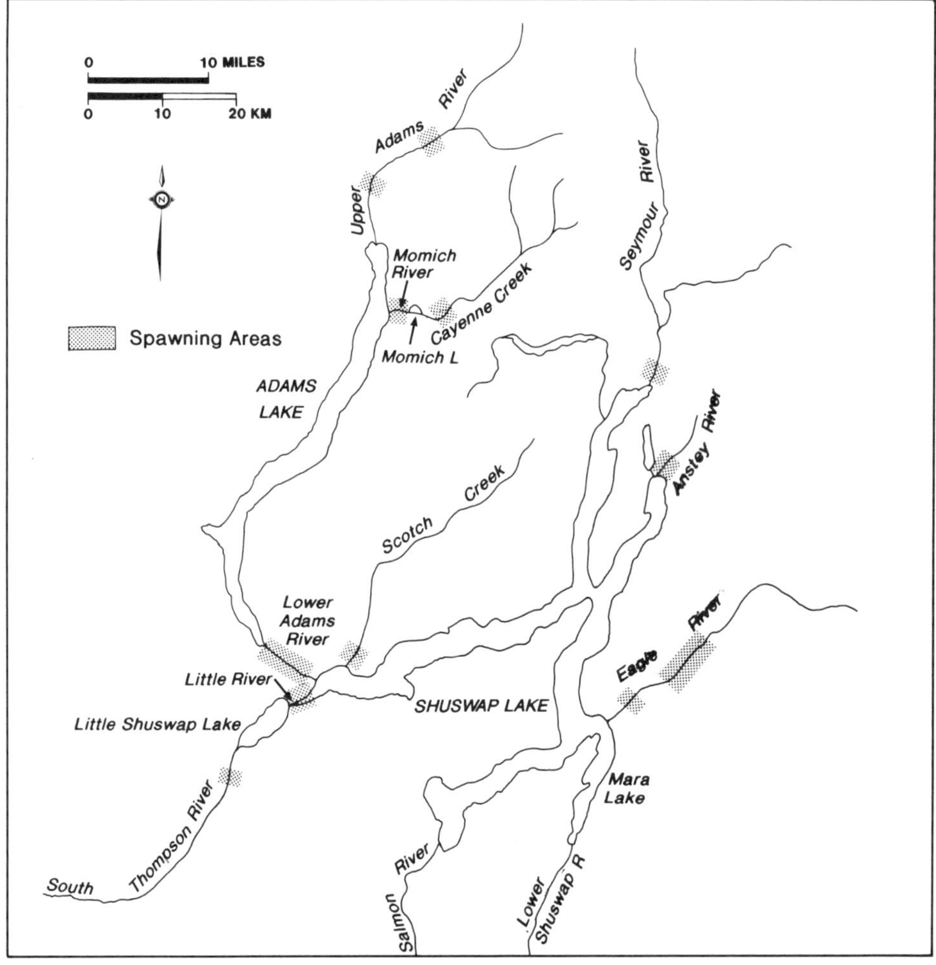

Figure 33. Part of the South Thompson River watershed.

rearing experiment. Hatching and growth was accelerated by using warmer water from Cultus Lake in an effort to increase survival of fry following release into the lake. In the spring of 1985, 771,000 fry were released near the mouth of the upper Adams River, 380,000 from the river plant, 360,000 from the incubation box and 31,000 from the Cultus Lake rearing. Fry (300,000) from the incubation box were held in floating pens at the mouth of upper Adams River and fed for four to six weeks before being released. Therefore it is possible that the present upper Adams population has genetic input from Seymour, Taseko, Momich and upper Adams River sockeye. By 1988, the escapement to upper Adams River had increased to 7,169 fish (PSC 1989).

The source and start-up of the Momich River populations present an enigma. Surveys were made of upper Adams River in 1902 and very few salmon were seen. J. Crawford also visited the Momich River and:

> I examined the Mo-mich (sic) River, a small stream near the head of the lake. The Indians had informed me that the Sockeye (sic) spawned in this stream. We found no salmon there, but could see where a great many had spawned last season. The beach was full of spawning beds made last year (State of Washington Department of Fisheries and Game 1902).

Sockeye were first observed in this system in recent years in 1960 and the spawning escapement increased rapidly on the cycle year to a record 5,854 in 1984 from 823 in 1964 (Figure 34). Only one other cycle (1985) contains any spawners increasing to 83 fish in 1985 from 11 in 1977. The Commission did not make egg plants to the Momich system. It seems very unlikely that the Momich race, because of its small numbers, could have survived the effects of Hells Gate and lower Adams splash dam whereas the great run to upper Adams River was destroyed. Is it possible that returning adults from the Seymour River egg plants into the upper Adams River strayed to Momich? This will likely never be determined for certain, but it remains a possibility. The first substantial transplants to upper Adams River on the 1960 cycle occurred in 1952 and 1956. In 1960 only one fish was seen in the upper Adams but several hundred were observed in the Momich/ Cayenne system and the population has continued to increase. The mouth of the Momich River is about six miles downlake from the mouth of the upper Adams River. Thus, there is a shorter migration distance to the Momich spawning grounds. Perhaps from homing instinct or energy requirements, the returning Seymour-origin fish found the Momich River met the necessary criteria for spawning in 1960. Genetic studies of the current runs in the system in relation to the donor populations would be of interest, but may still not provide the answers to these questions. Whatever the reasons, the important point is that the transplants have been successful and the Commission was able to create a run in the area.

Effects of the 1964 Chilcotin River slide provide some additional information on the possibilities of straying when sockeye are faced with stresses caused by migration blockages, siltation or other physical anomalies. Upstream migration was affected for about six days at this slide. Scale analyses identified about 1,000 Chilko-origin sockeye below Mission Dam on Bridge River near Lillooet, far below the mouth of the Chilcotin River. In addition, about 1,600 Chilko sockeye were identified in such distant rivers as the Horsefly, Stellako, and the Middle

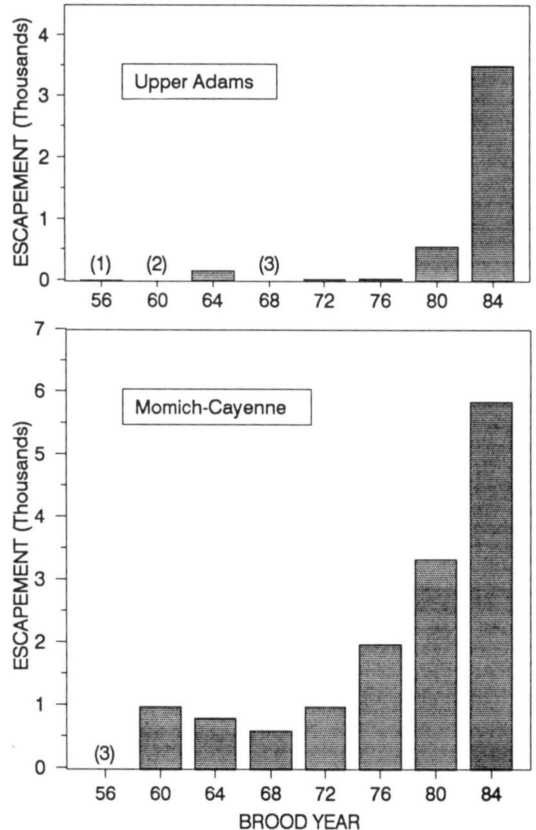

Figure 34. Early timed sockeye salmon escapements to two tributaries of Adams Lake on the 1912 cycle, 1956-1984.
Notes: (1) Only nine fish seen.
**   (2) Only one fish seen.**
**   (3) No observations were made.**

and Tachie rivers of the Stuart Lake system. These observations and the many observations in earlier years of sockeye spawning in tributary streams and rivers below the Hells Gate blockage prove that sockeye stray when normal migration is impeded. With respect to the few sockeye returning in the 1950s and 1960s from mainly Seymour River donor eggs placed in the upper Adams River, it is possible that the extra energy required by returning sockeye to ascend the lower Adams River (an expenditure not required by Seymour sockeye) may have resulted in most of those fish selecting Momich River instead of the upper Adams River.

In 1960, there were several hundred spawning sockeye observed in several locations of the North Thompson River between

Kamloops and Raft River. Neither the Commission staff nor local residents had previously observed sockeye in this area. A total of 316,000 Raft River eggs had been planted in Barriere River, downstream from North Barriere Lake (Figure 1), in 1956. Is it possible that other fish (in addition to the 146 fish which moved on to Fennell Creek) destined for Barriere River in 1960 did not home into the local area where the eggs were planted, and instead spawned in the North Thompson River? Sockeye escapements in the North Thompson River on two cycles now range from about one to two thousand fish. Did the upper Adams River run establish its roots initially in both Momich/Cayenne Creek with the several hundred fish in 1960 and in the upper Adams River in 1960 with just a very few spawners? Escapement to the upper Adams River increased to 162 fish in 1964 and 823 sockeye were estimated to have returned to the Momich River in 1964. Straying may have continued for several more cycles.

Upper Adams River sockeye pass through the commercial fisheries slightly earlier than the Chilko population (Figure 35). The Chilko population is now dominant on the 1912-1984 cycle whereas the original upper Adams River run appeared to be dominant on the 1913 cycle. The upper Adams River run is now establishing itself on the 1912 cycle and should be compatible with the well-established Chilko race. It would be expected that, eventually, the upper Adams River race will be more abundant than the Chilko population. The 1984 Fraser cycle, which has historically been the smallest of the four cycle years, will increase significantly as the upper Adams run continues to increase. The potential now exists for complete stock rehabilitation and utilization of the rearing capacity of Adams Lake.

Because the Commission considered this as the most challenging and potentially rewarding area, the greatest effort in the transplant program was in the Adams Lake system. Significant contributing populations were also estab-lished through transplantation efforts at Portage Creek, Fennell Creek, Eagle River, Scotch Creek and middle Shuswap River. Each of these races now is self-sustaining and produces significant adult returns.

A hydro-electric plant on the Barriere River, tributary to the North Thompson River, was taken out of service in 1952 by the British Columbia Power Commission. It was reported that the Barriere River supported a sockeye run of considerable numbers in earlier years. With the breaching of the diversion dam on the main Barriere River and the storage dams on North and East Barriere lakes (all of timber construction and all breached in 1952), the passage of residual sockeye was possible, allow-ing continued restoration of the sockeye runs to that system. One portion of the system which appeared to offer promise for supporting a sizeable run was at Fennell Creek, a tributary of North Barriere Lake. The Fennell Creek race has an interesting origin. The first and only egg plant was in 1959 with a plant of 490,000 eggs from the nearby Raft River. There were no sockeye observed in the creek prior to 1958, when five were seen and in 1959, when 27 were counted. The 1963 escapement increased to 439 fish and no additional transplants were made on that cycle. By 1979 (1959 cycle) the run had increased to 15,590 spawners. In 1960, eggs from Raft River origin were planted in the nearby Barriere River. There had been 23 natural spawners in the Barriere River in 1960 but none in Fennell Creek. The escapement in 1964 to Barriere River was 85 sockeye and an additional 146 sockeye (presumed from the egg plant) went further upriver to Fennell Creek. Without further assistance, these sockeye produced an escapement in 1968 of 954 fish and by 1984 this cycle escapement had in-creased to 11,021 fish. For the same creek, one cycle (1959) began primarily from an egg plant whereas the other cycle (1960) resulted from straying from nearby Barriere River. In both cases, the donor stock was from the Raft River.

At Scotch Creek, close to the lower Adams

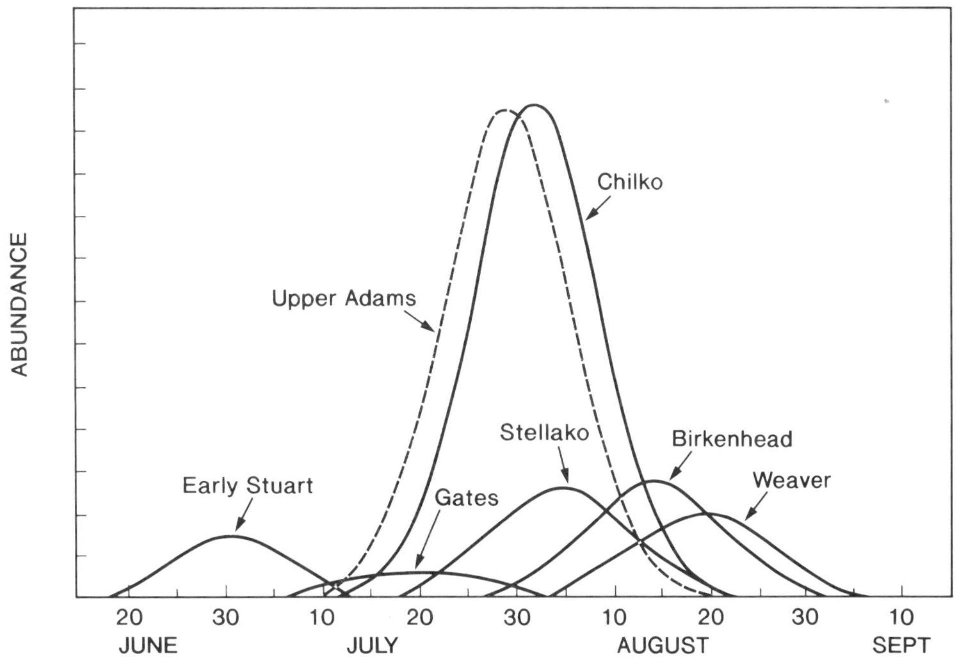

Figure 35. Estimated timing in Juan de Fuca Strait of the potential upper Adams run in relation to other sockeye populations on the 1984 cycle.

May 8, 1946. This B.C. Electric power dam on Barriere River was no longer being used so it and another at the outlet of North Barriere Lake were removed in 1952, which opened up potential sockeye spawning areas. An excellent run to Fennell Creek above North Barriere Lake developed as a result of this work.

River, no early-run sockeye population existed on the dominant Adams River cycle (1962) except for seven spawners in 1962. A total of 1,023,000 eyed Seymour River eggs were planted in Scotch Creek in 1962 and the returning escapement in 1966 was 459 fish. The migration distance to Scotch Creek is much less than that to Seymour River. By 1978, escapement was 2,000 and in 1986, a reported 26,624 early-run sockeye had reached the Scotch Creek spawning grounds (PSC 1988a).

Other races have also increased as a result of egg plants. The middle Shuswap River had no late-run sockeye present in 1954 and had been barren for many years. From about 1.4 million eggs of lower Adams River origin planted in 1954, the escapement increased to 3,000 by 1974 and in 1986, 80,529 fish were present *(ibid.)*.

Anstey River is tributary to Shuswap Lake and has an early-timed sockeye run. The first sockeye, four of them, were observed in the river in 1949. No egg transplants were made, but by 1978 almost 900 sockeye were counted in the river. This was the largest escapement of any year observed. The escapement in 1986 was reported to be 7,080 fish *(ibid.)*. It appears this race succeeded on its own but it took many cycles to do so.

Eagle River is another early-run sockeye population located adjacent to Shuswap Lake near the entrance to lower Shuswap River. The first year that sockeye, 11 of them, were observed in this river by the IPSFC was also in 1949. Eggs from Seymour River were planted in Eagle River in 1958 and again in 1962. In 1962 only 167 sockeye spawned in the river and about 200 spawned there in both 1974 and 1978. However, in 1982 the escapement increased to 1,642 and by 1986, an escapement of 7,138 fish was recorded *(ibid)*. Based on the length of time it took for several of the egg plants to produce significant returns, it appears that, in some areas, the Commission's expectations for early returns were too high.

Newly established stocks may be unstable for several cycles until populations become adapted to their local spawning and rearing environments. This was the case for Skagit River (Washington) sockeye which were transplanted to Lake Washington. Over seven generations were required before there was a significant increase in production. The Horsefly River race, too, went through many cycles of low production, aberrant timing and severe pre-spawning losses before it reached greater stability and greatly increased production in the 1980s.

HATCHERIES

During the Commission's tenure little success had been reported in successfully rearing sockeye and pink salmon in hatchery environments. Many years of earlier experiments for sockeye at Cultus Lake in the 1920s and 1930s concluded:

> …from a statistical analysis of the data obtained it is readily apparent that artificial propagation exhibits no significantly increased efficiency, in point of seaward-migrating young sockeye, over natural production. (Foerster 1938)

In 1959 the Commission was concerned about high fishery exploitation rates on less abundant stocks migrating through the fishing areas at the same time as races of high productivity. An example was the Pitt River run, which suffered from low fry production caused by frequent flooding and scouring of its spawning areas. An experimental hatchery was built in 1960 adjacent to the upper Pitt River. Some of the findings obtained at the Quesnel Field Station, such as protecting incubating eggs from light, were incorporated into the operation of the hatchery. While these procedures were followed for the next few years, the hatchery environment could not produce a normally-timed fry emergence. Hatchery fry developed faster and reached the emergence or "swim-up" stage several weeks earlier than wild fry, resulting in smaller and weaker fry not optimally timed with temperature and feeding conditions in the lake environment. The hatchery program was terminated and in-gravel incubation, using an upwelling-flow facility, was substituted.

Several attempts have been made to rehabilitate the Cultus Lake race. Research was conducted to re-evaluate whether or not juvenile sockeye of certain stocks could successfully be reared to the smolt stage. If this could be done, facilities could be established elsewhere in the watershed to develop new populations and strengthen less productive populations. For several years, starting in 1966, studies were conducted at Sweltzer Creek Field Station. Problems were encountered with disease, the most serious of which was Infectious Haematopoietic Necrosis (IHN). In the first years, very few fish survived to be released as smolts, but by 1972, 12,000 sockeye smolts were released and in 1974, 1,446 adults (12 percent) returned. In 1974, 67,807 Cultus Lake smolts were released and 656 adults (one percent) returned to the counting fence at Cultus Lake. Wild smolts had a 5.3 percent survival to adult returns. It is probable that survival from smolt-to-adults for released fish was four percent. The 1972 smolts were three times heavier at release time than their wild counterparts and the 1974 migrants were twice as heavy. These investigations suggested that rearing sockeye to the smolt stage was not a profitable or beneficial method of enhancing the stocks. Haig-Brown (1980) said: "It is important to emphasize that hatcheries do not and cannot replace the productive yield of natural stocks". In addition, he said: "It is urgently important to protect natural stocks".

Recent investigations in Alaska by the Alaska Department of Fish and Game appear to offer hope for successful production of hatchery-reared sockeye salmon (Dr. B. Allee, personal communication), but this work has not yet produced large numbers of adult sockeye as of this writing. Large pink salmon production had been achieved in the 1980s through hatchery programs in Prince William Sound

(ibid.). Hatcheries, however, can cause serious problems for maintenance of wild stocks (Goodman 1990).

EGG PLANTS FROM MORIBUND SOCKEYE

When pre-spawning mortalities were prevalent in many areas in the 1960s and early 1970s, the Commission was concerned about the great number of eggs lost each year on the spawning grounds. Experiments were conducted to evaluate the viability of eggs taken from moribund, unspawned females as well as the viability of sperm from moribund males. It was found that by being selective, based on subjective but experienced examination of dying fish, fertilization was successful and fry were normal in all respects. Large numbers of eggs were taken from dying females, fertilized and planted in the Horsefly area and in Gates Creek channel to augment production. The survival at Gates Creek from 300,000 planted eggs to the fry stage was 80 percent. But at McKinley Creek (Horsefly area), only 20,000 fry were obtained from a plant of 6,347,000 eggs. The techniques used in this operation were probably unsatisfactory and the location of the plant may have been a negative factor. Since pre-spawning losses had greatly diminished by the 1980s, there was no longer any need to continue these efforts.

EVEN-YEAR PINKS

The absence of pink salmon in the Fraser system in even-numbered years is intriguing. An approach to establish an even-year run was suggested by Washington State authorities as early as 1920, using eggs from stocks of even-year pinks from other British Columbia rivers and Alaskan streams (*Canadian Fisherman* 1921). Research and technological developments suggested other possible methods of establishing an even-year pink run in the Fraser. Several approaches were considered by the Commission and the one chosen involved growth retardation of pure Fraser River stock by one year with photo-period control to delay sexual maturity until the fish were three years of age (IPSFC 1975).

In 1974, 45,000 fry from the Seton Creek spawning channel were taken to the Sweltzer Creek Field Station and 30,000 to Bamfield Station, the marine biological facility on the west coast of Vancouver Island operated by western Canadian universities. All but 60 fish were lost at the Bamfield Station due to a disease (Vibrio) outbreak. The survivors were transported back to Sweltzer Creek. By the end of 1975, only 19 of these fish remained alive. At the end of 1975, only 1,400 pinks remained at Sweltzer Creek. Various problems plagued the experiments—kidney disease, fungus infections and heavy copepod parasitization. In the fall of 1976, 32 three-year-old female pinks were spawned and the eggs fertilized with milt from 84 three-year-old males. Approximately 12,000 eggs were recovered. Unfortunately 55 percent did not develop. Because of high mortality at hatching, only 2,000 fry were obtained for further rearing. The fry grew well until October 1977, at which time all died from a bacterial gill infection.

Another group of juvenile pinks, produced from eggs of the 1975 brood-year spawning at Seton Creek, was reared at Sweltzer Creek Field Station. A better diet, supplemented with vitamins, was used and the fish received prophylactic antibiotic injections in an attempt to control kidney disease, but they still suffered heavy mortalities in their third year and all died from kidney disease prior to spawning. At that point the program was terminated. Although transfer of large numbers of eggs from other sources might be practical for incubation in the Seton Creek spawning channels, any success in establishing a viable, even-year pink salmon population would have to be compatible with sockeye populations migrating through the fishing areas at a similar time on the even-years. One of the even-year cycles (1986) now contains the dominant Adams River sockeye run; however, up through 1913, the dominant lower Adams River sockeye run and the Fraser pink run occurred in the same year.

SPAWNING CHANNEL CONSTRUCTION

The first studies that led to production-type spawning channels were conducted at the Quesnel Field Station in the early 1950s. It was found that eggs should not be exposed to light during incubation and that sockeye fry survival rates in channels were four to eight times greater than in the wild if proper waterflow and gravel were used.

The Canadian Department of Fisheries constructed a small pink salmon spawning channel in 1954 at Jones Creek, a tributary to the lower Fraser River (Hourston and MacKinnon 1956). This was the first channel built in the Fraser system and was constructed as mitigation for a power development on the creek. The channel is still in operation and has served as a successful prototype for other channels. Salmon spawning in this type of facility enter the channel and spawn under a natural setting where discharge and gravel size are controlled to maximize egg-to-fry survival.

PITT RIVER

The results of earlier studies, some by the IPSFC at the Quesnel Field Station and by other agencies on hatchery evaluations, led the Commission to construct an incubation channel for sockeye salmon at the Pitt River facility. Studies in the hatchery at this site and further investigations at the Sweltzer Creek Field Station showed that the incubation of sockeye eggs in hatchery trays resulted in smaller and lighter fry than when the eggs were incubated in a gravel substrate. Without gravel or some other means of support, alevins fall onto their sides and then expend much of their yolk-sac

energy in swimming and trying to right themselves rather than in growth. The gravel incubation channel was a combination upwelling and stream-type incubation area designed to eliminate the adverse effects of the hatchery environment. In 1963, the first year of operation, 2,967,000 eyed-eggs were planted and 2,250,000 fry emerged the next spring, a survival rate of 75.8 percent. The channel has been in operation each year since. Channel fry emergence coincided with the emergence of wild fry, eliminating serious effects of timing differences due to the accelerated development of fry in the hatchery. In all respects, the channel fry were similar to wild fry. From 1963 brood year through 1976, egg-to-fry survival averaged 81.5 percent. Eggs planted varied annually from 2,133,000 to 4,648,000. For brood years 1963 to 1971, the channel had produced a total catch estimated at $1,892,947 landed value (Cooper 1977). It has continued to produce similar numbers since. The capital cost of the facility was $74,142 in 1961-1963.

Knowledge gained in experiments at Quesnel Field Station and the Sweltzer Creek laboratory was used to build a successful production facility for 3 million sockeye eggs on Seven Mile (Corbold) Creek, a tributary of Pitt River above Pitt Lake. Beginning operation in 1963, the gravel incubation bed has an area of 6,400 sq. ft.

March 2, 1967. Pitt River is subject to severe flooding and erosion. Egg-fry survival is so low that the egg incubation facility on Seven Mile Creek was considered to be essential for maintaining the commercial value of the sockeye resource.

October 4, 1961. A spawning channel was built adjacent to the upper part of Seton Creek to compensate for spawning area that was destroyed by construction of the Seton Creek hydroelectric plant. The channel is 2,918 ft. long and 20 ft. wide at the gravel surface. It began operation in 1961.

September 27, 1961. A portion of the upper Seton Creek pink salmon spawning channel.

In view of the success of the first spawning channel, a much larger channel was built in 1967 on Seton Creek as an enhancement project. Built on leased Indian Reserve land immediately adjacent to Seton Creek, it is 9,486 ft. long by 20 ft. wide at the gravel surface.

SETON CREEK

The Commission's first spawning channel for pink salmon was the upper channel in Seton Creek. Although it was initially regarded as an experiment, it has been in operation each odd-numbered year since 1961. The facility was to offset the loss of about 30,000 square yards of prime spawning area flooded by the dam placed at the outlet of Seton Lake in the mid-1950s. The capital cost of the channel in 1960-1961 was $32,259. For brood years 1961-1973 it was estimated that the channel produced a pink salmon catch with a landed value of $618,630 *(ibid.)*. For all brood years from 1961 through 1983, the average egg-to-fry survival rate was 49.8 percent. The numbers of spawners in the channel ranged from about 4,000 to 14,000 fish. Since larger pink salmon runs returned in the late 1970s and 1980s, the landed value in the fisheries from 1963 to 1985 was much larger.

The success of the upper Seton Creek spawning channel prompted the Commission to consider and recommend a second and larger channel in the lower Seton Creek area. In addition, there was an obvious need to increase pink salmon production in the Fraser River system. By 1965, plans were drawn, approved, and the funding provided to construct another channel. It was ready in 1967. The lower channel has capacity for 21,000 spawners, more than three times that of the upper channel.

Pink salmon stocks throughout the Fraser watershed were generally at low levels of abundance from 1957 to 1961 and total escapement to all areas averaged only about 1.5 million fish each year. For the same time period, the escapement to the general area of Seton Creek averaged 46,300 fish each year. In just a few years, the returns to Seton Creek increased dramatically and from 1979 to 1983 the average annual escapement was 613,000 fish. Total pink salmon escapement to all spawning areas for those same years averaged 4,227,000 fish. During the period from 1957 through 1983, total escapement of all stocks increased by a factor of 2.8 whereas the Seton Creek area escapement during the same period increased by more than 13 times. In 1979, the total area escapement reached a record 713,000 pink salmon. The spawning channels at Seton Creek have had a very positive impact on total production and it is likely the run was enhanced beyond the original capacity of the natural spawning grounds prior to power development in the mid 1950s.

The number of spawners in the lower channel ranged from 14,900 to 33,900 and fry production from 9.0 million to 22.6 million, averaging 15.7 million each cycle year from 1967 to 1985. Egg-to-fry survival rates for brood years 1967 to 1983 averaged 56.2 percent.

Benefits derived from the lower channel in its first four brood years (1967-1973), in terms of value to the fishermen, was $1,419,311 (*ibid.*). The channel cost $218,665 to build in 1966-1967. The benefit/cost ratio estimated at 1967 dollars was 5.7. Not included in those calculations is the value of salmon produced by spawners that originated in the channel but spawned elsewhere in the Seton Creek area. Benefits received from brood years 1975 through 1985 were not calculated but they would have been substantial.

· · · · · · · · · · · · · · · · · · · ·
WEAVER CREEK

The Weaver Creek sockeye run is a late-season run returning to the creek near Harrison Lake in the lower Fraser River. Fry from Weaver Creek move downstream to Harrison River and then migrate upriver to Harrison Lake. The population was in jeopardy because of highly unstable spawning grounds due to flooding and scouring of spawning beds—a consequence of extensive logging of the watershed. For this reason, a first-of-its-kind sockeye spawning channel was constructed and placed in operation in 1965. The channel is 9,614 feet (1.8 miles) long by 20 feet wide.

During its first 20 years of operation through to the 1984 spawning, the egg-to-fry survival rate varied from 25.1 percent to 89.5 percent and averaged 64.2 percent. A severe flood in upper Weaver Creek in the winter of brood-year 1977 (the year of the low 25.1 percent survival) deposited large amounts of silt, sand, gravel and logging debris in the channel. Efforts were made to clean the channel but survival never exceeded 57.3 percent following that flood. Survival in brood years 1965-1976 before the flood averaged 74.3 percent. For the years following the flood (1977-1984) survival averaged only 49.0 percent. In comparison, in certain years, the survival of sockeye fry from natural spawning in Weaver Creek proper was as low as less than one percent.

The number of sockeye spawners using the channel was controlled at a gate where fish were counted as they entered. In some of the early years when runs were depressed, such as in 1968, only 2,176 sockeye entered the channel. However, this situation soon changed as production increased rapidly and in some years escapement far exceeded the need (for both the channel and creek). In 1982, a record of 57,932 sockeye spawned in the channel. The total sockeye escapement of 295,474 to Weaver Creek in 1982 was the largest on record. Sockeye that were not allowed to enter the channel spawned in the creek. Some of the excess escapement from this race was experimentally harvested in the Harrison River. The canned product was deemed by industry as less than desirable for a large-scale venture even though much of the product graded favorably.

To appreciate the impact this channel has had on the size of the runs, consider that the total escapement each year to Weaver Creek and the channel during the first four years of channel operation (1965-1968) averaged only 14,400 spawners but by 1982-1985 the escapements averaged 108,900 each year— a 7.6 fold increase.

It was estimated that fry production from 1965 to 1968 from spawners in Weaver Creek proper averaged 1,775,000 annually. Fry production from the channel averaged 6,416,000

April 9, 1963. Weaver Creek, like many other coastal streams that have been logged, has a wide, abraded flood channel. Egg-fry survival was measured at about 6% compared to about 50% in the spawning channel.

The Weaver Creek spawning channel began operating in 1965 and after three cycles of operation there was a large surplus of spawners despite a heavy commercial fishery. Used primarily by sockeye, large numbers of chums and a few pink salmon, the channel is 9,614 ft. long by 20 ft. wide at the gravel surface.

fry each year during these same first four years of channel operation *(ibid.)*. With greatly increased escapements to both the creek and the channel from 1973 to 1976, annual fry production from the creek increased by a factor of 2.2 to 3,911,000 fry while the number of fry produced in the channel for the same period averaged 37,587,000, almost six times more fry than in the 1965-1968 era. From brood years 1965 through 1976, creek fry production averaged 2,788,000 annually, while the channel fry production averaged 19,712,000 each year. From 1965 through 1984, the channel produced an average of 26,433,000 sockeye fry each year. One of the prerequisites for channel construction was that channel-produced fry had to be the equivalent of wild fry. Examination of the channel fry confirmed this (Mead and Woodall 1968).

Total production of Weaver Creek sockeye, including commercial catches and escapement, averaged 84,500 sockeye annually before the channel was constructed (Cooper 1977). Production from the channel alone for the first eight brood years (1965-1972) was estimated at 203,800 sockeye annually. Following this period some very exceptional returns occurred. In 1982, total Weaver Creek production was a record 960,000 sockeye, more than double the previous largest run, which was also channel assisted. About 95 percent of the total return in 1982 was due to the spawning channel. The commercial catch in 1982 was 659,000 fish, more than the total number of Weaver Creek fish caught during all of the 11 years prior to channel construction .

The landed value of the commercial catch for the channel production of the first eight brood years (through the 1976 return) was calculated at $5,305,482 *(ibid.)*. The huge returns in the 1980s would be additive to these values. The channel was constructed in 1964-1965 at a cost of $280,725. Annual costs to operate the channel averaged $41,000. The benefit/cost ratio exceeds eight. The large returns from the 1978 brood-year spawning in

the channel alone had a value of $6,000,000 to fishermen.

The Weaver Creek run was near extinction in the 1961-1964 period, as evidenced by an annual average total run of only 9,000 fish. The large returns in the 1970s and 1980s, culminating in 1982 with a record return of 960,000 fish, clearly indicate the remarkable success of the facility. This leaves no doubt about the importance of preserving each race of sockeye, regardless of how depleted it might be. Chum salmon and pink salmon have also benefited from the channel and have contributed to the fisheries.

. . . .

GATES CREEK

The fourth spawning channel constructed by the Commission also involved a sockeye salmon run that had been reduced by human activity. As cited earlier, the Seton Creek power project had an adverse impact on local salmon runs. In addition, much of the natural spawning grounds at Gates Creek had been lost earlier due to logging and other encroachments in the watershed. The channel was built in 1967-1968 and is 6,200 feet long by 20 feet wide.

Another serious problem with the Gates Creek population was the recurring loss of sockeye due to pre-spawning mortality. For brood years 1968 through 1976, losses of spawners in the channel averaged 32 percent each year and was as high as 63 percent in 1971. Similar losses were recorded in the wild.

The annual average escapement to the creek and channel during the first four years of channel operation (1968-1971) was only 1,424 adult sockeye. By 1982-1985 average adult escapements had increased to 10,445 fish.

Fry produced in the channel during the first four years of operation (1968-1971) averaged 1.886 million annually. From 1981-1985, annual fry production averaged 6.405 million. As mentioned earlier, escapements increased

221

The Gates Creek spawning channel, which began operating in 1968, was built adjacent to Gates Creek near Anderson Lake on leased Indian Reserve land. Flowing from right to left, the channel has a silt settling basin just downstream from the intake in Gates Creek. A barrier fence across Gates Creek directs sockeye into the channel, which is 6,201 ft. long by 20 ft. wide.

March 15, 1968. A pipeline was placed to a depth of 200 ft. in Anderson Lake to provide an emergency water supply for the channel as well as for temperature control during the spawning period and ice control during winter months.

at a much higher rate than did fry production during the 17 years of channel operation. Part of the reason was lower egg-to-fry survival rates from brood years 1979-1984—only 58 percent compared with 78 percent during the first six years of operation. Pre-spawning mortalities also seriously reduced egg-deposition and fry production.

In comparison to the Weaver Creek channel, increases in sockeye fry production from the Gates Creek channel have progressed much slower and at a much lower level. From 1981 to 1984 at Weaver Creek, average annual fry production was 38,100,000, about 5.9 times greater than that achieved at Gates Creek. It follows that adults produced by the Gates Creek facility would be fewer, the landed value from commercial catch would be lower and the benefit/cost ratio would be low. The channel cost $315,452 to build in 1967-1968. The total adult return from brood years 1968-1972 produced a commercial catch with a landed value of $844,463. The benefit/cost ratio was 1.87. The largest commercial catch produced by the Gates Creek channel was 95,000 sockeye in 1976 with a landed value of $609,000 (*ibid.*).

The potential increase of the Gates Creek sockeye run had been delayed by pre-spawning losses and low escapement levels. In addition, Gates Creek fry migrate directly to Seton Lake and do not use Anderson Lake for rearing. It is possible that a lower survival rate is intrinsic in this race, compared with other populations, because of this aspect of the race's life history. Transplantation experiments utilizing donor races to produce fry which rear in Anderson Lake might be rewarding. Planting of Chilko River sockeye eggs in the upper portion of Portage Creek would be an interesting test since Chilko fry would be expected to migrate upstream and rear in Anderson Lake.

Apparently, in earlier years, sockeye fry reared in Anderson Lake for a year. In 1903 Babcock observed and described in detail, the behavior of large schools of sockeye yearlings migrating from Seton Lake. For the migration from Anderson Lake he said: "In addition to the movement of yearlings from Seton Lake, there was observed a large movement from Anderson into Seton through Portage Creek. The schools issued from Anderson Lake in the same manner as from Seton" (Babcock 1904a). Commission investigations did establish the presence of large numbers of kokanee that may, in some manner, be related to the earlier anadromous sockeye stocks which utilized the rearing opportunities of Anderson Lake.

It is reasonable to assume that maintenance and present run status would not have been possible without the channel and there is good reason to believe the runs can be increased even further. However, it is expected that progress will be slow until major problems facing the population are overcome. Smolt migratory problems at the Seton powerhouse need to be resolved. Large increases in production can be achieved with careful enhancement, environmental restoration and maintenance, and good fishery management. The continued enhancement of this race is important because of the population potential and its early timing into the Fraser at a time of relatively low abundance of other stocks.

NADINA RIVER

The fifth and final spawning channel was for the late Nadina race that spawns in the upper reaches of the Nadina River (Figure 29) about 680 miles from the Fraser River mouth. Natural expansion of this race was restricted by a limited spawning area. Estimates were made that Francois Lake could support additional fry from Nadina River to produce a catch of 5,154,000 sockeye every four years (IPSFC 1972a). This lake is currently the most underutilized lake in the Fraser system and has enormous potential.

The channel is 9,759 feet long by 20 feet wide, very similar in size to the Weaver Creek facility. It was constructed from 1970-1973 at a total cost of $761,159. During its first four years of operation (1973-1976) the channel had an annual average escapement of 5,576 spawners. By 1981-84, escapement to the channel increased by a factor of 2.3 to an average of 12,574 spawners each year. This was a disappointment in view of the 15,540 average return to the channel in the previous four years. Maximum escapement into the channel was 40,577 in 1979. From 1973 to 1985, an average of only 11,311 spawners used the channel, considerably below its design capacity.

Fry production during the first four years of operation averaged 6.213 million each year (1973-1976) For the last four years (1981-1984) fry production averaged 10.005 million annually or an increase of only 1.6 times that obtained for the earlier period. Fry production for all 12 years of channel operation averaged 8.586 million each year.

In its initial four years (1973-1976) the channel functioned well and egg-to-fry survival rates averaged 71.5 percent or as expected. However, for reasons not clearly understood, from 1981 to 1984 the average egg-to-fry survival declined to 41.2 percent. For all years, egg-to-fry survival averaged 54.3 percent.

Analysis of the commercial catch value attributable to the Nadina channel has not been completed; however, the total estimated late Nadina run in 1977, the first return year from channel production, was a record 300,000 sockeye. This would have yielded a commercial catch in excess of 200,000 fish with a value to the fishermen several times that of the initial channel cost. But total production of late Nadina sockeye in 1981 was only 148,000 sockeye and an estimated 146,000 of these came from the channel. In 1983, the total return was 85,400 fish, of which 74,300 sockeye came from the channel. The desired production of the late Nadina channel was not reached during the first 13 years of its operation. In spite of this, it is expected that the channel will be a significant contributor to the commercial catch.

In 1972, the Commission proposed that a spawning channel be built for the early Nadina River race. It was estimated that at full capacity the channel would produce a catch of 4.65 million sockeye every four years. It is unfortunate that from an escapement of 30,000 in 1957, the returns in all years have declined. No fish returned in 1982. The escapement in 1983, 1984 and 1985 was only 1,337, 806 and 18 sockeye respectively. The race is in desperate need of an incubation-spawning channel facility.

In a relatively short time span—up to 1976—all four spawning channels (Weaver, Gates, Seton) and one incubation channel (Pitt River) had given a total landed value to fishermen of $10 million for Fraser sockeye and pink salmon. The cost to build the five channels was $921,243. Large additional benefits from all six spawning facilities have been received by the fishing industries of both countries from the mid-1970s to the present time but the value of these benefits has not yet been calculated. The benefit/cost ratio, based only on the years of channel operation up to 1976, averaged 5.66 *(ibid.)*. The numbers of fish produced by the sockeye channels in any year did not exceed ten percent of the total run.

An important attribute of spawning channels for long-term stability of the runs is that the genetic integrity and diversity of the wild stocks is maintained. This vitally important biological consideration is generally ignored under commonly used hatchery practices.

STREAM IMPROVEMENT

The Commission successfully developed a gravel cleaner for use in spawning channels. It was tried in the natural spawning grounds but boulders buried in the river interfered with its operation. It was not designed to handle the larger rocks found in rivers. Further tests indicated that gravel cleaning in natural streams could be achieved by using a track-mounted Gradall excavator. The top foot of gravel was removed and then replaced back into the stream while the silt and sand were washed downstream. In 1973, 32,000 square yards of gravel in the Horsefly River were cleaned by this method. Subsequent sampling indicated high egg survival in both cleaned and uncleaned areas of the river. Another 4,000 square yards of gravel was cleaned in McKinley Creek below the McKinley Creek temperature-control structure. It was believed that about a three times greater egg-to-fry survival rate could be achieved to offset losses due to pre-spawning mortalities. Thousands of sockeye spawned successfully in the cleaned gravel areas.

Experiments to clean gravel in natural spawning streams continued in 1975 in the Nadina River when a length of about 700 feet of the spawning area of the early Nadina race was cleaned using a new technique. A track-mounted excavator using a screened bucket was used to dig gravel from the streambed. The bucket was then vibrated to drop material finer than 1/2 inch through the screen into the hole from which the gravel had been excavated. Cleaned material larger than 1/2 inch was placed on top of this fine material, producing a 12 to 16 inch layer of clean, permeable gravel. Only 481 early Nadina sockeye reached the spawning grounds in 1975 and unfortunately

August 6, 1973. The Nadina sockeye spawning channel began operating in 1973. It has a very reliable gravity water supply from Nadina Lake. Despite exposure to severe winter weather, it operates very successfully. The channel is 9,759 ft. long by 20 ft. wide at the gravel surface.

A gravel cleaner invented by IPSFC engineers was regularly used to remove algae, egg detritus, silt and sand from spawning channels to ensure that they continued to provide high intragravel flow, which resulted in good egg-fry survival. The cleaner was also tested for cleaning natural spawning grounds but a vibrated screen-backed bucket proved more effective.

this was not enough to provide any spawning in the cleaned gravel area.

In 1976 another 1,000 square-yard section of early Nadina spawning gravel was cleaned the same way. Gravel in the 1975 cleaned area was sampled in 1976 and found to have no fines less than 1/2 inch. Further surveys of the 1975 cleaned area were conducted in 1977 following two annual high-runoff flows. One-half the length of the cleaned area had reverted back to its original condition with fines and low permeability. In view of the small number of spawners and the substantial cost of gravel cleaning, the Nadina River program was terminated.

Cleaning of natural spawning gravel was not attempted in other areas, partially due to the high costs involved and limited success, but also because by the late 1970s many natural spawning populations were showing dramatic production increases. Gravel cleaning in all of the spawning channels continued on an as-needed basis.

The Adams River sockeye run had declined drastically in the dominant-cycle years of 1962 and 1966 following the 1958 record-breaking run. Engineering surveys and studies of spawning distribution suggested that fry production from 1958, 1962, and 1966 escapements was reduced because of a natural change in flow distribution between two major river channels where sockeye spawned. Adequate water was available during spawning but by winter low-flow periods, 90 percent of the flow entered the left bank channel, leaving only 10 percent in the right bank channel. The lack of water resulted in exposure and increased egg mortality. Corrective action in 1970 consisted of excavating the river bed at the point of flow so that at least 40 percent of the total flow entered the right bank channel during the winter. This correction was done in following years until by 1978 the lower Adams run had regained abundance.

By 1968 the Commission had recorded significant stock abundance increases. These were due to fishways and scientific management of the stocks. The Commission also visualized great potential for salmon enhancement through the use of spawning channels, since only minor success had been achieved with transplantation programs and because hatcheries offered very little hope. Further stock rehabilitation was necessary. All channels constructed through 1968 were built to mitigate human activity on the environment. Early results were promising enough for serious consideration to expand the technique. Further production could be achieved in areas with limited space on the spawning grounds or where other factors were limiting natural production. Many lakes in the watershed have greatly underutilized fry-to-fingerling rearing areas. Thus, the groundwork was laid in the late 1960s for the Commission to apply its findings on a much larger scale in accordance with its terms of reference under the Convention.

In late 1971, following extensive examination of the current status and stock requirement, including spawning areas, rearing areas, lake productivity, etc., the Commission submitted to the governments a proposed program for restoration and extension of the sockeye and pink salmon stocks of the Fraser River (IPSFC 1972a). The program called for the construction of nine spawning and incubation channels for sockeye and three spawning channels for pink salmon. The 12 channels would cost $14 million to build at 1971 prices. When fully operational they would produce an average annual catch of 5,664,000 sockeye and 4,098,000 pink salmon every two years. The projected annual landed value of these catches was $14.7 million, with the market value of the processed catch at $30.9 million (1971 prices). The benefit/cost ratio would be 9.5 to 1. Total sockeye catch after completion of the chan-

April 14, 1970. The lower part of Adams River divides into two channels, both of which are heavily spawned. In 1970 it was noted that minor erosion at the head of the island formed by these two channels had caused a severe reduction in flow to the right bank channel, which resulted in excessive redd exposure in this heavily spawned area.

Aerial view of a flow deflector, built of quarried rock to a maximum height of six feet, placed in Adams River to equalize the flow distribution in the two channels shown in the upper photo.

nels, including both naturally-produced and channel-produced fish, would exceed 10,000,000 fish annually. The pink catch would be 8,910,000 fish every two years. The projects were to be phased in over a 16-year period.

The two governments reacted to the proposal quite differently. The United States government and the country's fishing industry quickly gave strong support. Congressional appropriation for the United States share ($7 million) was obtained in 1972 following an effective lobbying effort by U.S. industry. In view of the problems encountered in obtaining Commission-supporting funds in the United States in previous years, this was a major achievement.

The Canadian government in a surprising move, stated that it was prepared to provide all the funds for the program and that it no longer wished to participate on an equal-share contribution basis, as provided in the Sockeye-Pink Salmon Convention. Therefore, it was clear that both governments approved of the program in principle and recognized the Commission's ability to restore the runs. The Commission petitioned the governments for several years in an attempt to get the program underway, but they did not give the required joint approval. The Canadian government would not proceed further under the existing treaty on a cost-sharing basis with the United States for capital construction projects.

Canada's position was not fully consistent with regard to Article III of the treaty. Although Canada had given its approval of the program, it was apparent it did not deem it desirable to provide its share of the funds in light of the ongoing U.S.-Canada negotiations concerning salmon interceptions.

The program did not proceed. For the first time in the Commission's history, one of the Parties failed to appropriate funds for "… restoration and expansion of the stocks," even though it had approved a Commission recommendation. In practical terms, Canada's action meant that if the Commission was to follow its mandate of extending the sockeye

and pink salmon resource, the Commission must pursue other avenues of approach. A national posture had never before been such a major issue affecting the work of the Commission. After several frustrating years and as a result of these developments, the Commission embarked on a program to increase escapements of certain selected stocks, where increased production potentials existed, by providing optimum escapements of the major races. Strong effort was directed to increase the escapement for sockeye races such as the Horsefly River, lower Shuswap River, Gates Creek, Pitt River and Nadina River. In essence, the development program was replaced by a major emphasis to increase escapements. In addition, pink salmon escapement was to be increased to all Fraser River spawning areas. In doing this, other smaller sockeye and pink races co-migrating through the fishing areas would also receive increased escapements. Associated with these would be increased protection for coho and chinook, a fact not publicized. There would be an initial cost to fishermen in terms of reduced catches but it was believed that these short-term losses would be more than compensated for in later years. Successes generated by this approach became very evident within a few years.

By 1981, a review concluded that the channels proposed earlier for the Horsefly River, upper Adams River, Ankwill Creek, Harrison River, Chilliwack River and Chehalis River would not be necessary to achieve production targets. In other words, it appeared that ten years later only four spawning channels, at a cost of $5.5 million in 1981 instead of the $14 million in 1972 for 12 channels, would be required to meet the goals. The Commission did not make a formal request to build the four spawning channels because of the previously stated government policy.

There was a large saving in capital costs because channels were not built. Increased natural production of both sockeye and pink salmon stocks was significant. However, the potential of the Stuart Lake and Francois Lake systems was not developed, the precarious status of the early Nadina race still exists, and the pink salmon runs of the environmentally damaged Chilliwack-Vedder system were left in in jeopardy. This represents a significant long-term financial loss to the industry as well as risking the extermination of these extremely valuable populations.

CHAPTER 14

The Commission's research and management findings were published in 26 Bulletins, 42 Progress Reports and 48 Annual Reports. In addition, there were more than 30 Technical Administrative Reports completed. Most of the reports were submitted to the governments (Appendix I). Many results of the Commission's findings have been published in other journals as well. A record of some of these publications can be found in the 1973 to 1985 annual reports.

No attempt is made here to present a full and complete review of Commission investigations because most findings are reported elsewhere. A wealth of data remains in the Commission's library and there are files to be analyzed and published. In 1985 the Commission prepared a list of 26 additional reports which could be considered for publication.

FRESHWATER AND MARINE INVESTIGATIONS

Sockeye fry and fingerling in the Fraser system exhibit different migration patterns, ranging from very simple to complex. The fry of most races, upon emergence from the spawning gravel (which takes place at night), drift downstream to a nursery lake. Other sockeye fry such as those of the Chilko River race, migrate upstream to their rearing lake for a year of feeding and growing. Weaver Creek sockeye fry migrate through small Morris Lake, down Morris Creek and then up the Harrison River to Harrison Lake. Fry from sockeye spawning in the Harrison River migrate to sea during their first year and do not rear in Harrison Lake. This unique behavior of the Harrison River race was first identified by Gilbert in the early 1900s. Sockeye fry have been observed rearing in side-channels in the lower reaches of the Fraser River. Substantial numbers of silvery-looking small fish were seen surfacing in the lower Fraser River in December 1960 by a Canadian Department of Fisheries biologist. One migrant sockeye was captured, and scale analysis indicated it had not completed one year in freshwater (a "zero" check) and had 15 circuli (Roos 1961). Returning Harrison River adult sockeye in 1963 had an average circuli count of 13.8. It would appear likely that those fish observed on December 14 and 15 in 1960 were from the Harrison River race and were on their way to the lower Fraser River estuary or Strait of Georgia. Seaward migration of Fraser River sockeye smolts normally takes place in April and May.

Cultus Lake sockeye fry emerge from beach-spawning gravels in their rearing lake. Gates Creek sockeye migrate down Gates Creek to Anderson Lake and then directly through and down Portage Creek to Seton Lake where they feed and grow for a year. Birkenhead River fry migrate downriver from the Birkenhead River spawning grounds to Lillooet Lake, where some reside for a year before exiting down Lillooet River to migrate through Harrison Lake and River and into the Fraser River. Other Birkenhead fry go directly to Harrison Lake for one year. Lower Adams River fry migrate directly to Shuswap Lake, but some are swept down Little River to Little Shuswap Lake and eventually, after some growth, migrate back up Little River and into Shuswap Lake. Brannon (1972) showed that these migration patterns are inherent. Even though sockeye fry of the various races were tested for their migration preferences in a foreign water source, they migrated in the direction they would have, had they been in their native area.

Some sockeye fingerlings make complex downstream migrations, such as the middle Shuswap River sockeye. These fish migrate from their Mable Lake nursery area, down the lower Shuswap River, through Mara Lake, a complex migration route through Shuswap Lake involving three 180-degree turns, down Little River, through Little Shuswap Lake, down the South Thompson River, through Kamloops Lake, down the main Thompson River to the Fraser River and out to sea. Most smolts exit the lakes at night, an apparent survival advantage.

The Chilko River sockeye population was chosen in 1948 for long-term studies to measure egg-to-fry, fry-to-smolt, and smolt-to-adult survival rates. The Commission continued these studies until 1985. The results laid the foundation for much of the Commission's ability to understand not only the life-history of sockeye salmon but also the delicate balance between productivity of a species and its environment. Moreover, the basis for many of the Commission's later decisions for protection and extension of the resource depended on knowledge gained from these investigations. Survival rates and production data are summarized in Appendix K.

A permanent field-station has been maintained at Chilko Lake for almost 50 years. Production of fry from the Chilko River spawners and their survival were measured by recording the unique upstream migration of the fry along both banks of the river using photographic techniques. Smolts were also enumerated each year as they migrated out of the lake and began their seaward journey. A weir or fence, placed across the river early each spring, guided the migrants through open but narrow gates. Photographs of migrants passing through the gates were made at timed intervals throughout the migration and total smolts migrating out of Chilko Lake were calculated.

Egg-to-fry survival rates ranged from five to twenty-seven percent. Fry population estimates ranged from three million to sixty million fish. Fry-to-smolt survival was usually relatively high, averaging around 50 percent and ranging from 35 to 70 percent. By the 1980s large numbers of fry were being produced by increased lake spawning. Thereafter, fry to smolt estimates of survival were not accurate. The greatest variations in the whole-life dynamics of this race occurred in smolt-to-adult survival, where a range from one to twenty-two percent was detected. Smolt out-migration estimates ranged from as few as one million to a maximum of thirty-six million. Studies on predation by Dolly Varden *(Salvelinus malma)* on

sockeye fry and fingerlings indicated that this species was the primary predator in Chilko Lake. Estimates of the total adult return were based on racial analyses of commercial catches and enumeration of spawning sockeye in Chilko River and Lake. Since scales were not available from the Fraser River Indian food fishery, other techniques were employed.

A major program similar to the Chilko research effort was established for pink salmon and started with the 1961 brood year. Fry-enumeration programs were initiated at the Harrison and Chilliwack-Vedder rivers spawning areas using standard fry-trapping techniques, but were discontinued after a few years' data were obtained. An innovative approach to enumerating all of the pink salmon fry migration downstream in the Fraser River began at Mission, in the lower Fraser River area, in 1962 and has been operating every year that pink salmon fry migrate down the river (Vernon 1966). To circumvent the impact of tidal fluctuations on river flow at Mission, a pontoon-mounted trap was propelled upstream over fixed transects at a relative velocity of 2.5 ft/sec for 15-minute periods. The bottom of this "scoop" trap was 40 inches below the surface. To determine numbers of fry migrating below that depth, a deeper trap device, fitted with conical nets, was fished to depths of 13.5 feet. A stationary trap was used to collect pink salmon fry at greater depths.

As was the case with sockeye at Chilko Lake, variations in pink salmon egg-to-fry survival rates were lower than those recorded for fry-to-adult in the marine environment. Freshwater survival rates for brood years 1961 to 1983 averaged 13 percent (ranging from 9 to 18.7 percent), while marine survival varied from 0.8 to 5.5 percent, averaging 3.2 percent. This was quite similar to the marine survival recorded during fourteen years of investigations at Auke Bay, Alaska—3.8 percent (range 0.2 to 10 percent) from 1971 to 1984 (Taylor *et al.* 1986). The range of Fraser egg-to-fry survival varied by a factor of 2:1, while the fry-to-adult marine

The Chilko River spawning area extends 3.75 miles downstream from the outlet of Chilko Lake. As shown in the lower photo, taken in late September during the sockeye spawning period, Chilko River is a broad, beautiful stream. The upper photo, taken on April 6, 1949 just prior to fry emergence, shows that some of the spawning area is exposed during the winter low-flow period. Sockeye fry migrate upstream into the lake and return downstream on their seaward journey one year later.

Starting in 1949, a fence was installed every year just below the outlet of Chilko Lake to count the sockeye smolts. The fish were counted by sequential-firing cameras placed in towers over white flashboards at each "V" in the fence. The first tower was being erected when this picture was taken in the spring of 1953. Each tower was enclosed in tarpaulins to exclude light and to minimize wind riffles.

Ross Stewart capturing Chilko fry for enumeration. Since velocities in the central part of the river are high, fry migrate upstream in lower velocities along the banks where they are easily captured. An improved enumeration technique was later developed using sequential photography.

survival range varied by a 6.9:1 ratio. Estimates of total fry populations ranged from 143.6 million to 590.2 million.

Estimates of Chilko sockeye smolts past Chilko Lake and pink salmon fry migration passing the Mission site were used extensively for predicting returning adult abundance. Many environmental and oceanographic data were analyzed and related to marine survival rates through the years.

The Adams River sockeye run was severely depressed in the 1960s and early 1970s. A major research effort began investigating reasons for the low production. Plankton studies continued at Shuswap Lake, as well as at other lakes in the watershed, to determine the density, species composition and relative sizes of plankters that served as food for sockeye fingerlings. Shuswap Lake was of particular importance because of the low standing-crop of plankton in the 1970s, compared with densities recorded in earlier years. There was concern that the large subdominant sockeye runs of 1967 and 1971, similar in size to the dominant runs of 1966 and 1970, were affecting the food supply of the fry produced by the dominant-year escapement.

In 1971, additional studies at Shuswap Lake focused on the timing of emergence, the distribution of fingerlings and their pattern of feeding migrations in the lake. A newly- developed echo-integrator developed by the Fisheries Research Institute at the University of Washington was also used to determine the numbers of juveniles in the lake (Figure 36). This work was done under contract to the University and for several consecutive years population estimates of fry-to-fingerlings were obtained. Water temperature, clarity, electrical conductivity (as a measure of lake productivity), and other limnological parameters were obtained at Shuswap; growth and survival of juvenile sockeye during the summer were concurrently monitored.

In conjunction with these studies, a comprehensive research program to examine the complete life history of the sockeye of the dominant (1974) and subdominant (1975) cycles began in 1974. Information was gathered on the number of eggs arriving at the spawning grounds, the numbers successfully deposited, the number of fry produced and the number and condition of fry surviving to the pre-smolt stage. An intensive study on fry loss to predators in the lake was also conducted. Based on results of previous investigations, it was expected that fry survival to the early winter stage would be much lower for the subdominant cycle than the dominant cycle. Contrary to these expectations, the rate of predation on the fry in 1976 was similar to the rate on 1975 fry. There was no evidence of compensatory or depensatory mortality, despite a five-fold difference in the number of sockeye spawners (Williams *et al.* 1989b). Acoustic estimates of fingerling abundance in several other Fraser lakes were routinely conducted for several years (Figure 37). Results were related to known egg-deposition, fry abundance and survival, and zooplankton abundance to assess the productive potential of the lakes. Useful information on numbers of juveniles to predict subsequent run sizes was also obtained.

Research on the feeding behavior of sockeye fry was conducted at Cultus Lake for two years starting in 1976. The purposes of the study were to identify food consumption and preferences of sockeye fry and the effect of varying fry densities on cropping of plankters. The cladocerans *Daphnia* and *Bosmina* were selected by the fry over calanoid and cyclopoid copepods. The percentage of cladocerans in the total plankton stock available for feeding fry was not affected by major changes in sockeye fry abundance.

Plankton and zooplankton studies were conducted for many years in several Fraser River lakes. These established the feeding preference and cropping by sockeye fry and provided many insights into lake productivity and differences in carrying capacity of the lakes (Goodlad *et al.* 1974). The Fisheries Research

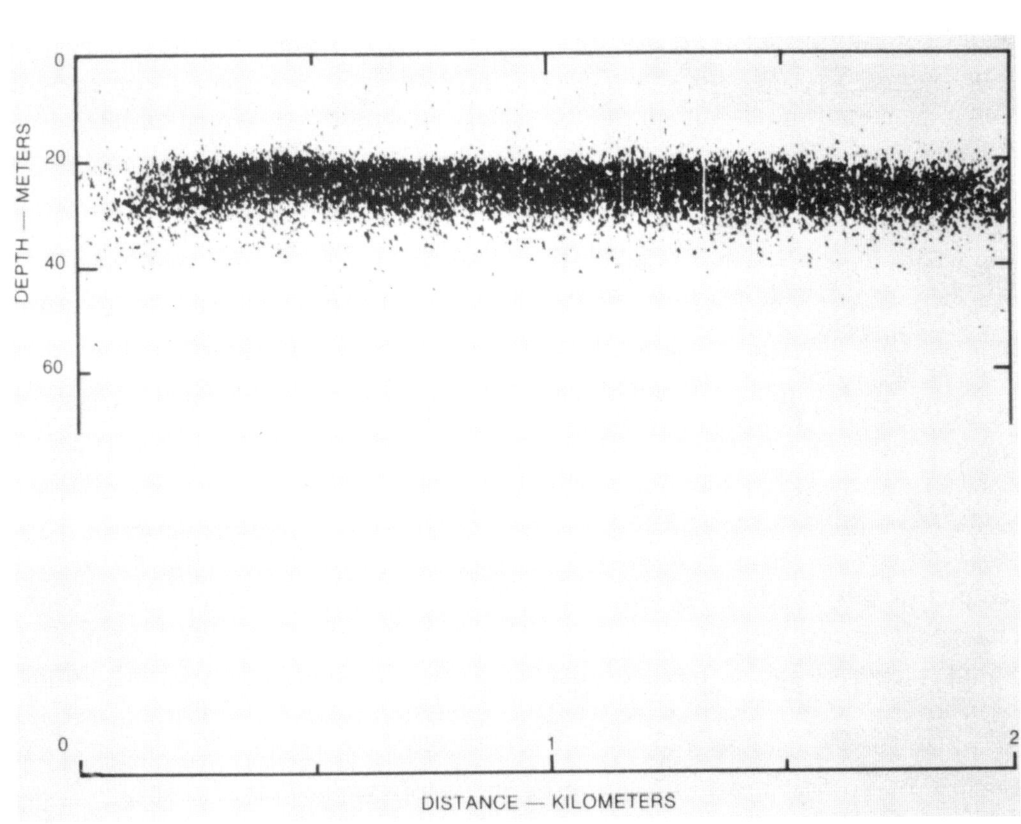

Figure 36. Echogram showing juvenile sockeye salmon in Shuswap Lake, October 18, 1979.

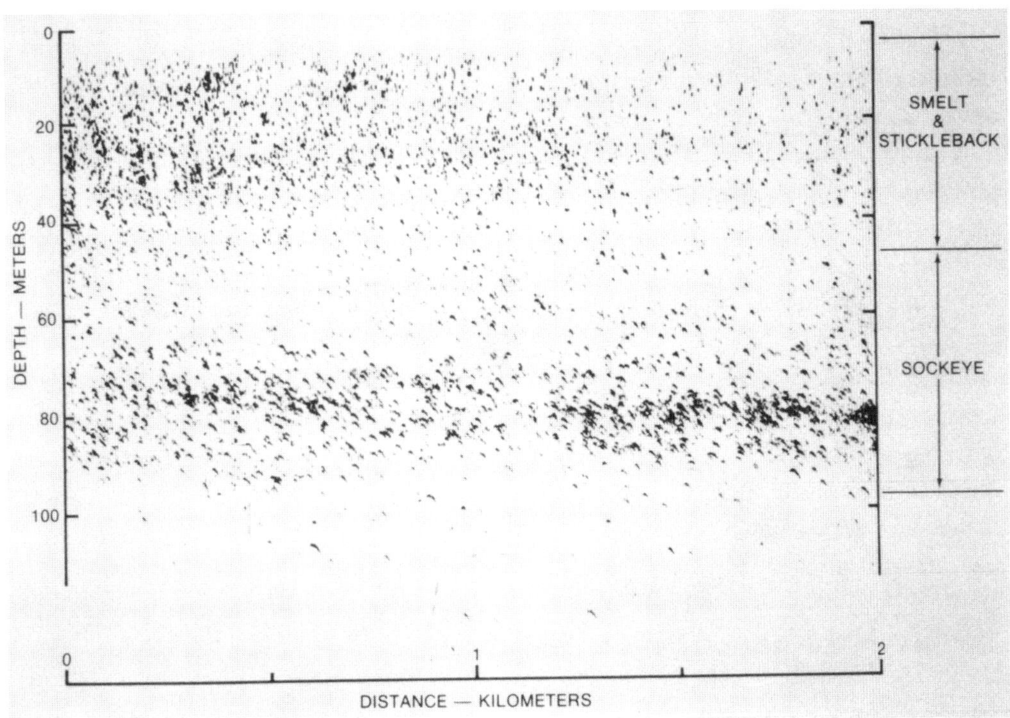

Figure 37. Echogram showing separation of sockeye salmon from smelt and stickleback in Harrison Lake, November 26, 1979.

Institute of the University of Washington and the Commission conducted the evaluation of a dual-beam echo-sounder at Cultus Lake to measure target strength of sockeye fingerlings. The enumeration of sockeye fry in the lakes could be simplified by this new approach. Comparison of the new technique was made in 1983 with the standard single-beam method. Comparable estimates were obtained but the new sounding technique took only one-half the time to obtain the population estimate (Johnson and Burczynski 1985).

Research continued to provide improved methods for forecasting the size of returning adult sockeye and pink salmon. This involved extensive analysis and correlation of environmental data with returns.

A pink salmon study was designed in 1983 covering selected biological parameters to establish baseline information required to protect Thompson and Fraser River pink salmon. A second railroad track was being built by Canadian National Railway along migration routes and spawning grounds located in the Thompson and Fraser rivers. The study, using highly specialized expertise from several disciplines, indicated that pink salmon metabolic rates are significantly higher than sockeye. Pink salmon have a 30 percent higher energy expenditure at the same swimming speed (Williams *et al.* 1989).

MARINE STUDIES

A joint program to determine genetic differences between Fraser sockeye stocks and other coastal sockeye stocks was conducted in 1982 with Commission and National Marine Fisheries Service scientists involved. Eye muscle, heart and liver tissues from early Stuart, Chilko, Birkenhead, Weaver, Adams and lower Shuswap populations were analyzed with electrophoretic techniques. These techniques were later applied to Fraser River pink salmon in attempts to identify these stocks and separate them from other pink salmon populations. Genetic stock identification (GSI) could have some application in certain years in separating Fraser stocks which migrate at the same time and have similar meristic characteristics.

In 1985, an effort was made to develop acoustical methods to enumerate salmon stocks delaying off the mouth of the Fraser River. The large area involved, unknown movements of the salmon schools, multiple and unknown fish species present, and target identification were just a few of the problems encountered. A major effort will have to be expended before valid information on the abundance of sockeye in that area is obtained. Greater emphasis in providing accurate estimates of late-season stocks delaying off the Fraser River could be rewarding.

Predator studies in the freshwater environment suggested that most of the mortality following emergence is related to predation (Williams *et al.* 1989). The limited information available from marine studies suggests that predation by Pacific lamprey (*Lampetra tridentatus*) and porpoise (*Phocoenoides dalli*) is low, undetermined for river lamprey (*Lampetra ayresi*), highest for sharks, and unknown for bird and fish predators (Williams and Gilhousen 1968, Roos *et al.* 1973, Gilhousen 1989). Gilhousen (1989) showed that significant mortality in the sea is likely due to predation and he concluded "that the vast majority of salmon lost during both freshwater and marine life are killed by predators."

CHAPTER 15

Indian Fisheries

UNITED STATES DECISIONS AFFECTING TREATY INDIANS AND THEIR FISHING RIGHTS

In 1854 and 1855 the United States government entered into a series of treaties with certain Indian tribes in what is now Washington State. The Indians agreed to release their interests in certain lands in exchange for their "right of taking fish at usual and accustomed grounds and stations … in common with all citizens of the Territory," and for other considerations. More than one hundred years later, the United States, as trustee for several Indian tribes, brought suit against Washington State in Federal District Court to obtain an interpretation of the treaties and to seek an injunction requiring the State to protect the Indians' rights to salmonids passing through the tribes' fishing grounds.

From 1974 to 1979 several court decisions were issued involving District Court, State Courts, Court of Appeals, and the U.S. Supreme Court. The first judgement occurred on February 12, 1974 when United States District Court Judge George H. Boldt rendered a controversial determination regarding Indian fishing rights in *United States v. Washington* (United States District Court 1974a). Indian fishermen with treaty rights were given the opportunity to take up to 50 percent (a ceiling) of the harvestable number of fish that may be taken by all fishermen—excluding ceremonial and subsistence catch. The decision contained only minor reference to the International Pacific Salmon Fisheries Commission and comments were limited to its role in management of the Convention Waters fisheries involving fishermen of Canada and the United States. Shortly following the decision in early 1974, Judge Boldt, in answer to questions posed by the State of Washington, stated:

> … that treaty right tribes fishing in waters under the jurisdiction of the International Pacific Salmon Fisheries Commission must comply with regulations of the Commis-

sion (United States District Court 1974b).

The District Court's conclusion was later upheld by the Ninth Circuit Court of Appeals which stated:

> … Congress sufficiently indicated its intent that all persons, including Indians, be subject to Commission regulations.… (United States Court of Appeals 1975)

The state petitioners in *United States v. Washington,* supra (No. 75-588) suggested to the Court that federal court actions would conflict with IPSFC regulations. The United States in its Brief in Opposition (to petition for writ of certiorari) in *United States v. Washington,* No. 75-588, pp. 21-22 stated:

> Both the District Court (384 F. Supp. at 411) and the Court of Appeals (Pet. App. 49-50) specifically held that all persons, including treaty-protected Indians, are subject to the regulations of the IPSFC. Thus there is no question of the decree impinging upon the authority of the IPSFC…As we have demonstrated, both courts below specifically recognized the supremacy of IPSFC regulations. If any subsequent actions interfere with the regulations of the Commission they are not sanctioned by the decision under review. (Brief for the United States in opposition 1975)

Finally, in 1978, after considerable petitioning to the court, following several years of discontent among the participants, the Supreme Court of the United States agreed to hear the case. Its decision was given on July 2, 1979. In essence, the District Court's decision and those of the Court of Appeals were upheld except that the Supreme Court found that:

> … the District Court erred in excluding fish taken by the Indians on their reservations from their share of the runs, and in excluding fish caught for the Indian's ceremonial and subsistence needs. (Supreme Court of the United States 1979)

It seems clear from the onset that the U.S. government believed that the IPSFC would apportion the 50 percent share of the United

States catch to U.S. treaty Indian tribes. This did not occur.

The foregoing is an overview of the major legal events between 1974 and 1979. The involvement of the courts, governments (federal and state), IPSFC, Indians and fishing industries during this period was extensive. The Commission was not a participant in the legal discussions from 1970 to 1974 and was drawn into the events unwillingly from 1974 and thereafter. Further explanation of its role and position is necessary. It was a period of animosity, divisiveness, and confusion.

IPSFC - A RELUCTANT PARTICIPANT

In 1970 the Commission received requests to make special accommodations for Indian fisheries within U.S. Convention Waters, primarily for the Makah Tribe located in the western end of the Strait of Juan de Fuca at Neah Bay, Washington. One such request from the Makahs was to open the fishing area each season at an earlier date to conform, with respect to the timing of fish migration, to openings in other areas of Convention Waters. The request was approved by the Commission in 1970 which allowed for earlier openings in specific areas for all fishermen (IPSFC 1971).

In 1971, discussions took place between the Department of State and the Secretary of the Interior on the matter of special treaty Indian fishing rights.

In a letter dated January 28, 1971, to the Secretary of State, the Undersecretary of the Interior stated the position of the Department of the Interior with regard to Indian fishing rights:

> We do not believe that proper recognition of the Indian treaty rights by the International Commission is in any way in conflict with the objectives of the international conventions or with the obligations of the United States thereunder. So long as the overall conservation objectives are not imperiled and the proper allocation of the resource between the respective countries is maintained, the international conventions are not concerned with what particular group of citizens gets what particular share of a country's entitlement to the common fish resource or generally speaking, what particular type of gear is used to effect that harvest. These are matters primarily for determination by the country involved. (Russell 1971)

A reply to this letter, dated May 10, 1971, from Donald L. McKernan, Department of State, Coordinator for Ocean Affairs and Special Assistant to the Secretary for Fisheries and Wildlife, stated:

> The Department of State concurs with the Department of the Interior's reading of the applicable law pertaining to the fishing rights of Indians in the United States. We agree that the United States can properly uphold its treaty fishing obligations under international agreements establishing the commissions while at the same time recognizing the treaty rights of the Makahs and any other affected Indian tribes. Accordingly, the State Department supports the position that where Indian fishing is affected by regulations approved by the United States as a participant in an international fishery commission, those regulations or their implementation should be based upon a recognition of the special interests and treaty rights of the affected Indians. (McKernan 1971)

This indicated the U.S. government was giving serious thought to the Indian issue in the early 1970s.

The United States Commissioners were subsequently notified of the positions taken by the Department of State and the Department of the Interior. It does not appear that Canada had been consulted in the matter up to that point.

The question of special fishing rights for the Makah Tribe was placed before the Commission again in 1971 and pressure was exerted upon the United States Commissioners in 1973 by the Department of State to give special fishing privileges to these specific fishermen. In a letter approving the Commission's proposed 1973 regulations, the Department of State included the statement:

It is understood that in the implementation of these regulations by the State of Washington due consideration will be given to the special treaty rights of the Makah Indian Tribe and of other Indian tribes which may be affected. (McKernan 1973)

Thus, United States Commissioners (in effect, the Commission) were asked to give special considerations to treaty Indians prior to the court decision in 1974.

Commission members and staff were invited to a meeting held in Washington, D.C. in July 1974, to discuss with representatives of both governments the obligations of the United States with respect to treaty Indians as a result of the Boldt decision and the regulatory needs for the future. The discussions were exploratory in nature. A senior Canadian government official summarized the Canadian position at the meeting as follows:

… division of catch between Indian and non-Indian fishermen in the United States should be achieved by supplementary regulations or other measures by the United States outside of, but consistent with Salmon Commission regulations… . The division of the catch between Indian and non-Indian fishermen within the United States part of the Convention area should be achieved solely through domestic United States regulations and not by authorization of special arrangements through the Commission. (Lucas 1975)

In addition, it was agreed that:

… proposed changes in IPSFC regulations would not reduce the regulatory options available to the IPSFC to meet the needs of Canadian fisheries (i.e. action on the U.S. side to accommodate Indian fishing would not require disadvantageous adjustments in Canadian fishing times and areas). (Hunter 1978)

None of the participants could have realized the difficulties that were to be encountered in the ensuing 11 years when efforts by the U.S. government to effectuate treaty Indian fishing rights continued. From the beginning, Canada was not favorably inclined to have the Commission directly involved in the domestic allocation of 50 percent of the U.S. catch to treaty Indians.

A major difficulty in the treaty Indian matter was the fact that the Convention of 1930, as amended in 1957, contained no provision whatsoever for dealing with Indian rights in either country. In fact, the Convention did not address Indian fishermen. From 1946 to 1974, commercial Indian fishermen in both the United States and Canada abided by and fished under the same regulations for each country as non-Indians with no distinctions being made for either group.

Effort was put forth by the U.S. government on behalf of U.S. treaty Indians immediately following the Boldt decision, as evidenced by a statement from the White House in 1974. A memorandum to Robert Schoning, Director of National Marine Fisheries Service stated:

It will be up to the skill and good will of the Commissioners, their staff, and the Indian representatives to work out together proposals which can then be persuasively presented to the Canadian Commissioners. (Patterson 1974)

The United States was faced with the problem of having treaties with two different signatories, Canada and various Northwest Indian tribes, on the same subject. It was soon apparent that the treaty provisions with Canada would impede the United States' obligations to treaty Indians.

The Commission discussed the matter at length and concluded that it could not assume responsibility for the domestic allocation of catch between different racial or ethnic

242

groups in either country. The U.S. District Court apparently believed the Commission had the authority to implement the special fishing privileges that would be required to give the Indians the opportunity to harvest 50 percent of the available catch. Nevertheless, in late 1974 "the Commission concluded that it was not a proper matter for consideration by the Commission" (IPSFC 1975).

The merit of the Boldt decision was never a subject matter for Commission deliberations. That decision was wholly a United States domestic matter. How either Party allocated its national share of the catch was not included in the Commission's mandate. The Convention did, however, recognize the existence of various gear-types and the Commission's regulatory scheme since 1957 had been to control the segregation of net gear-types to provide by regulation (setting fishing times) reasonable distribution of the harvest among the various gear user groups—not ethnic groups.

Possible changes in the salmon convention arising from the need to implement the Boldt decision were expressed in an October 1, 1974 letter from William L. Sullivan, Jr., Acting Coordinator of Ocean Affairs, Department of State, to Donald R. Johnson, a Commissioner and the Northwest Regional Director of the National Marine Fisheries Service. The letter, in part, stated:

> ... Moreover, great care must be exercised in pursuing any particular action in the IPSFC, since the question of basic revisions in the salmon treaty, or even its continued existence, has existed for some time. It would appear that the likely results of any revision of the treaty or its abrogation would be less salmon, perhaps very substantially less, for the American fishermen, Indian and others alike. We believe this result is to be avoided, and that any revision in the salmon treaty must be the result of a deliberate decision by the United States Government and not an inadvertent result of any particular move to meet an immediate need. (Sullivan 1974)

Perhaps neither government was prepared to amend the existing treaty to facilitate implementation of the U.S. treaty Indian fishing right because of ongoing negotiations relating to a new salmon treaty. Because there was no amendment to the treaty giving the Commission the authority to divide the U.S. catch equally between treaty Indian and Non-Indian fishermen, the Commission was destined for a long and difficult period of confrontation. The United States government would attempt for several years to accommodate treaty Indian fishing rights within the regulatory framework of the Commission.

However, as early as 1974, U.S. government officials recommended that an amendment to the Sockeye-Pink Salmon Convention be obtained to provide a seventh Advisory Committee member for each country. It was reported that the United States government would appoint a treaty Indian. The apparent purpose of the appointment would be to facilitate and assist treaty Indians in securing their court-mandated share of the catch. The presence of an Indian Advisory Committee member would not have resolved the larger issue of treaty Indian fishing rights under the existing Convention. Perhaps the recommendation was made to answer some of the demands of treaty Indian fishing groups. Strong representation was also made in 1974 for the replacement of one of the U.S. Commissioners with a treaty Indian.

From the outset, it was clear that the governments did not want any reference to Indians in the IPSFC regulations. The official summary record of a 1974 meeting stated: "<u>The changes would not require the Commission to recommend specific regulatory actions for Indians alone</u>" (Summary Record 1974) (Emphasis added).

Fishing regulations by the Commission for 1974 were put in place as in previous years and the percentage of the total U.S. Convention Water sockeye catch taken by treaty Indians remained very low (1.4 percent). This was because nothing special was done on behalf of the tribes to increase their court-mandated share of the catch through extra fishing time or the amount of Indian fishing gear.

The Commission's minutes of January 24, 1975, record the affirmation of a Commission policy similar to statements previously made regarding the treaty Indian situation:

> 9. The question of United States Indian fishery rights in Convention Waters was discussed. The Commission decided that this was not a proper concern for the International Pacific Salmon Fisheries Commission, whose areas of responsibility are defined by the Sockeye Salmon Convention.

Before the 1975 season, meetings were held in an attempt to develop regulations acceptable to all parties that would increase the treaty Indian share. The U.S. Department of State formally instructed the U.S. Commissioners of their responsibilities in proposing and adopting regulations:

> … sufficiently flexible to ensure that domestic decisions on the Indian fishery can be implemented.… Throughout the talks it was the Canadian position that the adjustments desired by the United States could be accommodated through domestic action within the United States without altering the agreed Commission regulations. (Sullivan 1974)

The Canadian government proposed that the 1975 season be opened under Commission regulations and that certain days opened under Commission regulations could later be amended by U.S. domestic regulations as treaty Indian-only fishing days. This would have had the effect of increasing the treaty Indian share of the U.S. catch but without extra fishing time for the United States, an imbalance in international catch would have resulted.

The Canadian government was purported to be concerned about recognizing the treaty rights of U.S. treaty Indians in Commis-sion regulations because of possible problems with Indian tribes in British Columbia.

At a similar time, the Commissioner of Indian Affairs, U.S. Department of the Interior, wrote the Director of IPSFC stating in part that:

> … full consideration must be given to the rights of the treaty Indian fishermen when regulations are drafted by the International Pacific Salmon Fisheries Commission. (Thompson 1974)

The Department of State and its attorneys and various Indian groups were also telling the Commission what it should do but none of these parties pointed out which Article of the Convention would give the Commission the authority to do so.

The United States Congress soon became interested in the escalating skirmish. Both senators and representatives expressed concern over the matter. In a letter to President Gerald R. Ford, Congressmen Joel Pritchard, Brock Adams and Floyd V. Hicks said:

> What we question is the action of officials in the Executive Office of the President who are indicating to U.S. Commissioners of the International Pacific Salmon Commission that they are obligated to provide special Indian fishing rights to the fish under the jurisdiction of the International Commission.
>
> This intervention in Commission affairs is weakening the position of U.S. Commissioners in negotiations with their Canadian counterparts and could seriously undermine U.S. rights to the Fraser River salmon run.
>
> It is clear, since the Commission was not even a party to the suit, that the decision does not require it to provide for special Indian fishing rights in its regulations and that the Court recognized this fact explicitly. Any action by members of the Executive Branch to require such a result would be an intervention in a judicial process, a cause for further unnecessary hardship to

non-Indian fishermen, and an interference with the independent judgement of Commissioners. (Pritchard *et al.* 1975)

The letter went on to ask the President to:

> … determine if such action is being taken by members of your staff and, if it is, to have it terminated.

A letter from Assistant Secretary McCloskey of the State Department to Senator Henry Jackson in early 1975 said in part:

> Because salmon under regulation by the IPSFC fall within the purview of the court's decision in U.S. v. Washington, this Department has attempted to ensure that those regulations do not prevent implementation of the court's decision. This Government, of course, cannot and has not attempted to direct the IPSFC, an international organization, to take any specific form of action in this regard. (McCloskey 1975)

Confusion and inconsistencies appeared before the 1975 season got underway. The Commission developed regulations for 1975 and submitted them to the governments. They did not contain special considerations for treaty Indians. William L. Sullivan, Jr., Department of State, presented an affidavit to the Western Washington District Court on July 21, 1975, following an unsuccessful meeting between the Commission and both Parties:

> The United States and Canada were unable to reach agreement on acceptable accommodation which would permit either gear flexibility for the Indians within the open days for fishing under the Commission's regulations or extra fishing time over and above that allowed by the Commission in State Area 2… . The United States has taken steps to temporarily withdraw its approval under the Convention of those parts of the Commission's regulations which seek to allocate the open fishing periods among various types of gear. (United States District Court 1975)

This action was taken in response to a court order issued by Judge Boldt on July 17, 1975. These changes were contrary to the regulations previously adopted by the Commission and approved by the United States on April 11, 1975.

The eighth point of the affidavit stated: The United States continues to object to that part of the Court's order which seeks to give extra fishing time not allowed by the Commission's regulations in State Area 2. Such an order is in contravention of the basic regulatory scheme elaborated by the Commission. *(ibid.)*

A difference of opinion between the two branches of the U.S. government added to the confusion. The Department of State's objection to the court-ordered, five-day-per-week fishery in Area 2 (later identified as Areas 4B, 5, and 6C) is interesting in view of later actions taken by the United States.

The 1975 fishing season was marked by several court orders involving both federal and state judicial authorities. Injunctions and stays were issued with the Washington State Department of Fisheries being heavily involved and being tossed to and fro between federal and state court decisions. The U.S. Department of State was seeking to direct a solution to the stalemate. The Washington State Department of Fisheries was unable to give treaty Indians special fishing times because of State Supreme Court rulings (Petty 1979). No satisfactory or agreed-upon solution to the problem was achieved. A very detailed and comprehensive review and record of the 1975 season is available and has been termed: International Disagreement and Unilateral Action by the United States *(ibid.)*.

The Department of State sent a letter to the Commission Chairman on July 21, 1975, outlining the United States government's action as described in the Affidavit of W.L. Sullivan (Clingan 1975a). The Commission responded in letters dated July 22 and July 25 to the Department of State expressing its strong dissatisfaction with the government's involvement in the regulatory process:

… The Commission must register the strongest possible protest of this action by the State Department.… The current state of confusion concerning the Commission's regulations in United States Convention Waters must be ended quickly,… . (Hourston 1975)

The Department of State responded by admonishing the Commission for its previous letters complaining about U.S. government actions. In a letter to U.S. Commissioners from the Department of State it is stated:

… The United States Government rejects the purported protest, which exceeds the authority of the Commission. The Commission is the creation of the two Governments and is responsible to them; the Governments are not responsible to the Commission. (Clingan 1975b)

The letter from the Department of State also said:

… The United States Government does not consider that there is any confusion about the implementation of the regulations.

The United States Commissioners were also chided by the Department of State for approving the letters sent to the Department of State protesting U.S. government actions altering Commission and government approved regulations. The 1975 season achieved little towards accommodating or assisting treaty Indian fishermen in their quest for significant improvement in catch. Their share of the U.S. sockeye catch was only 3.6 percent and of pink salmon, 3.0 percent.

At the end of the 1975 fishing season the Director of the National Marine Fisheries Service summarized the federal government's predicament as follows:

In attempting to find a cure for these State problems, it was discovered that the federal government had no authority under which it could promulgate regulations which would provide for Indian treaty fishing. Rather, in the context of the IPSFC fishery,

it was found that the federal government could only enforce regulations as adopted by the IPSFC and approved by the United States, and could not make allocations between Indian and non-Indian fishermen. Long hours of research and discussions seem to have made it clear that such residual federal authority will have to come from new legislation. (Schoning 1975)

The "new legislation" was never put in place during the Commission's tenure.

Long before the 1976 fishing season, extensive negotiations were initiated in an attempt to avoid the bitterness and confusion that existed in 1975 and again to seek real gains in the Indian catch. Months of discussion between the two governments and the Commission finally brought about an agreement in the wording of the Commission's regulations which, it was believed, would provide meaningful results for treaty Indian fishermen. The gear regulations for 1976 would be qualified by inclusion of the words: "to the extent permissible under the laws of the parties" (United States Department of State Aide Memoire, January 14, 1976).

The United States government instructed its Commissioners to adopt this proposed regulatory clause (Ridgeway 1976a).

In a letter to the three United States Commissioners from legal counsel representing the purse seine and gillnet associations, strong opposition was expressed to the 1976 clause:

… By including such language in its regulations, the Commission would be abrogating its duty to regulate the fishery by allowing "the laws of the parties" to determine what its regulations are. This is clearly unauthorized by the Convention. (Mikkelborg and Mijich 1976)

On several occasions the Commission and staff discussed this at length and strong differences of opinion were expressed over inclusion of the clause. The Commission sought advice from its counsel and was advised not to approve such a request. Counsel concluded: "… would seem to me to nullify the whole purpose of the Convention for it would permit unilateral action by either country" (McIntosh 1976). Nevertheless, in March, 1976, a required majority of the Commission approved the clause. This was an extremely agonizing decision for the Commission.

The regulations as issued, would limit all-citizen fishing to the periods open as specified for each gear-type and provide treaty Indian fishing with any gear at any time within the entire open period. The result of this effort was only a slight improvement in the Indian sockeye catch to 6.7 percent of total U.S. catch. Because the offshore coastal troll fishery was essentially unregulated by the Commission and open for seven days per week, greater fishing time was initially given to Indian fishermen in 1976 because of the newly adopted clause. The Commission was caught off-guard because it believed only net fishery times were being considered. The federal court stepped in and ordered a five-day-per-week fishery for treaty Indian fishermen. Both the Commission and the United States government protested and the order was subsequently amended to three days. Even at that, this order exceeded the preseason regulations in that it provided fishing times outside the limits of Commission regulations. Thus, the earlier agreement with Canada (that there would not be fishing outside the times set by the Commission) was no longer being fulfilled.

After numerous meetings and scores of ideas and recommendations during 1974, 1975 and 1976, the Indian percentage of the sockeye catch had increased to only seven percent in 1976 because additional, separate fishing times for Indians given by the United States were insufficient.

The District Court and Court of Appeals had ruled that treaty Indians must comply with Commission regulations. At the same time, the U.S. government maintained its position insisting that the Commission agree to regulatory suggestions they believed would help implement the Court's decisions. On the other hand, the Canadian government would not agree to have the Commission directly involved in issuing regulations that would give special treaty Indian fishing privileges. The Commission likewise was adamant that it did not have the authority to enable compliance with the Boldt decision and reiterated that the domestic allocation of the catch was not a proper concern for the Commission. Unclear also was the limited and disputed role of Washington State in enforcement and management responsibilities in regard to treaty Indian fishing.

Meanwhile, three fishing seasons after the Boldt decision, anger and frustration were building in the Indian community, even toward the federal government. This is expressed in a letter to Rozanne L. Ridgeway, Department of State, from M. D. Morisset, Legal Counsel for the Lummi and Makah tribes. He pointed out that the tribes required three to five nights of fishing per week if they were to increase their catch. He stated:

> Despite these continued requests, there seems to have been nothing more than a continuous and voluminous outpouring of mumbo-jumbo from the various departments of the government involved, all of which have accomplished virtually nothing in practical terms for the Indian Tribes… . Thus, we have been instructed once again to prepare possible litigation against the United States and the individuals involved. (Morisset 1976)

By the end of 1976 all groups were disgruntled with the state of affairs because there was not a single, unified federal decision or legislative authority in place to implement the Court's decision. Repeated attempts to circumvent this issue had been unproductive and had exacerbated an already explosive situation.

United States Commissioners were, on many occasions, pressured in opposite directions by federal authorities and fishing interests concerning action taken or not taken with respect to the treaty Indian fishery problem. A letter from A. Ziontz, attorney for several Indian tribes, to United States Commissioners Tollefson, Johnson and Saletic and Washington State Fisheries Director D. Moos stated that it is: "… no proper concern of the International Pacific Salmon Fisheries Commission as to how the resource is allocated by the United States among its citizens" (Ziontz 1975).

Ziontz, continued: "… the Commission continues to pay mere lip-service to these policies and to promulgate regulations which simply ignore Indian treaty fishing rights." The letter also pointed out that both Commissioners Johnson and Tollefson, in separate depositions, did not feel that the Boldt decision was binding on them as Commissioners from a legal point of view. The letter concluded: "… It would clearly appear that there is a sound basis in law to bring actions against individual U.S. Commissioners for their illegal conduct." No such action was taken.

The Commission wrote both governments of its disagreement over the events which took place in 1976 and was told by the U.S. Department of State that: "… this subject is not properly within the purview of the Commission!" (Ridgeway 1976b). The letter further stated:

> … We urge the Commissioners to be guided in their activities by the terms and responsibilities set forth in the Convention for the Protection, Preservation, and Extension of the Sockeye Salmon Fishery of the Fraser River System, May 26, 1930, as amended.

For three years both governments had been leaning heavily on the commissioners to take action exceeding any authority provided by the Convention in order to accommodate the U.S. court decisions. The special wording incorporated in the 1976 regulations: "… to the extent permissible under the laws of the Parties…" was done so only after strenuous urging from both governments.

The Canadian government was not pleased with the outcome of the 1976 fishing season because of the confusion which arose over interpretation of the special regulatory clause incorporated in those regulations.

As in the previous years, diplomatic negotiations for the 1977 season resumed long before the fish were due to return. Various proposals were made to Canada by the U.S. Department of State. In an Aide Memoire from the Canadian government to the United States government on November 19, 1976, a review is given of the meeting of officials of Canadian and United States governments held in Ottawa on October 1, 1976:

> At the meeting under reference, United States officials enquired whether Canada could consider instructing its Commissioners to the International Pacific Salmon Fisheries Commission to approve, in the Commission's Regulations for 1977, a five-day fishing week for Indian fishermen in United States Convention waters.
>
> The Canadian authorities have reviewed the United States request, and regret that it is not possible to accede to it. In the opinion of the Canadian authorities adoption of such a regulation would be inconsistent with the provisions and objectives of the Convention and ultra vires of the Commission.

The United States government proposed that the Commission should include in its regulations special days for treaty Indians only. The Canadian government clearly stated that such action would be beyond the legal authority of the Commission. Commission staff were asked to comment on the United States proposal for the 1977 season contained in the Aide Memoire to the Canadian government of January 14, 1977. Commission Director, A.C. Cooper stated that "the initial reaction of the staff to this proposal was incredulity that it should be proposed as a responsible solution to the problem being addressed." He pointed out that "the understandings of the July 18, 1974 meeting in Washington have not been acted upon by either government in terms of amendments or additions to the existing convention, and they therefore are not part of the Commission's official terms of reference." He further said "The Commission should not be the vehicle for implementation of regulations which neither state nor federal governments have uncontested recognized authority to implement." He termed the fishery hours proposed for reef nets as "ridiculous".*

A Commission meeting was held in Vancouver, B.C. on February 4, 1977, involving the Commission, staff, senior government representatives of both countries and the Commission's Advisory Committee. The purpose was to develop regulations agreeable to all that would substantially increase the U.S. treaty Indian catch in 1977. To gain additional hours of fishing for treaty Indian fishermen, it was proposed that the Commission adopt regulations that scheduled fishing for reef nets in U.S. waters during the night. Under these regulations, if the waters were open to one type of gear for non-Indian fishermen, they would be open for treaty Indian fishing with all types of gear. By giving the reef nets fishing time at night, the Indian gillnetters (on some nights) would be fishing without the competition of non-Indian gillnetters and their catches would increase. However, the fishing time for non-Indian reef nets at night would be of no value to that gear as, first of all, they need daylight to see fish entering the net and secondly and more importantly, the strong tides necessary for reef net operations during the summer fishing season occur only during the day. The reef net advisor, as well as other advisors, the Commissioners, and Commission staff were all surprised by the suggestion. The recommendation was rejected. This illustrated the extent of the U.S. government commitment to accommodate U.S. treaty Indian fishing opportunities.

The Commission was unable to reach agreement at that meeting. Another was held on March 25, 1977. Just prior to that meeting

---

* Comments of Director of IPSFC, A.C. Cooper, on the United States Proposal in portion of Aide Memoire received January 25, 1977. 4pp.

U.S. Commissioners were again instructed by the Department of State to vote for adoption of government proposals (Ridgeway 1977). In another Aide Memoire from the United States government on May 26, 1977 to the Canadian government, the Department of State said:

… the Department believes that the maximum prospect of avoiding difficulties between the two Governments in this extremely sensitive and urgent matter would be to agree to a joint interpretation of the Convention excluding the concerned Northwest U.S. Indian tribes from coverage. Failing this, the United States would consider achieving the same result as was agreed in that understanding through the exercise of its authority under Article VI of the Convention, as amended, not to approve the regulations of the Commission with respect to those Indian tribes. The United States would undertake to regulate the fishing of these Indians pursuant to domestic authority in order to ensure their fishing is not in excess of the amount that had been envisaged in the previous understanding.

The United States views this problem as one that could best be solved in the long run in the context of an overall Pacific salmon agreement. The Department remains ready to resume negotiations on this important matter at the earliest possible time.

It appears that the United States government was ultimately looking for relief through a new treaty with little intention of amending the current Convention.

The Canadian government was quick to respond with its Aide Memoire on May 31, 1977:

The Canadian authorities wish to stress again that they consider the problems being encountered in the United States in fulfilling treaty obligations to certain Indian tribes to be matters of domestic United States concern, in which the Government of Canada ought not to be involved and which, moreover, cannot affect the rights and responsibilities of the parties under the Convention. It is understood that this view is shared by the United States authorities.

The Canadian authorities have examined the proposal put forward by the United States for a joint interpretation of the Convention that would exclude the relevant Northwest Indian tribes from Convention coverage. They believe that this proposed approach is not appropriate to the domestic nature of the problem in question, even putting aside other considerations, such as conservation of the stocks, that might arise if such action were to be contemplated.

The United States authorities have stated that their final alternative course of action is for the United States not to approve the regulations of the Commission with respect to United States Treaty Indian tribes. The Canadian authorities have serious doubts as to the appropriateness of such selective disapproval or approval of regulatory proposals under the terms of the Convention, and would reserve their position in this matter. In any event, the Government of Canada would wish to ensure that the Indian fishery was conducted and regulated in a manner consistent with the objectives of the Convention and the rights and obligations of the parties thereunder.

The Canadian authorities wish to reiterate again their desire to help facilitate the resolution of this matter in the same spirit of mutual understanding and co-operation which has marked the fisheries relations of the two countries within the framework of the IPSFC. Like the United States, Canada considers that an appropriate long-term solution should be found and would be prepared to seek one in renewed negotiations towards a comprehensive Pacific salmon agreement.

Canada seized the suggestion put forth by the United States to resolve the issue within a new Pacific salmon agreement. Finally, on June 1, the Commission officially met again. Following strenuous and unanimous objection to the proposal by Advisory Committee members from both countries and the subsequent withdrawal of support by the Canadian government, U.S. government proposals for 1977 were not agreed to by the Commission. The regulations agreed to by the Commission and finally submitted to the governments were much the same as were submitted to the fishing industries in December of 1976. The clause recommended by the U.S. State Department in 1976 was not included and no special accommodation was given to U.S. treaty Indian fishermen.

The United States proceeded to take matters into its own hands. The regulations as submitted by the Commission for 1977 were approved by the U.S. government, with the exception that United States Indians:

> … are entitled to exercise fishing rights by virtue of treaties with the United States in U.S. Convention Waters and are fishing in accordance with Federal regulations providing for the exercise of such fishing rights. (Brewster 1977)

A United States Department of Commerce memorandum of June 2, 1977 outlines the steps the government contemplated:

> Although Interior had earlier agreed on a tentative basis to an Executive Order which would have delegated to Commerce the President's authority to implement treaty rights, the assumption at the time was that federal regulations would be used to limit non-Indian participation in certain of the fishing times provided by IPSFC regulations. Now that the group to be regulated is the Indian fishermen, Interior is not willing to transfer its traditional responsibility for Indian affairs. (Powell 1977)

The Department of Interior was to be responsible for the promulgation of the regulations for treaty Indians but these regulations would require the concurrence of the Department of Commerce. The proposed relationship between the Departments of Commerce and Interior was outlined in a draft Memorandum of Understanding.*

The U.S. Department of the Interior, in which the Bureau of Indian Affairs is located, then promulgated the federal regulations for these Indians which generally tracked the Commission's regulations, except that the Indian fishery was allotted more fishing time than the Commission granted to all fishermen. Earlier starts or extra fishing time after IPSFC weekly openings had closed were permitted by the government outside Commission regulations. This unprecedented unilateral action returned resource management to dual management. This was contrary to one of the basic purposes for establishing the Convention. The status of the U.S. special Indian fishery was subsequently characterized as "A Continued Special Indian Fishery Unrecognized by Commission Regulations—An Inefficient and Risky Solution" (Petty 1979).

The Canadian government advised the United States by Aide Memoire on June 30, 1977 that Canada doubted the appropriateness of the 1977 action in terms of the Convention and reserved its position. The Commission had notified the governments that it considered fishing other than as allowed under the Commission's regulations would be "in contravention of the Sockeye Salmon Fisheries Convention" (Johnson 1977).

In response to the Commission's strongly worded letter of June 27, 1977, the Department of State issued a terse reply on August 1, 1977 (Negroponte 1977). This letter in part said: "… It is our view that domestic implementation of the salmon regulations is not a subject properly within the purview of the Commission."

The Commission, in response to the U.S. government's actions with respect to the regu-

lations, sought to clarify the exemption issue since the Commission regulations did not provide exemptions. Therefore, the Commission issued an emergency order on June 27, 1977, stating that its regulations applied "<u>to all citizens without exception</u>." This order was later held to be without force or effect by a United States federal judge in a lawsuit brought by a non-Indian fishing industry group contesting the validity of the United States action (United States District Court 1977). The decision of the federal court is interesting in view of jurisdictional authority. Petty (1979) stated: "The federal courts lack jurisdiction over an international organization and its foreign representatives."

The Canadian Aide Memoire of June 30, 1977 stated:

Of greatest concern to the Canadian authorities is the untenable situation arising out of the unsatisfactory relationship which is developing between the Commission and the two Governments as reflected in the Commission's emergency order and press statement of June 27. This appears to have been brought about as a result of the situation regarding Indian fishing and is threatening the effectiveness, perhaps even the very existence of the system established under the Convention. The Government of Canada urges the United States authorities to reconsider their approach to this matter and to take all necessary steps to ensure that their actions are fully consistent with the objectives and terms of the Convention and also to ensure the maintenance of a proper relationship between the Commission and the two Governments which is required for the continued effectiveness of the system based on the Convention.

With more fishing time granted directly to treaty Indian fishermen by the United States, the treaty Indian sockeye catch in 1977 increased to 19.2 percent and the pink salmon catch to 8 percent.

---

* DRAFT MEMORANDUM OF UNDERSTANDING between DEPARTMENT OF COMMERCE AND DEPARTMENT OF THE INTERIOR concerning REGULATION OF TREATY INDIAN FISHING FOR FRASER RIVER CONVENTION SALMON. S.J. Powell, June 1, 1977. 4 pp.

The Commission's 1977 recommended regulations and all annual recommended regulations through 1985 were approved by the United States, "… except as to U.S. Indians who are entitled to exercise fishing rights by virtue of treaties with the United States." Regulations recommended by the Commission following 1976 were, as usual, applicable to all fishermen; yet the United States had approved IPSFC regulations that applied only to "All-Citizen" fishermen. By exempting treaty Indians, the totality of United States regulations for sockeye and pink salmon fishing were not identical to those submitted by the Commission. The result each year thereafter was that regulations controlling fishing in U.S. Convention Waters were not Commission-recommended regulations. Treaty Indian fishermen were controlled by the regulations of the U.S. Department of the Interior and "All-Citizen" fishermen were controlled by the Commission's regulations. The treaty Indian regulations were published separately in the Federal Register by the United States Department of the Interior; the Commission's "All-Citizen" regulations were filed with the Federal Register by the National Marine Fisheries Service of the Department of Commerce. This action was taken by the United States government despite admonitions from the Canadian government.

In 1978, the two countries held deliberations to discuss proposed regulations. Canada rejected U.S. proposals which would have required specific IPSFC regulatory actions for Indian fishing. Kronmiller stated in that regard: "The political implications for the Canadian government are said to be unsustainable" (Kronmiller 1978).

Starting in 1977 the National Marine Fisheries Service and U.S. Coast Guard carried out the enforcement of regulations, rather than the Washington State Department of Fisheries. In addition, technical observers from the Department of the Interior and the National Marine Fisheries Service were invited into the normally closed in-season regulatory meetings of the Commission. This was done at the request of the governments in order that the Interior Department's schedules for the treaty Indian fishery could be formulated in light of staff reports made to the Commission and the Commission's decisions. With few exceptions, this system worked sufficiently well to achieve the major objectives of the Convention; i.e. division of catch between the two countries and obtaining the required escapement. However, a serious over-fishing problem on Early Stuart sockeye occurred in 1977 (IPSFC 1978).

Throughout this era of dual management, the Commission was very vocal about its feelings with regard to the United States involvement in management of the Fraser River sockeye and pink salmon runs. On many occasions the Commission, through its chairman, wrote both governments strongly expressing its disagreement with government intervention. The Canadian fishing industry, through its Commissioners, consistently protested the United States action, especially as it affected Canadian fishermen. For example, treaty Indian fisheries in United States waters on Saturday and Sunday would significantly reduce Canada's catch in the Fraser River on Monday.

Despite all of the rhetoric directed toward the Commission and its policy decisions regarding extra fishing privileges for treaty Indians, the Commission's staunch posture of not having authority to allocate domestic catches (the exception being the 1976 regulations) was later vindicated, to a certain extent, by a Chief United States District Court Judge's order in July, 1977. Judge Walter T. McGovern ruled as follows:

> (5) Article VI, as amended, does not give to the Commission the authority to direct the domestic allocation of the fishery allotted to the United States. How the latter nation divides its share of the fishery amongst its citizens is the business and the responsibility of the government of the United States and not of a commission composed in half by

citizens of another nation. (United States District Court 1977)

There appeared to be little interest directed at a critical examination of the Articles of the Convention in regard to the government's actions. Article IV did not provide for any sockeye or pink salmon fishery "in said waters" when the waters were closed by Commission regulations. There was no provision in the Convention for any government or other fishery agency to reopen Commission closed waters.

Article IV in part states clearly:

> … any order adopted by the Commission limiting or prohibiting taking sockeye (and pink) salmon in the waters covered by this Convention, or any part thereof, <u>shall remain in full force and effect unless and until the same be modified or set aside by the Commission</u>. Taking sockeye (and pink) salmon <u>in said waters</u> in violation of an order of the Commission shall be prohibited (emphases added). (Appendix A)

Even though it appeared that the government's decision to exempt treaty Indians from Commission regulations usurped the court's mandate that Commission regulations prevailed in this matter, the Commission's authority over fishing in the said waters was still officially in force.

Because an amendment to the Convention was not achieved, treaty Indian fishermen were deprived of opportunities to catch their court-mandated share of the salmon harvests. The United States approach in granting extra fishing time was cautious and gradual until the 1980s and it therefore took many years for Indian tribes to reach their 50 percent court-decreed share of the catch. In fact, it was not reached for sockeye until 1983 (53.3 percent) and for pink salmon in 1985 (50.9 percent).

UNITED STATES AND CANADIAN GOVERNMENT INTERACTIONS

From 1974-1985, particularly in the earlier years, there were many meetings between the two governments concerning U.S. treaty Indian fisheries in U.S. Convention waters. Commissioners were involved in many of these meetings.

The 1854-1855 treaties gave fishing rights to the U.S. Indians and these rights were re-affirmed and defined further by the District Court, Court of Appeals and by the Supreme Court from 1974-1979. As trustee of these rights, the U.S. government was obligated to ensure that these rights were upheld.

Clearly, the court's pronouncement involving the Commission's regulations placed the U.S. government in an extremely difficult predicament. The government was aware of its responsibility and obligations to Canada and to the basic provisions of the Convention. The United States government maintained that there had been no serious harm caused to the resource as a result of its independent actions outside the Convention. For the most part, the United States was very careful in its actions to protect the resource; however, that position could not be construed as imparting a legal status to the actions taken. In a 1974 memorandum summarizing a meeting of interested parties concerning the IPSFC regulations, a National Marine Fisheries scientist said:

> Finally, from a management standpoint, it would seem incumbent upon the Salmon Commission that they retain control on the extra day fishing for Indians in the same manner as for the entire fleet so that adjustments in the days fishing could be made throughout the season to achieve the Commission's goals for escapement and division of the catch. (Blondin 1974)

The Commission disagreed with the U.S. government involvement throughout 1974 to 1985 and their relationship during this period was strained. An amendment to the Convention giving the Commission the authority to divide the U.S. catch equally would have clarified the Commission's obligations. The new 1985 Pacific Salmon Treaty gives the Fraser River Panel of the Pacific Salmon Commission the authority to allocate catches in U.S. waters to treaty Indian and non-Indian fishermen.

On several occasions the U.S. government attempted to draw an analogy of its regulatory authority of treaty tribes in U.S. Convention Waters with the Canadian government's action on the Indian subsistence fishery in the Fraser River watershed. But there were basic and important differences between these fisheries. First, the United States treaty Indian fishery for Fraser fish took place entirely within the commercial fishing areas of the State. In Canada, the Indian subsistence fishery is almost exclusively above and outside the commercial fishing boundary. Another important distinction is that those Canadian Indian fishermen fishing commercially in the commercial fishing zones of Convention Waters (catch counted in division of catch) were subject to and abided by the regulations of the Commission (approved by the Canadian government) from 1946 through 1985. They were not identified by separate regulations within the framework of the Commission or by special designation by the Canadian government. The Indian subsistence fishery (catch not counted toward international division) above the commercial fishing boundary in the Fraser River at Mission was subject to Canadian government control long before and after the Sockeye Convention was ratified. And Canadian Indian fishermen fishing commercially in Convention Waters were not controlled by special Canadian government regulations during the Commission's tenure. Since 1977, fishing by U.S. treaty Indians was unilaterally put under the control of U.S. federal government regulations by the United States.

The Supreme Court of the United States in its decision on July 2, 1979 referenced that Canadian Indians were also exempted from Commission regulations:

> … Indeed, the Canadian Government has long exempted Canadian Indians from regulations promulgated under the Convention and afforded them special fishing rights. (Supreme Court of the United States 1979, 31)

These special fishing rights for Canadian Indians, however, pertained exclusively to subsistence fisheries located almost entirely upriver in a non-commercial fishing zone. Thus, the attempt to establish similarities between the U.S. treaty Indian fishery and the Canadian Indian subsistence fishery was inappropriate.

When the United States government became involved with the Commission's regulations in 1975, Canada was quick to respond with a news release on August 15, 1975, as follows:

> Vancouver: Canada has informed the United States Government of its objections to a recent United States action in rescinding certain regulations regarding salmon fisheries under the regulatory authority of the International Pacific Salmon Fisheries Commission (IPSFC). Canada considers this action to be contrary to the provisions of a bilateral convention… . The United States action in rescinding part of the IPSFC regulations and substituting its own in their place is viewed with serious concern by the Canadian authorities. (Environment Canada 1975)

When the United States government embarked on unilateral regulatory control of the treaty Indian fishery in 1977, the Canadian government issued a news release on July 7, 1977 as follows:

> Ottawa: Fisheries Minister Romeo Le Blanc today expressed concern over the conduct of salmon fishing in U.S. waters covered by the International Pacific Salmon Fisheries Commission (IPSFC… I am concerned that the U.S. action may not be consistent with the Convention…It was clear after two weeks' fishing that the situation presented inherent dangers for the Fraser River salmon runs. (Environment Canada 1977)

In 1978, The Canadian government commented on another United States proposal:

> The primary criterion used in assessing the United States proposal was whether it provided for the IPSFC to clearly and unmistakably regain regulatory control of all fishing for Fraser River sockeye salmon in U.S. Convention Waters as provided for under the terms of the Convention. Our examination confirms the preliminary views which I presented in Montreal, that the proposal would not meet this criterion but would serve to formalize a dual management system in U.S. Convention Waters which would be inconsistent with the terms of the Convention, and that having the IPSFC recognize that deviations from its regulatory regime would occur, would be no more than legal "cosmetics" to cover up this basic inconsistency.
>
> You are fully aware of the concerns expressed by Canada that the operation of two regulatory agencies governing fishing of one stock of fish, even when those agencies work in close but informal cooperation, contains latent dangers. We are particularly concerned that the ability of the IPSFC, to which both Governments have assigned the responsibility for management of Fraser River sockeye salmon to manage the fishery in accordance with the object of the Convention could be adversely affected, notwithstanding the stated intention of the United States Government to ensure fulfilment of its obligations under the Convention.
>
> … The IPSFC has, over the years, developed a rapport with the fisheries communities of both our countries. The fisheries community in Canada, at least, is insistent that IPSFC continue to carry out its management role, until new arrangements are negotiated, without reference to domestic problems in the United States. Serious departure from a position held unanimously by the fishing industry is obviously very difficult for the Canadian Government… . I remain hopeful that we can reach a solution to this issue that will permit IPSFC to resume and retain the regulatory control over the sockeye fisheries. (Hunter 1978)

In 1979 the Canadian government again expressed its concern with regard to United States intervention in the fisheries. The Minister of Fisheries sent the vice chairman of the Commission a letter saying, in part:

> … As you know, Canada and the United States appear to have different views concerning the legality of separate regulations which have been issued by the Government of the United States for treaty Indian fishing in U.S. Convention Waters in the last three years. (McGrath 1979)

These and other comments by the Canadian government clearly indicated the government's apprehension over United States' intrusion into the regulation and control of a portion of the fishery. Nevertheless, in spite of the numerous objections made by Canada up to 1979, the 1979 United States Supreme Court decision (p.32) contains the following footnote:

> Although the IPSFC has refused to accede to the suggestions of the United States that special regulations be promulgated to cover the Indian fisheries, we are informed by the Solicitor General that the Canadian Government has no objection to those suggestions, has unilaterally implemented similar rules on behalf of its own Indians, and has expressed no dissatisfaction with the unilateral actions taken by the United States in this regard (Brief for the United States at 40 n. 26). (Supreme Court of the United States 1979)

The highest court in the U.S. had been given misleading information. The "similar rules" were applicable to the subsistence fishery, not to the commercial fishery.

The Canadian government, and for that matter, the Commission also, received strong objections concerning the United States involvement. These comments, directed to the Minister of Fisheries came from the Fisheries Association of British Columbia (Maxwell 1977), the Fishermen's Union (United Fishermen and Allied Workers' Union 1977), and many members of the industry as well as Advisory Committee members, who also made public statements condemning the action (The Fisherman 1977).

In 1979 the Commission was invited to give its views to the U.S. Senate Committee on Commerce, Science and Transportation on salmon issues pertaining to management problems arising from U.S. federal court decisions. The Commission's statement said:

> … the Commission is concerned with the manner in which sharing   of catches with the tribes is being effectuated, since it involves fishing for sockeye and pink salmon by Indians in Convention Waters at times when the waters are closed by Commission regulations… The Convention does not give any authority for regulation by any other agency with respect to the catching of Fraser River sockeye and pink salmon in Convention Waters. In exchange of views between the United States and Canada in 1976 and in subsequent years, Canada advised the United States that Canadian authorities would not agree to any formulation which would permit either Party to allow fishing during periods when regulations promulgated by the Commission prohibit taking of sockeye and pink salmon with any type of gear. Nevertheless, the United States has proceeded in 1977, 1978 and 1979 to exclude the Indians from Commission regulations and adopt separate regulations under the Department of Interior which provide substantial fishing time for the Indians when the waters are closed under Commission regulations. (IPSFC 1979b)

In 1980, Canada commented strongly on United States actions during that fishing season at a meeting with U.S./Canadian government representatives:

> The Canadian side vigorously protested recent U.S. actions — which resulted in U.S. Treaty Indians being given extra fishing time after the Commission had closed the river. Campbell stated that Canadians were losing confidence in the Commission and were worried about its future. He stated there was a strong feeling on the Canadian side that the U.S. Treaty Indian fishery was out of control and indicated that Canada was seriously considering withdrawing from the Commission unless a satisfactory solution was found to prevent future incidents of this type.*

In the memorandum it was acknowledged that appointing an Indian to the Commission "would not alleviate the basic problem in that all Commissioners represented the U.S. not the Treaty Indians." "… It was agreed that the two governments would prepare virtually identical letters of instruction to their commissioners for the December Commission meeting." These letters were to contain instructions regarding the treaty Indian issues.

Following the meeting of government representatives, H.A. Larkins** (who was Northwest Region Director of National Marine Fisheries Service and a U.S. commissioner on the IPSFC) stated: "As it is now, everyone involved in Commission affairs pussy-foots around the 'U.S. Government Fishery,' ignoring it when possible and coming unglued when Indian fishing is perceived to have been contrary to Commission design." Larkins proposed a draft letter that both senior government representatives send to the Commission so that "it might help bring the Boldt decision out from under the Commission table, at least to the extent that the Commission staff could openly consult with us regarding technical aspects of alterna-

---

* Memorandum of Conversation, Department of State, Bureau of Oceans and International Environmental and Scientific Affairs, September 25, 1980.

** Memorandum of H.A. Larkins to Morris D. Busby, OES State Department. November 12, 1980.

tive domestic management measures for the Indian fishery." Larkins further stated "John Roos feels that a letter such as I am suggesting would allow the staff to more openly and directly assist the U.S. in walking the tight-rope between U.S./Canada and U.S./Indian commitments."

In late 1981 following the fishing season Canada again reiterated its opposition to United States involvement and actions in the regulation and management of the Indian fisheries in U.S. waters. In 1981 both governments sent identical letters to Commission Chairman, William G. Saletic. The Canadian government letter was signed by A. Campbell, Director-General, International Directorate, and the United States letter by Theodore G. Kronmiller, Deputy Assistant Secretary for Oceans and Fisheries Affairs, Department of State. The letters in part said:

> …in order to meet its obligations to certain treaty Indian tribes the United States has taken regulatory actions which are not identical to nor strictly consistent with the Commission's regulations. The Government of Canada has formally protested the regulatory action taken by the United States, and continues to believe that the exclusion of a group of fishermen from the purview of IPSFC regulations is illegal. The United States does not share this opinion. (Government of Canada 1981, United States Department of State 1981)

The letters requested the Commission's cooperation in the matter.

"The Commission will agree that a practical solution must be found to this difficulty and, to this end, both Parties to the Convention recommend and request that:

(1) The Commission and its staff, during their management deliberations, be aware of the rights of U.S. treaty Indian fishermen operating in Convention Waters; and

(2) To the extent practicable, the Commission staff provide technical advice to the appropriate U.S. Commissioner regarding the features of the U.S. domestic management regime (including in-season adjustments) which could satisfy U.S. Government obligations to its treaty Indian fishermen, while not jeopardizing the achievement of the objectives of the Fraser River Convention.

These two recommendations were almost identical to Larkin's previously referenced proposal a year earlier.

From 1982 to 1985 the Commission and its staff did attempt, as much as possible, to comply with the recommendations and requests contained in the 1981 letters. This action was taken by Commissioners and staff in deference to the resource and their commitment to protect it, not as an indication that their actions implied approval of continued U.S. interference in regulation of the fishery. Admittedly, there was a very narrow path to follow during those years between steps taken to protect the resource and what might be perceived by many as capitulation to the United States position. No significant long-term damage to the resource was inflicted and the basic objectives of the Commission were achieved.

In 1974, the United States government and U.S. Indian tribes were anxious to have a treaty Indian representative on the Commission's Advisory Committee. The two governments held discussions again in 1976 to amend the Convention to provide for a seventh member of the Advisory Committee. The amending Protocol was signed at Washington, D.C. on February 24, 1977. The Amendment was finally entered into force on October 15, 1980 (Appendix A). It was understood, at least on the U.S. side, that the seventh member would be a treaty Indian representative, although the amendment did not stipulate it would be an Indian. Thus, the only change in the amendment was the substitution of the number seven for the previous number six. The 1930 Convention provided for five advisors. In the

Aide-Memoire from Washington, D.C. of August 10, 1976, the U.S. government stated:

> The Canadian authorities would have preferred in principle to avoid making any amendments to the Convention at the present time and in the present circumstances. They appreciate, however, the concerns of the United States in this matter.

Input from the U.S. treaty Indian member of the Advisory Committee was limited from 1981-1985 because the United States government had exempted treaty Indians from Commission regulations. The United States appointee—an Indian Tribal representative—did not comment on Commission regulations, apparently because treaty Indians were under government regulations; however, the Indian advisor was in a position to convey to the tribes the basis of the Commission's decisions.

These were trying years for the Commission, years in which it was also struggling with other issues. Both the courts and the U.S. government seemed to take the position that as long as no harm was being inflicted on the resource, there was no serious problem. For those who had spent their entire careers in strong allegiance to the Convention, this was unacceptable. In this regard, the Commission, in a statement to Senator Warren G. Magnuson, Chairman of the Senate Commerce Committee, said:

> There is a tendency to view the matter in terms of the management problems created, the implied assumption being that if there are no problems, then the method of allocation is considered acceptable, regardless of the responsibilities under the Convention. (IPSFC 1979b)

INDIVIDUAL COMMISSIONER ACTIONS

At no time during the entire episode did any Commissioner or staff member receive any formal communication from either government showing, under the Convention, that the Commission was abrogating any responsibilities it was required to perform. Industry representatives, Commission members, staff and government officials on both sides were very concerned. Some Commissioners felt the government had failed to live up to its responsibilities under the Convention and, privately, all Commissioners expressed grave concern about the situation and its possible consequences.

The United States method of handling the treaty Indian fishing issue, as it affected the Commission, created a situation that Canadian Commissioner and magistrate Haig-Brown adamantly refused to accept. His strong opposition to United States government interference never wavered and by 1976, Dr. Haig-Brown had grown very disillusioned: "… He no longer looked forward to the Bellingham meetings of the Commission, except with hopeless irritation" (Metcalfe 1985).

Dr. Haig-Brown passed away suddenly in October 1976, while still a Commissioner. His dedication to the resource was appropriately recognized in 1978 with the establishment of the provincial Roderick Haig-Brown Conservation Area encompassing the Adams River.

Most vocal in rendering public complaints over United States' actions was Canadian Commissioner Richard Simmonds, who, for four years was very agitated by government intervention into Commission matters. He commented in 1978 on the situation that existed in 1977: "… the Canadian Commissioners are fed up with the politics being played… " and that "… we're ready to throw in the towel…" (*ibid.*). In addition, speaking of the 1977 regulatory system Simmonds said: "… that system was very unmanageable and very frustrating—it was very lucky we didn't do some serious damage to the resource" (*Bellingham Herald* 1978).

Simmonds resigned in protest in 1980. The treaty Indian issue also led to the resignation of United States Commissioner Ted A. Smits in 1985. The manner of these resignations was unprecedented.

Neither the state nor federal fishery agencies of the United States had adequately studied the biology or management of Fraser River salmon and were in no position to manage the sockeye and pink salmon fisheries unilaterally. Accordingly, it is fortunate that the U.S. Department of the Interior's regulations for treaty Indians usually tracked the Commission's regulations.

INVOLVEMENT AND ACTIONS BY OTHERS

The Northwest Indian Fisheries Commission was formed by the Indian tribes in 1974, soon after the Federal Court decision. The Indian Fisheries Commission was vocal in its opposition to IPSFC policies, as evidenced by remarks in various newsletters and publications. One such report was critical of the seven percent Indian harvest in 1976:

> This has been largely due to the IPSFC steadfast refusal to recognize treaty Indian fishing rights by allowing treaty Indians extra fishing time. (Northwest Indian Fisheries Commission 1977)

In the same report they acknowledged Judge McGovern's earlier referenced decision that clearly stated the IPSFC did not have the authority to direct domestic allocation of catch. Further illustrating the confusion and apparent misunderstandings, the newsletter continues:

> … The open opposition of IPSFC Commissioners to the U.S. accommodation of the Indian fisheries in 1977 is an assumption of authority beyond that granted by the respective Governments of the United States and Canada.

An observer from the National Marine Fisheries Service was allowed to attend regulatory in-season meetings to ensure that the U.S. government understood the rationale behind Commission decisions. In this way, the government could advise the Department of the Interior on regulatory changes necessary for the tribal fisheries. Apparently even this was not satisfactory as stated in the same above-referenced newsletter:

> The tribes do not believe that the National Marine Fisheries Service (under the Department of Commerce) is supportive of their interests and they are concerned that a representative of that agency is presently their only access to the IPSFC data.

In subsequent years, a Department of Interior representative was also permitted to attend the regulatory portion of Commission in-season meetings. This latter decision was the subject of lengthy and sometimes sharply divergent views among Commissioners and staff. In 1985, a tribal representative was present at the meetings in preparation for the 1986 season when the new Pacific Salmon Commission would be regulating the treaty Indian fishery.

On many occasions from 1977 to 1985, the treaty tribes, through various representatives, requested direct access to IPSFC technical data. The Commission consistently stated (with government concurrence) that the Convention clearly established that there was only one management authority and there was no other entity with which management responsibilities were to be shared. This was never a debatable issue. Therefore, the Commission instructed the staff that there would be no exchange of data or liaison between the Commission staff and representatives of the tribes or any other entity seeking management involvement.

The Northwest Indian Fisheries Commission wrote to John D. Negroponte, Department of State, stating in part:

> … the Tribes have not been permitted direct access to Commission meetings and technical staff information. This has created a major obstacle to their abilities to efficiently regulate their sockeye and pink salmon fisheries … . (Heckman 1977)

The tumultuous situation from 1974 to 1985 regarding treaty Indian fishing in United States Convention Waters was unfortunate. The United States government ultimately approached the problem from a unilateral point of view. The Commission in its public and private criticism of the U.S. government's actions, incited the wrath of the government. Indians were totally frustrated and lashed out at both the Commission and the U.S. government. In 1977, as their percentage of catch began to increase, the tribes became more supportive of the U.S. government. All the while, the State of Washington was incurring criticism from fishery associations and state and federal courts. In addition, State Court decisions differed significantly from Federal Court opinions. Even the Federal Court and different departments of the federal government were at odds with each other. The Commission attempted to maintain a hands-off policy, but was unwillingly drawn into the fray. The Canadian government generally supported the Commission's position, even stating that U.S. action was illegal, but at the same time had to maintain its diplomatic relationship with the U.S. government, and emphasized serious concern while acknowledging the need for achieving catch division and conservation goals. The sometimes-acrimonious debates which ensued resulted in dissipation of effort involving many individuals and organizations.

Recently, the sentiment towards the Commission's position on the treaty Indian issue was explicitly expressed (Yanagida 1987, attorney in the Office of the Legal Advisor of the U.S. Department of State). Yanagida stated that treaty Indian participation in international fishery management institutions had been precluded "largely as a backlash to *United States v. Washington*." The "hostility of non-Indian interests" was confused with the policy of adherence of the IPSFC to its terms of reference as mandated in the Convention and the frustrations this evoked from the United States government. In addition, Yanagida said: "Elements of the U.S.-Canada International Salmon Fishery (sic) Commission (IPSFC) were among the interests that opposed implementation of *United States v. Washington*." On the contrary, the Commission never took a position vis-a-vis the Boldt decision.

Yanagida, later in reference to "rebellion" brewing from non-treaty interests, stated: "The IPSFC continued to assist the rebellion." Reference was made that the two governments initially agreed to IPSFC regulations early in 1977 before the fishing season and then the IPSFC refused to adopt their proposal. It was further said: "Unabashed, the U.S. commissioners contravened the Department's directives... ." It must be pointed out that pre-season regulations always originated with the IPSFC and not with the governments. Moreover, the Department's directives would have instructed the Commission to contravene the Convention. In effect, the Commission was accused of defying the federal court along with the State fisheries agency and State Supreme Court as detailed in further remarks: "Because of the IPSFC's defiance of *United States v. Washington*, the United States had to resort to a scheme that excluded treaty Indians from the regulatory regime of the IPSFC. This was an imperfect solution for all." The concluding remark on the matter was: "Recalling recent IPSFC contravention of U.S. law with respect to Fraser River fisheries"—an apparent reference to the Commission's inferred refusal to submit to Department of State directives. Yanagida seems to suggest that it would have been permissible for the IPSFC to contravene a long-standing international Convention to accommodate U.S. domestic law that could not be legally implemented by either the state or federal government. Amongst others, Petty (1979) concluded that the United States did not have "the power to compel Commission's recognition of the treaty Indian fishery in its regulations" and said "Renegotiations of the Sockeye Convention could provide an opportunity to remedy this." In the end, this is what was finally done.

On many occasions from 1974 to 1985 the United States government said that it would not do anything to prevent it from honoring its commitment to Canada in achieving the two basic objectives of the Convention. What was lacking by the United States was the depth of biological understanding necessary to manage the resource scientifically while obtaining division of catch and the proper escapement. Thirty years earlier, Dr. W.F. Thompson, the Commission's first Director, made a pertinent comment:

> To divide the catch, so that political and economic difficulties in the functioning of the treaty may be avoided, is necessary but is not the purpose of the treaty. Either of these two, the details of regulation or the arrangements to share the catch, is administrative in nature and can be assisted by technical knowledge if necessary. But to the difficult task of understanding the needs of the fishery, all the skill and patience of purposeful scientific research should be brought. (Thompson 1945b)

Thus, from 1977 through 1985, a dual and not necessarily scientifically-based management regime prevailed in United States Convention Waters. The Convention was compromised by one of the governments. In spite of these obstacles, the Commission was able to stay on course.

CANADIAN INDIAN SUBSISTENCE FISHERY

The Canadian Indian subsistence fishery throughout the Fraser River watershed was in existence long before the sockeye Convention was signed in 1930. The Commission accounted for, but did not regulate this fishery. The Indian fishery was controlled by the Canadian Department of Fisheries. The Department specified the location of fishing, the gear to be used and the fishing time permitted. The salmon were for personal consumption only and not for commercial purposes. These fish were not counted into Canada's share of the catch between the two nations.

Accurate records for the early years of this fishery are lacking. Dr. W. E. Ricker stated that the Indian catch of Fraser sockeye in earlier years could have approached several million fish (Ricker 1987); however, he did not present evidence to support this estimates. Killick and Clemens (1963) present estimates for the years 1915 to 1940 which ranged from 20,000 to 150,000 fish each year. G.W. Hewes, an anthropologist, calculated the aboriginal salmon consumption based on estimated population and estimated per-capita consumption (Hewes 1973). From his data, an estimate of two to two and one-half million salmon could have been consumed by Fraser River Indians in the years prior to the first white contact (1770-1800). If approximately 80 percent of these fish were sockeye, between one and one-half to two million sockeye could have been taken for food each year by Indians living in the Fraser watershed.

If the big-year runs were in the order of 50 to 75 million fish and the off-year returns were perhaps up to 10 million sockeye, the Indian catch in the late 1800s would have taken up to 20 percent of the total off-year runs and four percent of the dominant big-year runs. Other estimates of the size of the big-year runs were in the order of 100 million sockeye (Ricker 1950) and some were even larger. With total runs of 100 million sockeye, the Indian food fishery would have been only one-half of the maximum rates given above.

Since modern-day commercial fisheries harvest near 80 percent of the total runs, it is unlikely that the Indian fisheries two hundred years ago had any measurable impact. Yet, the apparent animosity towards this traditional fishery surfaced at the Commission's first meeting with its Advisory Committee in 1938 when the Advisors recommended that the fishery be eliminated. No such action was taken. There were other earlier attempts to scuttle the fishery, such as in 1918 when a proposed treaty contained a clause which would have essentially rendered the ancient subsistence fishery above

Mission, B.C. ineffective by the prohibition of any fishing except with hook and line (Gilbert 1988). Almost all of the Indian fisheries take place above the commercial fishery boundary at Mission. The gears used are now mainly gillnets and dip nets. Mitchell (1925) describes the relative efficiency of Indian and non-Indian fishermen:

> When the Indian struck at a salmon with his spear he got that salmon, and when he had enough he quit; not so the white men. They got after them with pitch forks, iron spikes, or pieces of thick telegraph wire fastened on poles, and with gaffs. After wounding several they would get one pinned to the bottom then lost it trying to get it to a shore. For every ten they got, they injured a hundred that escaped.

Sockeye on or near the spawning grounds were also sought by non-Indians in earlier years.

From time to time the Commission has requested closures of the Fraser River Indian fisheries to protect specific sockeye races, such as the Chilko, Nadina and early Stuart runs. These have been requested in the mainstem Fraser River fisheries and in tributaries. Chilcotin Indian fishing was closed for several years in the 1930s and 1940s. In most instances, the Fisheries Department was responsive to the Commission's recommendations and closures were implemented as requested. In recent years, those closures have been preceded by extensive negotiations between Indians, Fisheries Department and Commission staff. On one occasion, the Commission considered implementing its own regulations to restrict certain fisheries, but this was not adopted.

In the 1970s and 1980s a more militant position was taken by many Indians about their fishing activities being regulated by the Department. Numerous confrontations took place. The issues were generally prompted by the desire for self-control and separate management. The selling of the fish was considered a right by many, just as their predecessors had bartered fish to the Hudson Bay Company in the late 1700s and 1800s. The Indians also wanted the right to set their own fishing times and catch limits. Resentment over fisheries control increased to the point where there were open confrontations between Department enforcement officers and Indians. Many threats were made and in some instances officers were warned by Indian fishermen not to check their nets for catch information. These problems continued to escalate and for safety reasons, during a three-week period in the summer of 1975, the Fisheries Service of Canada withdrew its fishery officers from night patrol in the lower Fraser Canyon area. Reports of increased illegal fishing and illegal salmon sales were prevalent during this time. The Commission was officially notified by the Department of Fisheries that it should make allowance for the removal of an additional 100,000 sockeye above the normal Indian food fishery catch.

Illegal fishing and sales by both Indians and non-Indians increased substantially in recent years. Fraser River Indian catch data from legitimate fisheries has always been difficult to obtain and somewhat suspect as to accuracy. These problems, plus known but unquantifiable illegal activities, made the validity of catch figures questionable. The problem of obtaining accurate catch statistics from this fishery persisted throughout the history of the IPSFC. Several methods have been used in the past to obtain catch data—fish counts, tag ratios and net sampling, and verbal reports from Indians. Defects in the catch-reporting system are serious and have been even more so in recent years as the reported catch has increased substantially.

The Commission and fishing industries of both countries became increasingly alarmed at the increasing Indian catch and alleged illegal activities. At times, it was reported that some of the subsistence catches were being combined with commercial catches legally caught below Mission. B.C. It was not possible

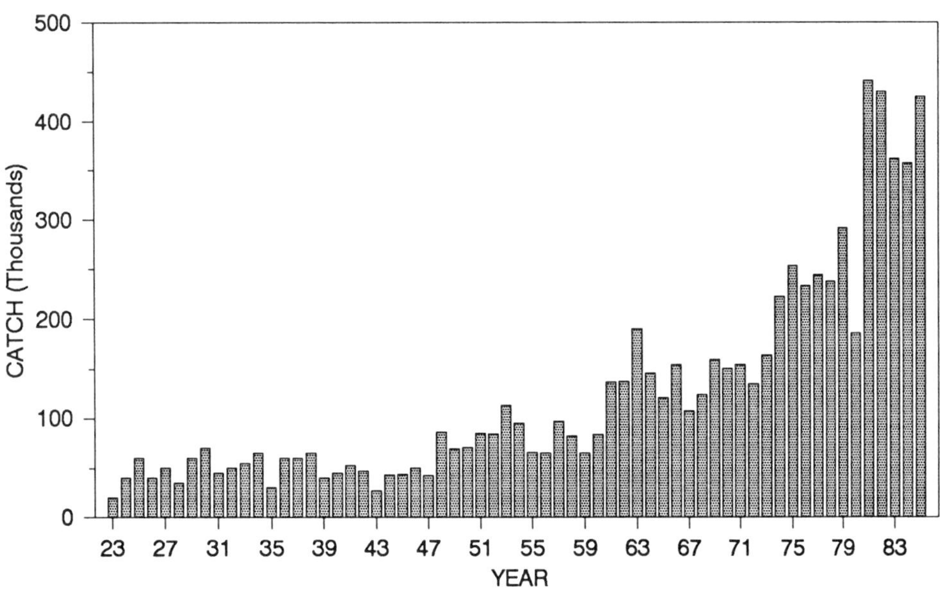

Figure 38. Indian food fishery sockeye salmon catches on the Fraser River, 1923-1985.

to validate the reports or quantify the amounts, but numbers were believed to be small. Some U.S. Commissioners and many others in the U.S., including industry representatives, felt that because the U.S. treaty Indian catch counted in division of catch, so should the Indian subsistence catch in Canada. This was not supported by Canadian industry and despite several requests, the Commission did not count the Canadian Fraser River Indian catch in division of catch.

Canadian industry representatives became increasingly adamant that limits should be placed on the Indian subsistence catch. They realized that additional closures were required in all Convention Waters fisheries to "pass-through" extra fish to be taken by the upriver Indian fisheries. Those fisheries were not on limits or quotas. The increasing Indian food fishery catch in the Fraser River system had an impact on the commercial catch in both countries. It was also increasing problems with the Commission's management goals to ensure the net escapement goals of the various stocks onto their spawning grounds. It was essential that the Commission staff obtain from Canadian fishery officers estimates of weekly catches by the various Fraser River fisheries below Lillooet. This information was useful in determining the relative magnitude of the harvests. These problems however, were similar to those experienced by the Commission in dealing with an excessive commercial fishing fleet.

The reported Fraser River Indian catches (Appendix L) (Figure 38) indicated that in the 20-year period before Commission management in 1946, the Indian catch held steady at about 40,000 fish each year (Roos 1984). As the runs increased, the catch in the Indian fishery also increased, but they were taking an even higher percentage of the total run (Figure 39). A record catch of 441,000 sockeye was reported in 1981 and the average catch from 1982-1985 was 356,000 sockeye. This represents an approximate ten-fold increase over the catches made before the Commission became involved in the fisheries. Yet the runs had not increased by a factor of ten. The percentage of the total Fraser sockeye run reportedly taken in the Fraser Indian catch increased to an average of about four percent after 1960 from about two percent from 1925 to 1960 (Figure 39). The Fraser River Indian catch from 1946 to 1955 was 3.4 percent of the Convention Waters catch

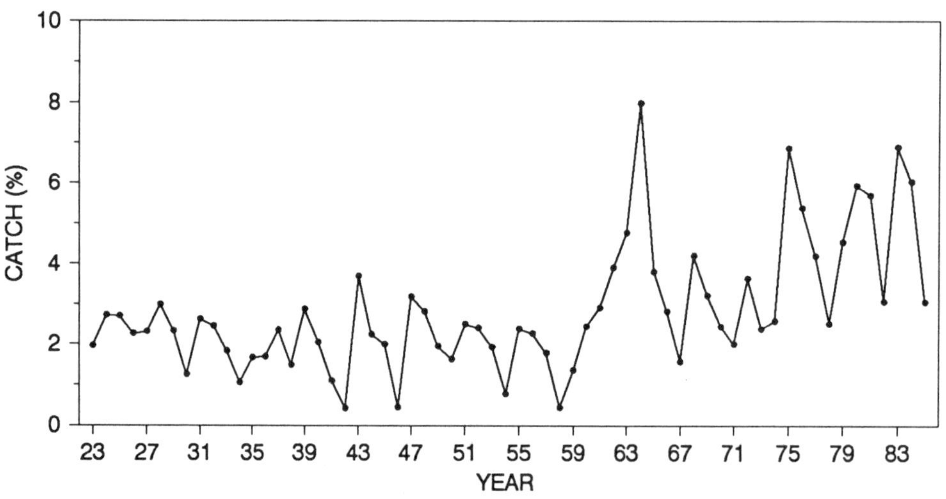

Figure 39. Proportion of the total sockeye salmon run taken by the Fraser River Indian food fishery, 1923-1985.

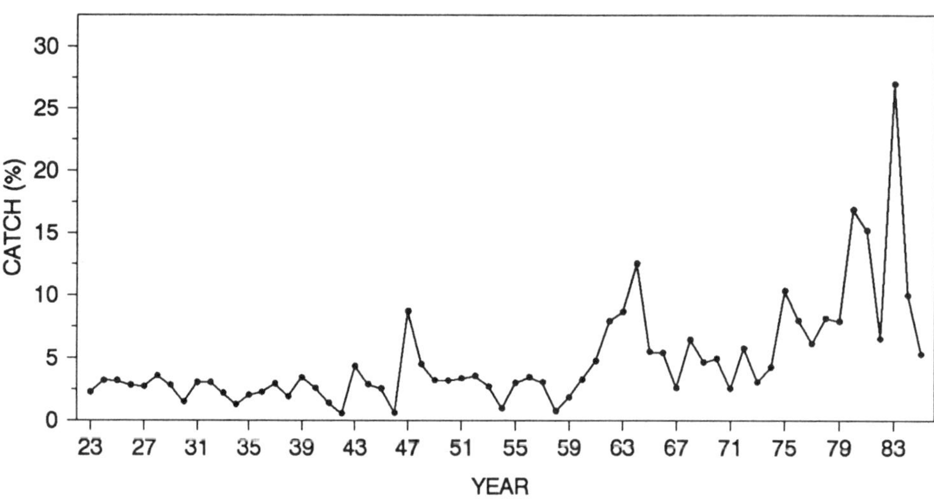

Figure 40. Proportion of the total Convention Waters commercial and Indian food fishery sockeye salmon catch taken by the Fraser River Indian subsistence fishery, 1923-1985.

(Figure 40) and 2.9 percent of total catch in all areas. From 1974 to 1985 the Indian catch averaged 11.1 percent of the Convention catch and 6.2 percent of total catch in all areas. Most of the increase has taken place on the summer-run stocks.

The Indian population increased by a factor of 1.6 from 1941 to 1981, whereas the number of fishing permits issued for the same general time period increased by 6.5 times. As a result of Canadian policy to give priority for harvestable fish to subsistence fisheries, fishing times have not been reduced in the Indian fisheries to the extent they have in the commercial fisheries.

With great catch increases, difficulties were encountered. This was particularly so on highly desirable races that are the first to migrate upriver, such as the early Stuart race. For example, in 1981 the Fraser River Indian fishery catch of the early Stuart race was 32 percent of the total catch of all summer-run races, whereas the early Stuart fish comprised only 16 percent of the total escapement of those races.

In spite of the fact that the early Stuart gross escapement passing Mission, B.C. in 1981 was estimated by echo sounding to be 335,000 fish, only 129,000 arrived at the spawning grounds.

Because of the great increase in fishing effort by various Indian bands in recent years, timely and accurate catch data from each major fishing area was very important but adequate information was not always available. Management actions to protect certain races in commercial fisheries below Mission were, in some instances, nullified by exploitation of these races by the upriver Indian fisheries. In-season adjustments of Indian fishing schedules were virtually impossible to achieve. Total Indian food fishery catch of Fraser sockeye from 1946 to 1985 was 6,734,000 fish.

Pink salmon catches in the Indian fishery were generally small and were taken incidental to other species. Because of poor quality, these fish were of little value to Indian fishermen. An estimated 615,000 Fraser pinks were taken from 1957 to 1985.

Management Strategies in the 1970s and 1980s.

The Canadian government's policy decision in 1971-1972 not to fund additional capital construction projects proposed in the $14 million development program was not anticipated by the Commission. The Commission therefore developed an alternate resource enhancement plan consistent with the objectives of the Convention. The escapements would be increased as quickly as possible to the extent practicable to accelerate the rehabilitation process. It was realized that there would be a short-term cost to the industry due to reduced catches but every effort was made to minimize disruption of fisheries and their catches.

The overall plan was to increase escapements to reach the optimum escapement of major races. It was recognized that some of the smaller, suppressed races would benefit concurrently from increased fishery protection and that production of all races should increase. For sockeye, a cycle-by-cycle racial approach was implemented as much as possible each year. A summary of the average adult escapements by four-year periods from 1974 to 1985 is shown in Table 5.

Sockeye escapements increased significantly between the 1974 to 1977 and 1982 to 1985 periods. While the increase to 2.018 million sockeye in the 1982 to 1985 period from an average of 1.172 million spawners in the earlier period may not seem substantial, significant

increases in total returns occurred concurrently with the increased escapements. The number of sockeye spawning each year from 1946 to 1985 is shown in Figure 41.

Even more dramatic increases in pink salmon escapement were achieved between the 1970s and 1980s. In 1973 and 1975, pink salmon escapements averaged 1.561 million fish, which was very similar to escapement levels achieved in the 1960s. By 1981/83 the average pink salmon escapement had increased by 292 percent to 4.560 million fish. In 1985 a record 6.461 million pink salmon escaped up the Fraser River.

For the most part, the Commission's objective of increasing total escapements was successful. The increases were beneficial in terms of catch to the fishing industries. Significant increases for individual races were meaningful as well. The Horsefly River sockeye escapement on the dominant cycle-year of 1985 increased to 1,135,000 fish in 1985 from 253,000 spawners in 1973 (Figure 42). Of great significance also, the Horsefly River race has established a subdominant population on the cycle-year following the dominant year run. In 1970, the escapement was only 1,350 fish but by 1982, 30,000 sockeye spawned in the river. Both of these cycle-year escapements represent great potential. It appears that an increase is developing on the 1983 cycle year, since the escapement increased to 2,036 spawners in 1983 from 201 fish in 1975.

The 1980-1984 cycle-year escapements for some individual races offer considerable hope for the future. For example, the Gates Creek escapement, which is dominant on this cycle, increased in three generations to 19,100 in 1984 from 8,500 in 1972. Of great importance is the upper Adams River increase to 3,502 fish in 1984 from only 31 spawners in 1972. Because of the large unutilized spawning area available in this river and the large rearing-area potential in Adams Lake, this run would be expected to be a major producer of sockeye by the turn of the century if adequate

TABLE 5. Average annual escapements by four-year periods for Fraser sockeye and pink salmon, 1974 to 1985.

Years	Sockeye (Million)	Years	Pinks (Million)
1974-77	1.172	1973&75	1.561
1978-81	1.553	1977&79	2.975
1982-85	2.018	1981&83	4.560
		1985	6.461

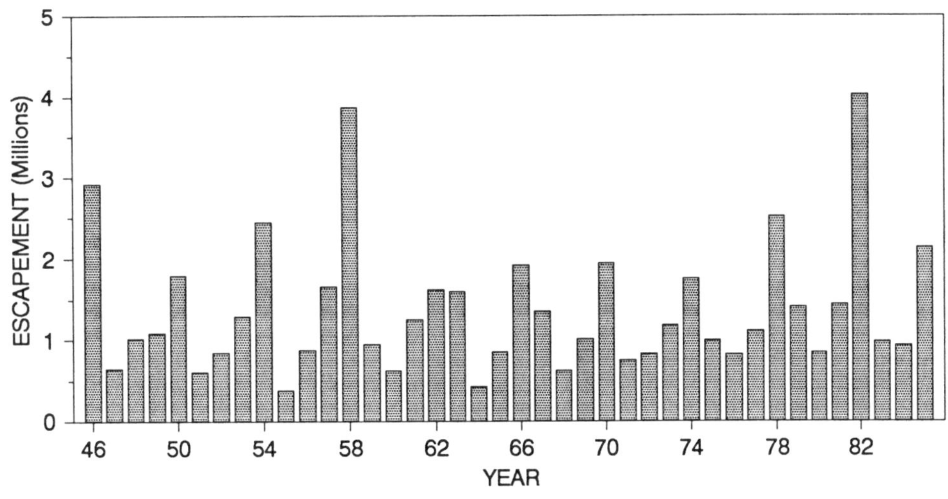

Figure 41. Fraser River sockeye salmon spawning escapements, 1946-1985.

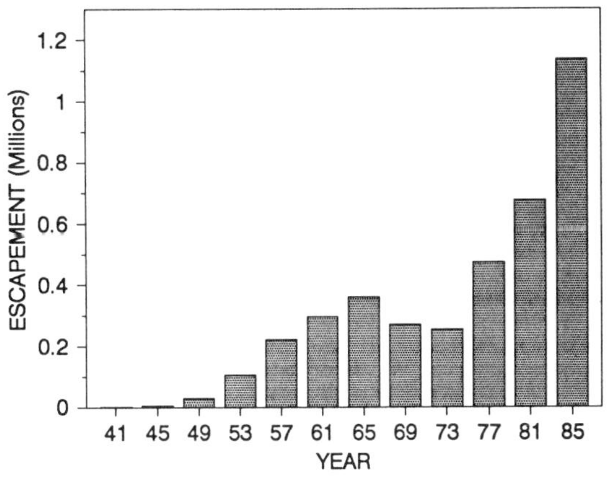

Figure 42. Horsefly River sockeye salmon spawning escapements on the dominant cycle, 1941-1985. Spawners were present in 1937 and 1941 but only about 200 and 1,000 fish, respectively.

regulatory protection is given. The Momich-Cayenne Creek population, in the upper Adams Lake watershed, increased to 5,900 in 1984 from 1,000 spawners in 1972. The potential for this stock, however, is limited by the meager spawning grounds available. Another close-by population is the Fennell Creek race which spawns in the North Thompson District. Escapement to this creek, which appears to be most abundant on the 1984 cycle, increased to 11,100 in 1984 from 1,900 in 1972. A very large increase in the Chilko Lake-South End population has also taken place on this cycle to 127,700 in 1984 from 2,100 in 1972.

Significant improvement of the off-cycle return to early Stuart streams has also been observed. In 1972, the escapement was 5,100; by 1984, 45,200 sockeye had spawned in the Stuart Lake region.

The 1983 cycle-year escapements have not shown increments similar to other cycles. This cycle-year is the subdominant year of return for the Chilko and lower Adams River populations. The Chilko River escapement increased to 332,000 in 1983 from 162,000 in 1971. The Chilko Lake-South End race increased to 55,000 in 1983 from 12,000 in 1971. But the lower Adams River escapement of 280,000 in 1971 declined to 202,000 by 1983. As noted earlier, overfishing in non-Convention Waters took place in 1983. The optimum escapement for the cycle is about three to four times greater.

The 1982 cycle-year escapements have shown some significant gains in recent years. The Weaver Creek escapements increased to an excessive escapement of 295,000 in 1982 from 11,000 in 1970. The Birkenhead River race has also produced well on the cycle; escapements have been good following the 73,000 spawners in 1970 to more than 100,000 in the intervening cycle years with 129,000 present in 1982. Significant improvement in the Portage

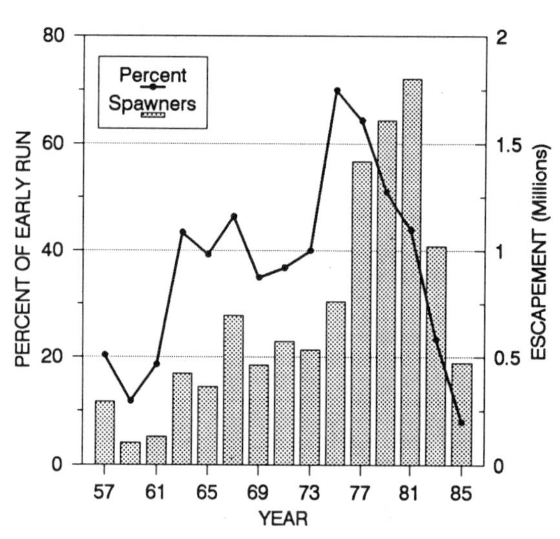

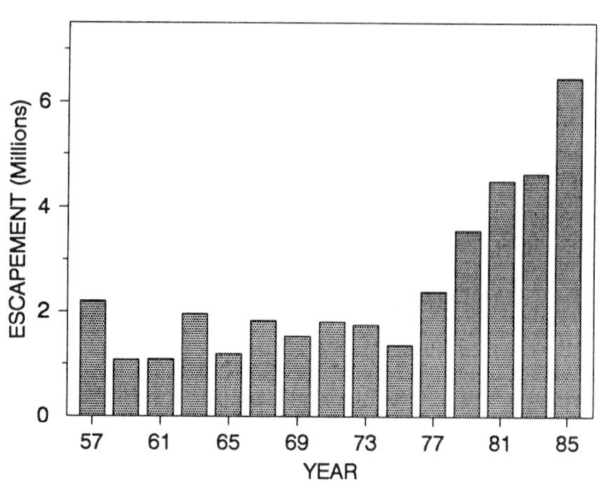

Figure 43. Number of pink salmon spawning above Hells Gate and the proportion of the total early run escapement spawning above Hells Gate.

Figure 44. Fraser River pink salmon spawning escapements, 1957-1985.

Creek population has occurred since 1970, when the escapement was 4,000 sockeye. By 1982, the escapement increased to 24,000 fish. Steady increase in the lower Adams-Little River escapements has also taken place, increasing to 2,310,000 in 1982 from 1,467,000 in 1970. Because of the large potential spawning area available, the increase of the lower Shuswap River spawning population to 514,000 in 1982 from 29,000 in 1970 constitutes a very significant step forward. In addition, an almost tenfold increase of the middle Shuswap River escapement took place when 40,000 sockeye spawned in 1982 as compared to 4,600 in 1970. On the other hand, the early Stuart escapement of only 4,600 in 1982 was a continuation of a decline on the cycle from a peak of 33,000 in 1970.

The switching of the dominant lower Adams River cycle from 1913 to 1914 (1986 cycle) may be viewed as beneficial from an industry point of view because it spreads total production of sockeye more evenly over the four-year period instead of an almost total dominance by the "big-catch" years that occurred prior to the Hells Gate slide. What effect this has had on the total productivity of all races is not known.

In regard to pink salmon management, a major objective of the Commission was to increase the early-run escapements above Hells Gate. Annual increases continued to occur and by 1981 about 1.8 million fish were recorded in the upriver spawning areas of Seton Creek and Thompson River. This was 43.8 percent of the total early run escapement or 40 percent of the total Fraser escapement. By 1985; however, high flow in the Fraser during adult migration and small size of fish resulted in only 468,000 pinks past Hells Gate. This represented only eight percent of the early run escapement and only seven percent of the total Fraser pink salmon escapement (Figure 43). These events were beneficial to the main Fraser River escapement which reached a record 5.3 million in 1985. But the escapement to Thompson River in 1985 was only 193,000 fish in comparison to 1.2 million in 1981. Clear evidence of blocked and difficult migration conditions in the Fraser Canyon in 1985 can be seen from an analysis of pink salmon escapements in the Fraser Canyon streams below Hells Gate. The increasing percentage of the total early-run escapement that spawned in tributary streams is shown in Table 6.

TABLE 6. *Distribution of early-run pink salmon escapement into Fraser Canyon streams, 1977-1985.*

Year	Canyon Stream Escapement	Total Early Run Escapement	Percent of Total Early Escapement
1977	9,245	2,203,902	0.4
1979	25,610	3,154,945	0.8
1981	43,234	4,097,269	1.1
1983	46,456	4,373,049	1.1
1985	169,858	5,886,698	2.9

These data indicate that proportionately, about seven times as many early-run pinks spawned in canyon streams in 1985 as compared to 1977. It is likely that this was due to the reduced size of fish, migration conditions, and volume of escapement. The spawning effectiveness of these diverted pink salmon is not known nor is it for the large numbers of fish that backed down river to the lower main Fraser River spawning grounds. The total Fraser River pink salmon escapement for each cycle year from 1957 to 1985 is shown in Figure 44 and the rapid increase in total escapement since 1977 is clearly evident.

A serious migration problem for pink salmon took place in 1985. Large numbers of pink salmon accumulated at several locations in the Fraser Canyon (IPSFC 1986). Fish were so crowded at times that some were forced up on the rocks and others pushed out into faster currents by the volume of fish. Several hundred thousand fish were estimated to be delayed at specific times. Congestion problems may have also occurred during large escapements in 1979 and 1981 and to a lesser extent in 1983. It is believed the migration passage problem in 1985 was because higher water levels made migration much more difficult. The problems encountered by the pink salmon were somewhat similar to those experienced by sockeye from 1913-1944.

There is also indirect evidence that the smaller individual size of the fish may have contributed to the migration problem in the canyon in 1985. With the large run of almost 19 million fish in 1985, the average fish weight was reduced to 4.8 lbs., one of the smallest average weights on record. Swimming ability is related to fish size—larger fish are more capable of swimming against high water velocities. In earlier years (1960-1970s), when the fish were larger and fewer in numbers, there probably were no migration problems. In 1985, the small size of the fish magnified the adverse effects of the migration difficulties. The effect of size on migration ability is evident from the sex ratios of tagged and untagged fish on the spawning grounds. Normally, more female than male pink salmon escape to the spawning grounds. This occurred in 1985 for the total Fraser River pink salmon run and particularly for those that spawned in the lower Fraser River; however, the 1985 escapement above Hells Gate, for the first time on record, was dominated by males. The likely explanation is that the males, being larger, were more capable of migrating through the high-velocity areas in the Fraser Canyon. This unbalanced sex-ratio was also recorded for sockeye in the 1920s and 1930s as a result of the blockages. Sex ratios for pink salmon at the lower Fraser River tagging site in 1985 were normal.

September 29, 1985. Difficult migration conditions, combined with an unusually large number of pink salmon competing for the limited space available for migration through the Fraser Canyon, prevented a large proportion of the 1985 pink salmon run from reaching the Seton Creek and Thompson River spawning areas.

The greater numbers of males implies that only some of the pinks destined for the spawning grounds above Hells Gate were successful. Further evidence was obtained from a small tagging program conducted in the Fraser Canyon below Hells Gate. Tags were applied to 900 pinks at Saddle Rock and at Little Hells Gate, two locations where apparent migration blockages were observed. Subsequent recoveries of tagged fish on the spawning grounds indicated that only 12 percent of the tagged fish reached the spawning grounds above Hells Gate. Some pink salmon spawned in tributary streams downstream, such as the Coquihalla River. But the majority of tagged fish (64 percent) moved back far downstream to spawn in the lower Fraser River or major tributaries such as the Harrison River.

Since tagging took place during a period of known blockage of fish, tagging was possibly selective for fish already delayed and stressed and therefore with a reduced chance of successful migration. It does illustrate that a significant blockage of upper river pink salmon occurred and that the blockage contributed to the sharp decline in upriver escapement. In addition, slightly more females than males were tagged, but nearly two thirds of the tagged fish recovered on the upper river spawning grounds were males. Similarly, from the enumeration tagging program in the lower Fraser River and recoveries at Seton Creek and Thompson River spawning grounds, there was a much higher successful migration rate for male tagged fish.

The canyon tagging results also provided information on the delay period in the area. Tagged fish were observed passing Hells Gate for up to 19 days after tagging. Normal movement would take only one or two days. The delay of pinks was studied for several days by tracking a small number of fish implanted with miniature radio transmitters. Fish were radio-tagged at Saddle Rock and Little Hells Gate. Delays of at least four days were observed in some fish at Saddle Rock and delays of one to two days were detected at China Bar Rapids below Hells Gate. Several radio-tagged fish disappeared during the study. They might have gone downstream and out of the study area, as did 64 percent of the fish tagged with plastic discs.

To examine stress, pink salmon were collected at two difficult migration locations: at Little Hells Gate and China Bar Rapids. Stress levels, as indicated by plasma-cortisol levels, in female pink salmon were significantly higher than observed in fish sampled in 1983. Also, stress levels of fish at China Bar Rapids were much higher than those at Little Hells Gate which was only three miles downstream, indicating a rapid increase in stress as the fish travelled this short distance.

It is possible that high stress levels observed in 1985 pinks may have occurred in some of the 1979-1983 fish when large numbers of migrating pinks also experienced congestion sand difficult passage at rapids in the canyon. Fish which successfully reach the spawning grounds but are highly stressed en route may be less productive because of improper egg development or inability to spawn normally. Also, because proportionately fewer female pinks reached upper spawning areas, potential egg deposition was reduced. This may partially explain the recent decline in the proportion of the Fraser pink salmon population migrating to the upper spawning areas even though total Fraser escapements have increased concurrently. Further stress studies might shed light on the decline of upper river spawners. Migration pathway improvements have been implemented by the Canadian government (P. Saxvik, personal communication) in an effort to increase and maintain the upper river pink salmon production.

Estimating the Losses

275

I t was fortunate that most of the proposed development projects discussed in Chapters 11 and 12 in the Fraser River watershed did not come to fruition. Impacts on the resource would have been added to those inflicted earlier in the early 1900s. It is doubtful the resource could have survived if some of the projects were completed. The Commission's actions in assisting to protect the resource were important factors in re-building and maintaining the stocks.

Events that created the physical and long-term losses caused by the Hells Gate blockages and the Quesnel and Adams River dams occurred more than seventy-five years ago. Yet the impact continues to the present day. Many stocks affected have still not reached pre-Hells Gate abundance levels. This is a constant reminder of the delicate balance between fish and their environment.

The quantity and value of fish lost have been enormous. Native Indians, the fishing industries of both countries, and the general public were deprived of a vast source of food and income for many years. It had been suggested that the white man should be responsible for paying the Indians for the lost production (Mitchell 1925). That suggestion was never followed through, but would certainly be more possible now than at that time. While no compensation was given to Indians, there was also no relief given to harvesting and processing industries. Moreover, there appears to have been little effort, if any, from the railways in offering assistance in the clean-up or restoration of the huge rock and gravel deposits caused by railway construction. There appears to be no record of requests, or rejection of such requests, made to the railways for compensation. Additionally, the railways were never charged for the effects of their actions. Although many believed the drastic decline of sockeye salmon in 1917 and subsequent years was due to the 1913 fish passage conditions and the 1914 slide, prevailing thought soon was that the river was restored, as far as fish passage was con-

cerned, and that the most critical problem was overfishing. Until the IPSFC investigations showed that Hells Gate was indeed the most important problem, it would have been difficult to prove that railway construction was responsible for the majority of the industries' woes after 1917. Similarly, neither mining nor logging companies were held accountable for damaging the Quesnel and upper (exterminating) and lower Adams River runs.

In effect, taxpayers of both countries have had to pay for extensive Commission investigations necessary to determine the cause of depletion, for the construction of the fishways, and for the remedial work required to alleviate the problem. In addition, there was a loss of income and food to Indians and to the fishing and support industries in both countries from 1915 to the present time.

No previous assessment of total losses to date has been made. Thompson (1945a) estimated the loss from 1917-1942 at "far in excess of $279 million" (in 1942 dollars). The Commission concluded in its 1948 Annual Report that the loss at 1948 values had reached more than $500 million. Another and more recent estimate of the total lost value of sockeye for all runs from 1917 to 1945 has been placed at $2.19 billion (1978 prices) (Ellis 1989). Sockeye production from 1945 to the present also suffered as a result of the rock dumping, slide and dams. A more detailed analysis of the loss is presented here. Separate reviews were made for both sockeye and pink salmon.

The annual commercial catches of Fraser River sockeye for 1893-1989 are shown in Figure 45 and the cyclic average catches are shown in Figure 46. The total reported sockeye catch for the four years preceding the slide (1910-1913) was 41,341,000 as given by Rounsefell and Kelez (1938). Gilhousen (1990 MS) and PSC have adjusted catch totals taking into consideration salted, shipped fresh, exported and wasted fish. This total for the same period was 42,055,000 sockeye. Production during these four years might have been low-

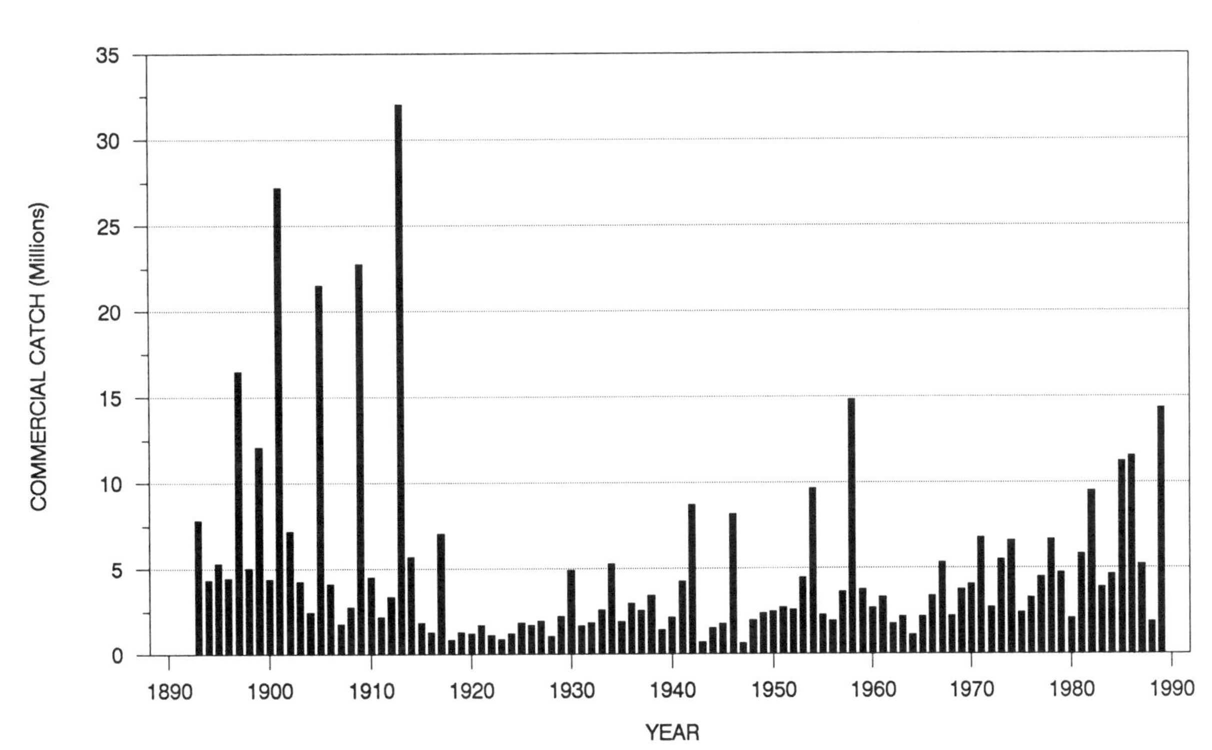

Figure 45. Annual commercial catches of Fraser River sockeye salmon, 1893-1989. From Rounsefell and Kelez 1938, Gilhousen 1990 MS and PSC.

ered somewhat by the effects of the Adams River and Quesnel Lake dams, and possibly further lowered by overfishing on previous off-year cycles. The total catch from 1898 to 1901 was 46,555,000 sockeye (Rounsefell and Kelez) and 48,690,000 fish (Gilhousen 1990 MS and PSC), suggesting a decline by the 1910-1913 period. Nevertheless, to arrive at a possible potential expected total catch from 1917 through 1985, for the 17.25 cycle-year periods covering 69 years it was assumed that the catch could have been maintained equal to that in the 1910-1913 period. The calculated total harvest during this period (using recent data from Gilhousen and PSC) would have been 725,449,000 sockeye under pre-Hells Gate slide conditions.

To determine production loss caused by human alteration of the environment, an estimate of the total actual catch of Fraser River sockeye from 1917 to 1985 is required. The difference between this catch and the projected total catch (725,449,000) would give the theoretical loss resulting from the rock dumping, slide and dams.

The total Fraser sockeye catch from 1917 through 1949* was 84,724,000 sockeye. All of these fish were considered to be of Fraser River origin although it is acknowledged that small numbers of other stocks would be included.

Total (all waters) Fraser River sockeye catch by both Canada and the United States from 1950 to 1985 was 160,787,000 fish (Appendix H_2). Therefore, approximately 245,511,000 Fraser River sockeye were taken from 1917 to 1985 and the calculated loss in catch resulting from environmental degradation is estimated at 479,938,000 fish. Increased Fraser River Indian fishery harvest reduced the commercial harvests by fishermen of both countries in recent years. In addition, increased escapements under Commission management would have reduced the commercial catches compared to pre-Commission regulatory prac-

---

* The time period not covered by Commission management.

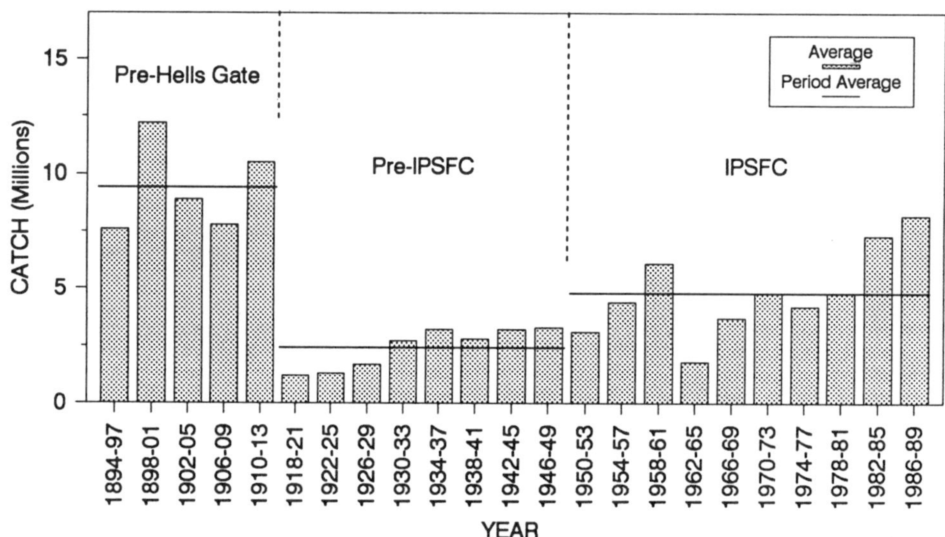

Figure 46. Comparison of average catches of Fraser River sockeye salmon in the pre-IPSFC period (the 1894-1913 pre-Hells Gate period and the 1918-1949 period before the effects of IPSFC regulation of the commercial fishery were evident) and the 1950-1989 period when the benefits of IPSFC fishway construction and commercial fishery management were obtained. Catches from 1914-1917 have been omitted because the brood year spawners were obstructed by the Hells Gate blockade in 1912 and 1913 whereas those in 1910 and 1911 were not. Catch information from Rounsefell and Kelez 1938, Gilhousen 1990 MS and PSC. 1986-1989 catches under PSC management.

tices. The loss at 1985 prices ($8.58 per fish),* would be $4.118 billion.

Estimating the loss of Fraser River pink salmon is more difficult than for sockeye. Prior to 1913, pink catch data were incomplete because these fish were just becoming economically important to industry and many fish caught were not recorded. Rounsefell and Kelez (1938) showed that for 1911 and 1913 the average yearly trap catch in Puget Sound was 6,450,000 fish. This was for the traps north of Deception Pass. All these fish, based on tagging studies, were considered Fraser origin. This same report gives average purse seine catch in Puget Sound of 457,000 pinks for 1911 and 1913. It is assumed that 75 percent of these seine fish were of Fraser River origin (343,000). The authors gave a catch of 6,177,178** Fraser pinks taken in the Fraser River from 1925 to

1933 during odd years. A minimum odd-year estimate for 1911 and 1913 at 1,235,000 pinks was assumed by using the average for 1925 to 1933 (these years affected by Hells Gate). The sum of these three area catch estimates is 8.028 million fish. For the 36 cycle years of returns from 1915 to 1985, the estimated total Fraser catch would have been 289,008,000 pink salmon. For reasons mentioned earlier, this would be considered a minimum estimate.

Inherent is the assumption that Fraser pink salmon production could have been maintained at that level. Based on run size averages from 1979 to 1985 of about 17 million fish and a 12 million catch each year, it appears that the 8.0 million average catch could have been maintained from 1915 to 1985. It is likely that an undisturbed Fraser pink salmon resource could probably have produced catch levels averaging at least 20 million fish. Thus, an estimated catch of at least 500 million pink salmon during the 1915-1985 period would not be unreasonable.

---

*   Using average value of sockeye in each country for 1985.

**   Catch from Strait of Georgia not included.

To determine the actual catch from 1915 (the first return year affected by Hells Gate) through 1985, heavy reliance was again placed on data provided by Rounsefell and Kelez. The trap catch north of Deception Pass from 1915 to 1933 was 19,695,991 pink salmon. The Fraser River catch from 1925 to 1933 was 6,177,178* pinks. The 1915 to 1923 Fraser River catch was estimated by using the ratio of the average yearly (odd year) trap catch north of Deception Pass from 1915 to 1923 compared with 1925 to 1933 applied to the 1925 to 1933 Fraser River area catch (6,177,178). The 1915 to 1923 average trap catch was 2,169,000 pinks compared with 1,770,000 pink salmon each odd year from 1925-1933. Thus, the estimated Fraser River area pink salmon catch from 1915-1923 is projected at 7,567,000 fish (all other factors being the same). The pink salmon catch by United States purse seines from 1915 to 1923 was 4,267,000 of which 75 percent were estimated to be of Fraser origin (3,200,000). Purse seines from 1925 to 1933 caught an estimated 23,958,000 additional pink salmon of Fraser destination. Few pinks were taken by U.S. gillnets and are not included in this analysis. Vernon (1958) calculated the total Fraser River pink salmon catch from 1935 to 1957 to be 71,012,000 fish. The 1959 to 1985 total Fraser pink salmon commercial catch in all areas was 93,506,000 fish. No estimates of Fraser River pink salmon catches in Johnstone Strait were made for the early years following 1913 up to 1957. The catch data/projections of Fraser pink salmon as given previously total 225,116,000 fish. Allowing for Johnstone Strait Fraser pink salmon catch in the years prior to 1959, perhaps about 240,000,000 Fraser pink salmon would have been taken from 1915 to 1985.

Thus, the estimated loss from 1915 to 1985 of production of Fraser River pink salmon caused by the Hells Gate situation is projected at about 260,000,000 fish which at 1985 prices ($1.75/fish)** would be $455 million. In total, the estimated combined loss of Fraser River sockeye and pink salmon would be $4.573 billion. The reduced production of other salmonid species caused by the slide would add to the total loss.

These estimates of loss as a result of human activities indicate the general magnitude of fish production lost to both countries. The financial impact in reduced catches to the industry has been substantial.

What would the loss have been if the two countries had not reached an agreement in 1937 and in 1956? It is likely that it would have been greater; however, it is possible (assuming reduced fishing exploitation) that certain sockeye stocks like the lower Adams River and Chilko River races would have been able to produce at abundance levels similar to the late 1930s and up to mid-1940s for many years. It would also have been possible for the Fraser pink run to produce runs similar to those of the 1930-1940s without production from above Hells Gate. It is likely however, that many other upriver sockeye races would have been destroyed.

Without the Commission, the two countries would probably not have been able to independently reduce fishing effort in their respective countries to both protect and enhance the stocks. Also, with lowered fish production, would the Fraser still be free of mainstem power development? The international agreements played a significant role in preventing power development on the Fraser River. A continuation of the low production of the 1930-1940s might have precipitated other uses of the Fraser watershed incompatible with protection and enhancement of the sockeye and pink salmon resource.

---

*   Catch from Strait of Georgia not included.

**   Using average landed value for each pink salmon in each country for 1985.

An Evaluation of IPSFC Contributions

STATUS AND IMPROVEMENT
OF THE STOCKS

Sockeye

P rior to the Hells Gate disaster in 1913, the average annual total Fraser River sockeye run from 1893 through 1916 was estimated to be about 11.4 million sockeye. The runs declined after 1913 to a four-year low of only 1.6 million from 1919 to 1922 and from 1917 to 1949 the annual average total Fraser run was only 3.3 million fish (Table 7). From 1950 to 1978, the Commission's efforts increased the annual runs to an average of 5.6 million (Appendix M)—an increase of 67.7 percent from the depleted period. Returns continued to increase, and from 1979 to 1982, average yearly run size was 7.8 million sockeye (Table 7). Further increases in Fraser sockeye runs took place from 1983 to 1986, with an average return of 10.2 million sockeye each year. In recent years, the cyclic average total run was approaching the pre-Hells Gate estimated returns (Table 7 and Figure 47). As noted earlier; however, considerable commercial catch waste, and under-recording of actual catch of salmon was reported in some of the big years prior to the Hells Gate blockages. Commission accomplishments must be evaluated in light of the possible under-estimates in total run size during the very early years.

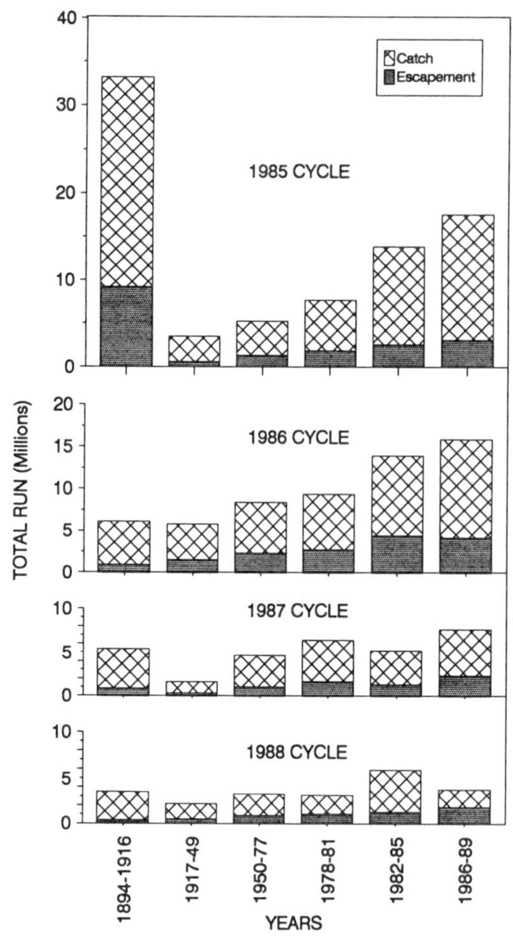

Figure 47. Comparison of average Fraser River sockeye salmon runs by cycle from 1894-1916 (prior to Hells Gate Disaster), 1917-1949 (prior to realization of IPSFC benefits) and 1950-1989 (when the benefits of IPSFC fishway construction and commercial fishery regulation were realized). From Rounsefell and Kelez 1938, Gilhousen 1990 MS and PSC.

. . . .

Table 7. Average total runs of Fraser River sockeye salmon, millions 1894-1989.[1]

1910-1982	Cycle	1911-1983	Cycle	1912-1984	Cycle	1913-1985	Cycle	Average
1894-1914	6.10	1895-1915	5.39	1896-1916	3.49	1897-1913	32.39	11.17
1918-1946	5.82	1919-1947	1.68	1920-1948	2.19	1917-1949	3.54	3.31
1950-1978	8.56	1951-1975	4.73	1952-1976	3.23	1953-1977	5.26	5.55
1982	13.99	1979	6.43	1980	3.13	1981	7.74	7.82
1986	15.90	1983	5.24	1984	5.92	1985	13.88	10.24
		1987	7.69	1988	3.76	1989	18.40	

[1] Catch data for 1894-1934 from Rounsefell and Kelez (1938) as adjusted by Gilhousen (1990 MS). Escapement estimates from 1894-1937 from Gilhousen (1990 MS). Catch estimates from 1935-1985 from IPSFC as well as escapements from 1938-1985. Data for 1986-1990 from Pacific Salmon Commission 1988a, 1988b, 1989 and 1990 as later adjusted.

If the eight-year period (1982-1989) is used as a measure of Commission achievements in sockeye run rehabilitation, the average run was 10.6 million sockeye compared to 3.3 million for the 1917-1949 period, indicating that IPSFC's management program increased the average total run size by a factor of about 3.2. The Commission's efforts have resulted in over two hundred percent increase in run size compared with estimated returns prior to its control (1917-1949). Total Fraser River sockeye runs from 1946-1989 are shown in Figure 48.

Commercial catches for the period 1894-1989 are given in Table 8. Prior to Hells Gate problems, the Fraser River sockeye commercial catch averaged 8.6 million sockeye each year from 1894 to 1916. During the period of depletion (1917-1949) the catch dropped to only 2.6 million each year. Under Commission management from 1950 to 1978, the average commercial catch of Fraser River sockeye increased each year to 4.1 million sockeye, an increase of 60 percent. Catches continued to increase from 1979 to 1985 even though major emphasis was placed on maximum escapements. The average annual catch during these seven years was 6.0 million sockeye, or an increase of 133 percent compared with the catches during the depletion period. From 1983 to 1986 (1986 under PSC regulations), the commercial catch averaged 7.8 million or 204 percent above the years of poor catches following the Hells Gate slide.

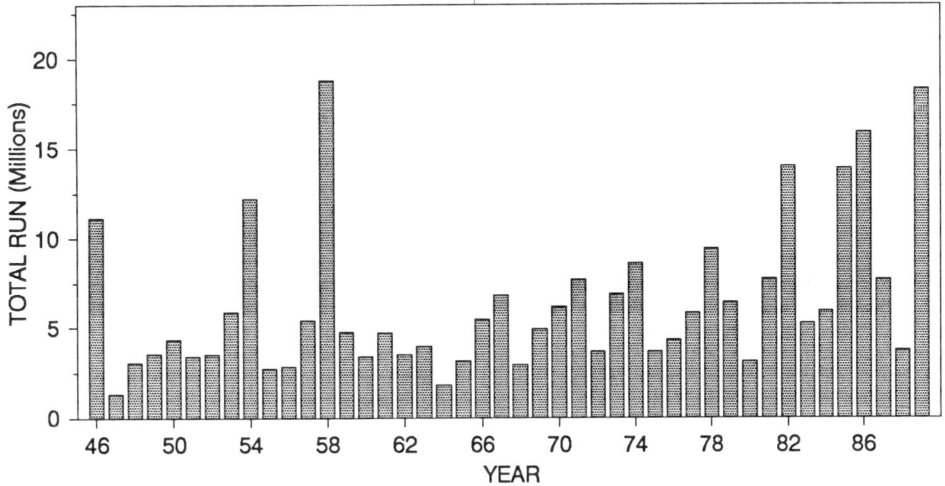

Figure 48. Total Fraser River sockeye salmon runs, 1946-1989. Data for 1946-1985 from IPSFC and 1986-1989 from Pacific Salmon Commission.

Table 8. Average annual commercial catches of Fraser River sockeye salmon, in millions 1894-1989.[1]

1910-1982	Cycle	1911-1983	Cycle	1912-1984	Cycle	1913-1985	Cycle	Average
1894-1914	5.14	1895-1915	4.57	1896-1916	3.12	1897-1913	23.99	8.56
1918-1946	4.26	1919-1947	1.30	1920-1948	1.74	1917-1949	2.93	2.57
1950-1978	6.18	1951-1975	3.64	1952-1976	2.36	1953-1977	3.91	4.10
1982	9.50	1979	4.73	1980	2.07	1981	5.84	5.54
1986	11.55	1983	3.89	1984	4.61	1985	11.24	7.82
		1987	5.20	1988	1.86	1989	14.35	

[1] 1894-1934 catch from Rounsefell and Kelez (1938) as adjusted by Gilhousen 1990 MS and PSC, 1935-1945 IPSFC Convention Waters only, 1946-1985 IPSFC, and 1986-1989 Pacific Salmon Commission.

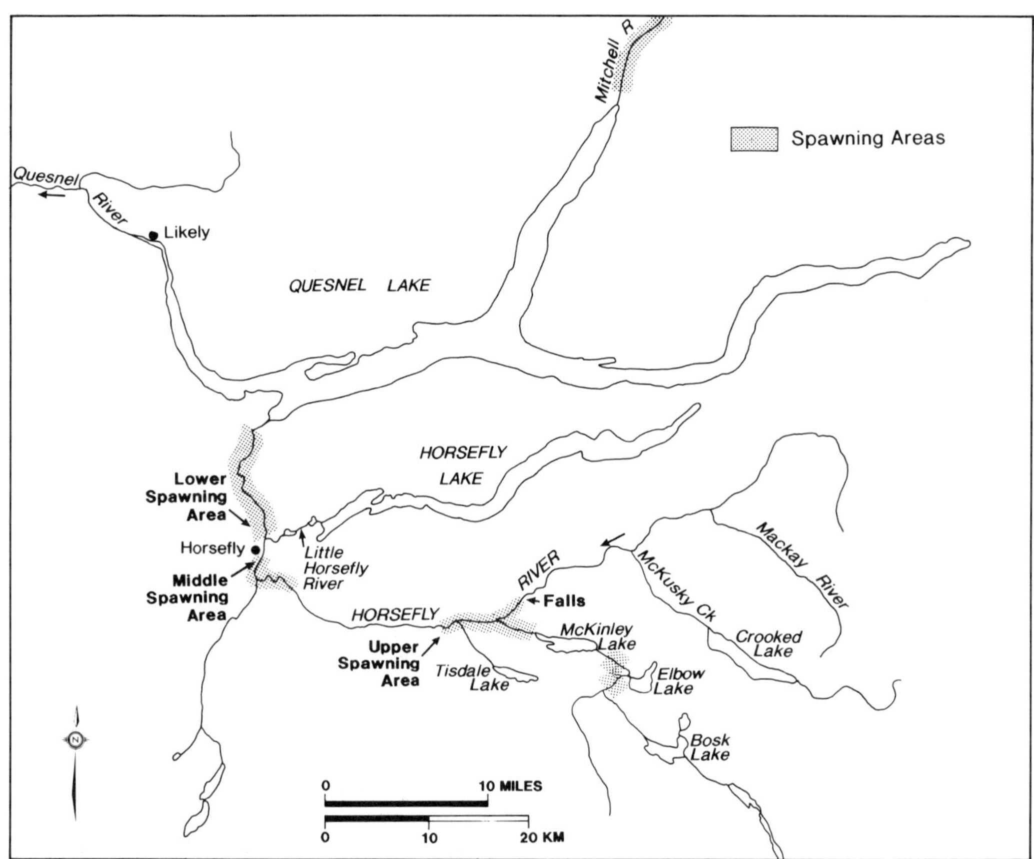

Figure 49. The Quesnel River watershed, showing location of sockeye salmon spawning grounds. McKinley Creek extends from Horsefly River to Elbow Lake.

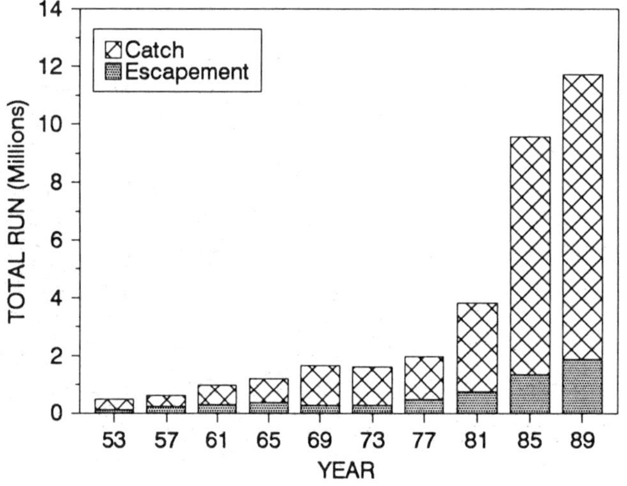

Figure 50. Quesnel area sockeye salmon dominant-cycle runs, 1953-1989 cycle. Data for 1989 from Pacific Salmon Commission.

Commercial catch value resulting from Commission regulatory actions in all fisheries from 1950 to 1985 was large. Canadian fishermen caught 100,158,000 (62.3 percent) Fraser sockeye with a landed value of $464.4 million (based on each year's value). United States fishermen took 60,630,000 (37.7 percent) sockeye valued at $254.9 million. In addition, during the same time period, 6,488,000 sockeye were taken in the Indian food fishery in the Fraser River system. These have an undetermined value.

One of the reasons for the Commission's success was the early concept and application of management by specific spawning race. Commission director, Dr. Loyd Royal, was at the forefront of pioneering this sound management strategy in the early 1950s. The most impressive increase has taken place for the Horsefly River/Mitchell River populations in the Quesnel Lake system (Figure 49). From a residual population of less than 1,000 spawners in 1941—from a total run of about 5,000 fish and a race on the edge of extermination—total production by 1985 had increased to a remarkable 9.6 million sockeye (Figure 50). It is quite possible that if the Hells Gate fishway had not been in place for the first year of operation in 1945 the Horsefly run might have been lost. The significance of this to both countries can be appreciated more fully when it is realized that the landed value of the commercial catch of Quesnel system sockeye in 1985 alone was about $68 million. The total 1985 return to the Quesnel area was the largest since 1913. In 1985 there was an investment of over one million spawners allowed to escape the commercial fisheries to provide larger returns in future years. In 1989, this run increased further to 11.7 million fish (Pacific Salmon Commission 1990) and the commercial catch in that year (1985 prices) was valued at $87 million. It is possible that the run now would be at an even higher level except for the serious problems with pre-spawning mortalities associated with early-timed runs in earlier years (1950-1970s).

Peak spawning times in the 1980s reverted to later dates (mid-September, about two weeks later) and this was beneficial. Redistribution of escapement (through commercial fishery protection of the latter parts of the run in several years) into the lower spawning grounds resulted in large run increases in the late 1970s and 1980s. Increased escapements in the 1950s and 1960s above those obtained would have been at a great cost to industry and of little benefit because of high pre-spawning mortalities.

Production has also increased in the Quesnel system on the subdominant cycle (1986). This will be beneficial in increasing the summer-run potential for catch on this cycle. Escapement had increased to 30,000 fish in 1982 from only 1,350 fish in 1970. In 1986 the Quesnel area return totalled 622,000 sockeye (Pacific Salmon Commission 1988a). The history of the dominant-subdominant run size relationship prior to the Hells Gate incident is not available. However, the 1986 return is promising.

Another success story is the tremendous increase in the Mitchell River spawning population at the far end of Quesnel Lake. In 1941, the total run was only 200 fish and a meager escapement of 41 fish arrived at the river. By 1981 escapement had increased to 66,000 fish from an estimated total run of about 260,000 fish. A record return was experienced in 1985 with a total run of about 1.5 million sockeye, of which 205,000 reached the spawning areas.

In 1973 an important milestone was reached when 1.4 million early Stuart sockeye returned, and similar number again in 1977. These were the largest returns on record, dating back to 1900, and very likely the largest ever to the system. These sockeye also benefited from fishways and selective protection through fishery closures. Its decline and rehabilitation process closely parallels that of the Horsefly River population. Early Stuart escapement in 1941 was only 6,500 spawners. Lake-rearing potential for early Stuart sockeye is great, but

August 11, 1945. The Horsefly River near the McKinley Creek junction contains good quality spawning gravel that supports a very large spawning population. From an escapement of less than 1,000 sockeye in 1941, the escapement increased to more than 100,000 spawners in 1953 and to over 1,000,000 in 1985.

August 23, 1961. Future Director John Roos, in charge of the Horsefly River sockeye enumeration in 1961, examining dead unspawned sockeye. An unusually high pre-spawning mortality (62%) occurred during this year as a result of the combined stress caused by early migration, high river temperature and columnaris disease.

A winter scene on Horsefly River in the fall of 1950. This river has a heavy discharge of frazil ice formed at a falls at the upper end of the sockeye spawning area and at downstream riffles. Soon after this photograph was taken, the river froze over completely. By cutting through the eight-inch-thick surface ice sheet at Black Creek, a short distance below the falls, a four-foot depth of frazil ice was found, extending to within a few inches of the stream bottom.

Severe winter weather is a major factor limiting the egg-fry survival of sockeye and pink salmon in the Fraser River watershed. Spawned gravel beds are exposed due to flow reduction and then the eggs and/or alevins are killed by freezing. Frazil ice dams sometimes divert flowing water away from spawned areas. Winter surveys showed that frazil ice diversions frequently occur in small northern streams such as Nadina River and the many streams utilized by the early Stuart sockeye population. Cold weather and deep snow also made travelling and working conditions dangerous for staff engaged in winter surveys, fishway maintenance and operation of spawning channels and other facilities.

spawning grounds are limited, with the exception of the Driftwood River. Under present circumstances, it is not likely that the early Stuart runs will reach the full potential available from Takla and Trembleur Lakes because of the spawning ground limitation. With assistance through spawning channels, this group of sockeye could be increased significantly. Large populations of kokanee (sockeye salmon that remain in freshwater) are also found in the system.

The late Stuart sockeye race is another population with a similar recent history to the Horsefly and early Stuart races. By 1941 escapement was reduced to 10,000 spawners. Migrating at a similar time and receiving the same protection through the fisheries as Horsefly River fish, the late Stuart run increased to 1.6 million fish by 1973. This was the largest return since 1913. The late Stuart potential is also limited by available spawning area and Stuart Lake has a much greater capacity to rear fry to the smolt stage than is now being realized. The subdominant runs to the Stuart System have greater potential for the future.

The resurgence of the lower Shuswap River (Figure 51) population is a success story

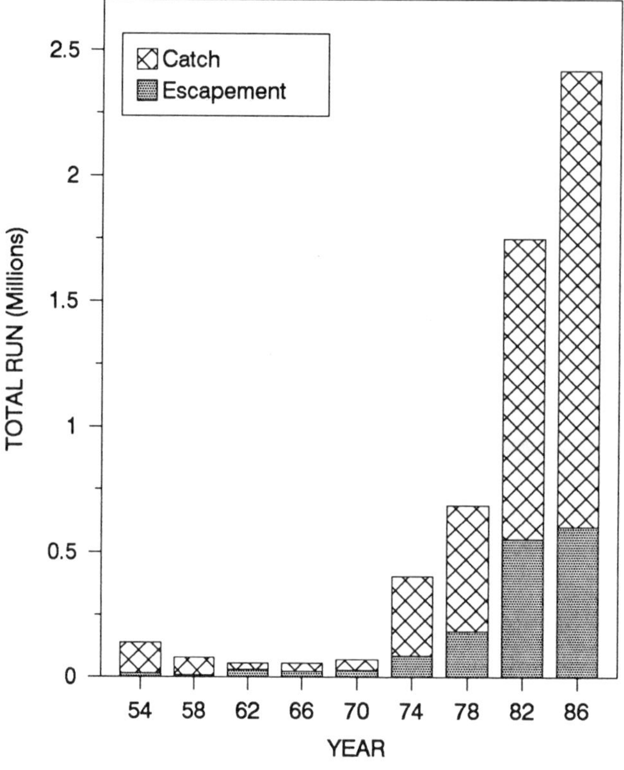

Figure 51. Lower Shuswap River sockeye salmon dominant-cycle runs, 1954-1986. Data for 1986 from Pacific Salmon Commission.

too. This race migrates at a similar time as the lower Adams River population and juveniles of both races merge in Shuswap Lake. In 1946, only 1,200 sockeye were estimated to spawn in the river. A remarkable increase in escapement has taken place since the very small escapement in 1946; 514,000 fish spawned there in 1982.* As shown in Figure 51, the total return in 1986, produced by that large parent escapement, was estimated at 2.4 million fish *(ibid.)*. This was probably the largest run to this river since 1913. The full potential of the lower Shuswap population is yet to be realized.

The middle Shuswap River (Figure 32) empties into Mable Lake, the outlet of which flows into the lower Shuswap River. The middle Shuswap River does not contain spawning area comparable to the lower Shuswap. It is restricted by Shuswap Falls and a dam. It is reported that from time to time the dam flushed out large quantities of silt (R. Stewart, personal communication); however, there is sufficient area below the falls for a sizeable spawning population. From only 50 spawners in 1950, the population increased substantially to a 40,000 escapement in 1982. Total return in 1986 was estimated to be 325,000 sockeye *(ibid.)*.

Other major races less affected by Hells Gate blockages were the lower Adams River and Chilko populations. With regard to the lower Adams River race, the 1942 and 1946 returns were substantial and estimated to have been between eight and nine million fish in each of those years. By 1954, the run totalled almost 10 million fish. In 1958, the return was a phenomenal 15 million. (Figure 52). This was the largest run to the Fraser system since 1913. The Adams race declined substantially following the record 1958 return and has only re-

---

* Some straying of Adams River sockeye (mentioned earlier in Chapter 7) might have taken place in 1982 as 32,000 late-run sockeye from either the Adams or lower Shuswap River races were observed in nearby Eagle River. The indigenous run to Eagle River has always been an early run but late-run "slop-over" stocks on the dominant years have on occasion been recorded. About 8,000 late-run sockeye were there in 1978.

October 7, 1946. Adams River sockeye are severely crowded during their upriver migration, especially in the Fraser Canyon when they seek refuge from the high velocities and severe turbulence.

The crowding of Adams River sockeye extends to the spawning ground, where the density of the spawners plus the waiting schools of unspawned fish is higher than in any other Fraser River population.

The Adams River spawning ground is so crowded that large schools of later arriving spawners wait their turn while ripening.

cently regained some of its former stature. In 1982 the total return of lower Adams River sockeye was eight million fish. It would be reasonable to expect consistent production of at least 8 to 10 million fish on the dominant Adams cycle, with an occasional return (similar to that of 1958) of around 15 million fish.

The Chilko sockeye population was fairly abundant when the Commission was investigating the Hells Gate blocks in 1940 and 1941. In fact, estimated escapements in both those years was about 300,000 fish, an amount approaching the optimum number of about 450,000 for Chilko River. The estimated total Chilko run in 1941 was probably around three million fish, a significant return for any Chilko run. This run has been viewed as the backbone of the summer-run fishery. Substantial production occurs on two of the four cycle-years. The highlight of Commission management of the Chilko River-Lake populations took place in 1984 when the total return was a record 3.8 million sockeye. The 1984 return was approximately 52 percent larger than the previous maximum run (under Commission management) of 2.5 million in 1960 (Figure 53). The production of the 1984 return was similar to that experienced in 1958 for the lower Adams River race. It is wonderful when it happens, but it is not the norm. Both the dominant and subdominant (1984-1983, respectively) stocks at Chilko are healthy and can be expected to continue to be major contributors to the fisheries. As mentioned earlier, the population spawning at the south end of Chilko Lake has also increased significantly.

The Birkenhead River population migrates through Harrison River, Harrison and Lillooet lakes and spawns in the Lillooet District. This migration route branches off from the Fraser River below Hells Gate and the race was not directly affected by the Hells Gate blockages. Production has increased significantly in the 1970s and 1980s, possibly because of the presence of large numbers of fry produced in the Weaver Creek channel rearing in

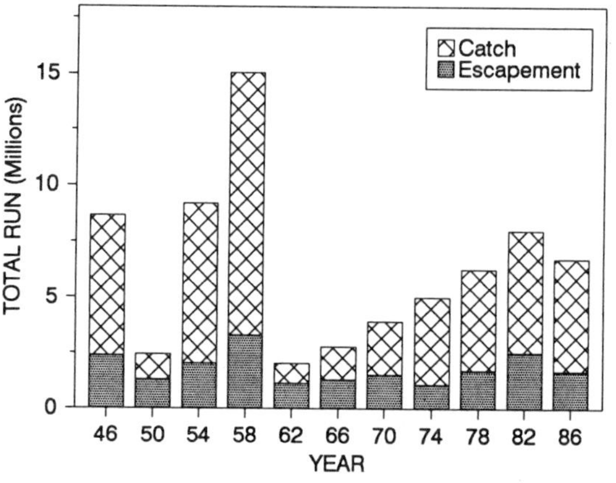

Figure 52. Lower Adams River sockeye salmon dominant-cycle runs, 1946-1986. Data for 1986 from Pacific Salmon Commission.

Harrison Lake along with Birkenhead River fry. There may be a buffering effect (from predators) beneficial to both races. The Birkenhead River run produced well on the 1982 cycle with a 1986 record run of about 1.3 million fish *(ibid.)*.

Weaver Creek sockeye escapements from 1960 to 1963 declined to 10,000 fish each year, only one-half that of the previous 16 years. Total annual production was less than 50,000 sockeye each year. To enhance this race, the Commission built its first sockeye channel in 1965 and the increase in production has been phenomenal. In 1982, a record 960,000 sockeye came back (Figure 54). In 1986 there was a return of 950,000 sockeye, providing an even greater commercial catch value *(ibid.)*. Juveniles from both Weaver Creek and Birkenhead River races rear in Harrison Lake. Both races appear to be dominant on the 1982 cycle-year.

Significant improvement has taken place in the production of early-run sockeye to the South Thompson District. From 1950 to 1953, the only race of any importance was the Seymour River population and the average annual escapement during that period was only 12,300 fish. As discussed earlier, the upper Adams River and Momich/Cayenne races have been firmly established as self-perpetuating populations. The upper Adams race has great potential and has increased to about 7,000

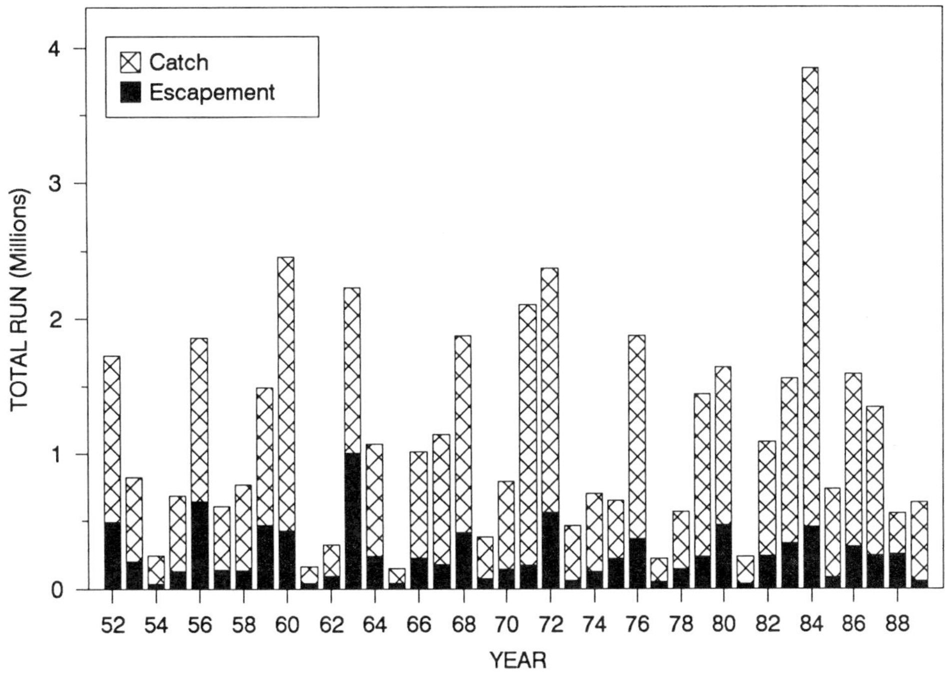

Figure 53. Chilko sockeye salmon runs, 1952-1989. Data from 1986-1989 from Pacific Salmon Commission.

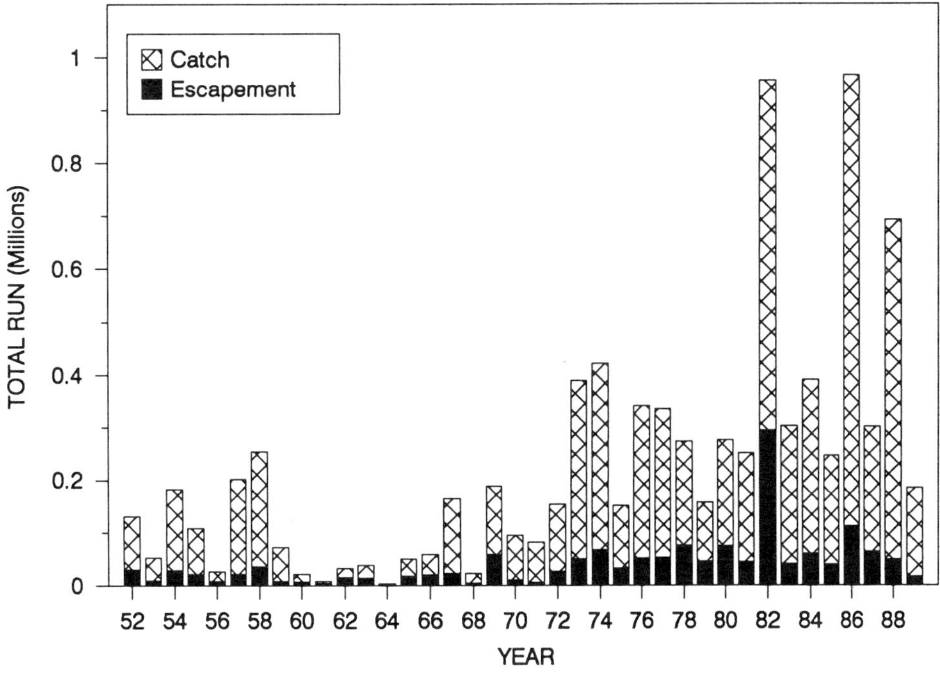

Figure 54. Weaver Creek sockeye salmon runs, 1952-1989. Data for 1986-1989 from Pacific Salmon Commission.

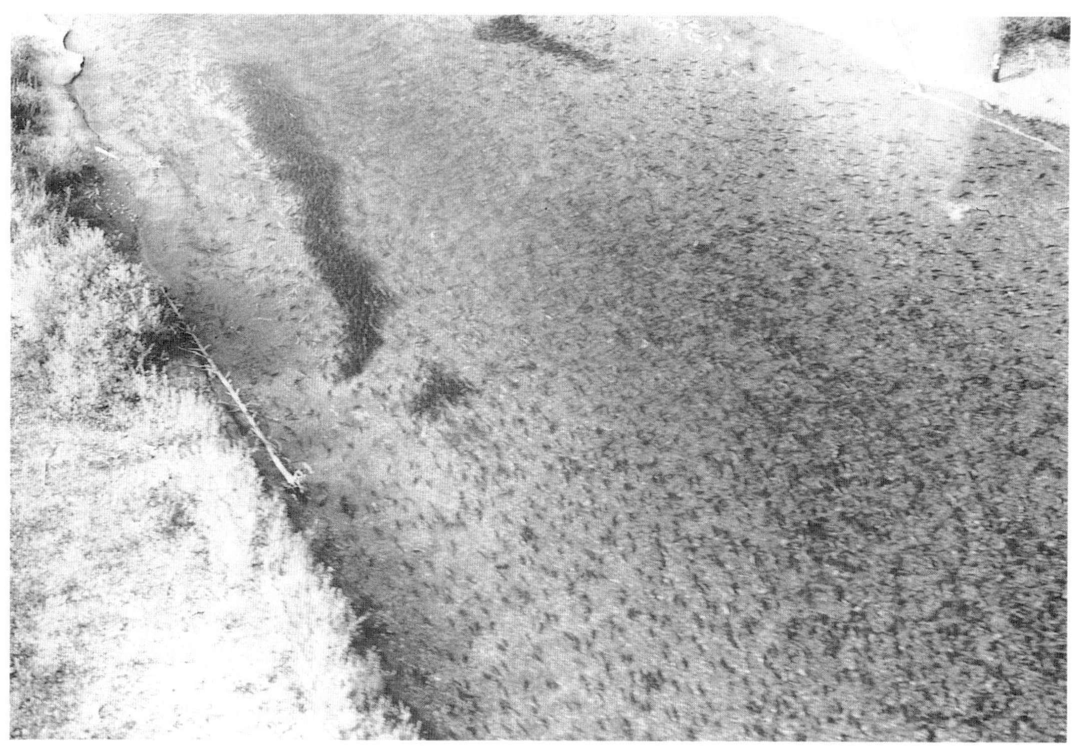

In dominant years, such as when this photo was taken in 1982, the Adams River spawning area is fully occupied by sockeye and dense schools can be seen throughout the river as well as in Shuswap and Little Shuswap Lakes.

October, 1982. Thousands of visitors are fascinated by the sockeye spawning activity in Adams River. Much of the river bank has been reserved as a park, called the Roderick Haig-Brown Conservation Area.

October 22, 1950. One little-recognized benefit of large salmon runs like the Adams River sockeye population is that they support major trout populations that provide a valuable recreational resource. The trout feed on sockeye eggs. Probably the main benefit to trout populations is the food source provided by newly emerged fry, rearing fingerlings and migrating smolts. Part of the Commission's work included studying the effect of trout predation on sockeye productivity.

spawners in 1988 from 560 in 1980 (PSC 1989). Other races in the Shuswap Lake system have increased substantially during the past 40 years with average escapements to all of these early races reaching 35,900, or about a threefold increase from 1982 to 1985.

Smaller races, such as those that spawn in Scotch, Fennell and Gates creeks, have also shown increases. As referenced earlier, restoration of the upper Adams River race bodes well for the future. It is the cumulative effects of enhancement and restoration, not only to the large races such as Stuart, Horsefly and lower Shuswap River, but also the added contributions of the Mitchell River, middle Shuswap River, Birkenhead River, and Weaver Creek populations and the minor races that in aggregate improve abundance of the whole Fraser River sockeye resource.

There were, however, some disappointments. The early Nadina River sockeye run had great potential for expanding sockeye production and utilization of the large rearing poten-

tial of Francois Lake. By 1957, escapement in the then-dominant cycle, was about 30,000 fish. Due to environmental problems, 1985 escapement on this cycle was only 18 fish. In 1982 no sockeye were present.

Total sockeye escapement reaching the spawning grounds has increased to 2,018,000 average annual escapement from 1982 to 1985 (Appendix N), (44 percent increase) from an annual average of 1,400,000 during 1946 to 1949.

Pink Salmon

Hells Gate blockage was the chief reason for the decline of sockeye and the diminished Fraser pink salmon runs. For sockeye, many of the upriver races were almost exterminated when the Commission became involved. Some races were even gone. For pink salmon, however, even though the conditions at Hells Gate prevented for the most part early-run fish from reaching upriver spawning grounds, the large early-run spawning area available in the main Fraser River below Hope was a buffer that undoubtedly produced large numbers of early-run pink salmon after 1913.

More recent pink salmon catch information relating to the Fraser pink runs are available starting in 1935 (Vernon 1958). The annual average catch of Fraser pinks in Convention Waters for the years 1935-1945 was 4.03 million fish.

There are no accurate estimates of the Fraser pink salmon run sizes prior to the Hells Gate blockage. However, from an analysis of the index of abundance data from Table 52 of Rounsefell and Kelez (1938) and the data on total catch of Fraser pinks from Vernon (1958) for the years 1935-1955, one could calculate estimates of the total Fraser run (using annual catch estimates of 500,000 fish in Johnstone Strait and 80% fishery exploitation) for each of the above odd-numbered years at about 8.2 million fish. It is assumed that large numbers of pink salmon had not spawned above Hells Gate from 1945-1953. The trap-catch index of Fraser pinks from 1927 through 1933 fell to less than one-fourth (24.3 percent) the 1907-1913 level. On this basis it is possible that the Fraser pink runs approached 34 million fish prior to 1915. Previously, estimates by IPSFC concluded that total runs for the Fraser of about 29 million could be realized from escapements of almost seven million (IPSFC 1972a). Thus, the Commission had a major task to restore the abundance of pink stocks to estimated potential run sizes.

The abundance of Fraser River pink salmon from the mid-1940s to mid-1950s increased despite the absence of major upriver production. Convention Waters catch of Fraser pink salmon from 1947 to 1955 immediately before the Commission assumed control in 1957, averaged 8.46 million pink salmon, ranging from 7 to 11 million (Vernon 1958). Vernon's analysis did not include catches of Fraser River pink salmon made in Johnstone Strait and these, therefore, are minimum estimates. It is also probable that with inclusion of escapements not enumerated during this time—total Fraser run size averaged over 11 million fish in the years prior to 1957 (1947-1955). This is a fair return considering the absence of spawning and production from above Hells Gate.

A few years after 1957 the pink returns declined. Total runs averaged only 3.2 million from 1961 to 1965. It is believed that much of this was due to poor environmental conditions. Whatever the reasons, the Commission was off to a less than auspicious start in management of the pink salmon stocks. It was not a comfortable situation for the Commission because both Parties and industries had previously expressed strong support for the Commission and wanted IPSFC to take over Fraser pink salmon along with sockeye.

For brood years 1957-1975, escapements were relatively consistent, averaging 1.58 million (1.1 to 2.2 million range) annually (Appendix N). The resulting returns from these escapements in 1959-1977 (Appendix M) averaged 6.27 million fish (range of 1.90 million to 12.97 million). In only one year out of those ten did a run in excess of 10 million occur, and that happened for brood year 1965 when freshwater egg-to-fry survival was one of the highest ever (18.4 percent, Appendix O). In contrast, when the Commission raised escapement goals, stocks responded in a positive manner. The average escapement for brood years 1977-1983 was 3.77 million, an increase of 139 percent compared with the period 1957-1975. The odd-year production from these increased escapements averaged 16.82 million fish from 1979-1985 (range - 14.40 to 18.86 million) (Figure 55). The low return of 7.1 million pinks in 1987 from the large brood year escapement was likely due to poor environmental conditions (Pacific Salmon Commission 1988b). It is also significant that return-per-spawner of 4.46 for the most recent period (1977-1983) was higher than the 3.92 fish-per-spawner for the 1959-1975 period, when escapements were smaller. A spawning ground limitation had not been reached.

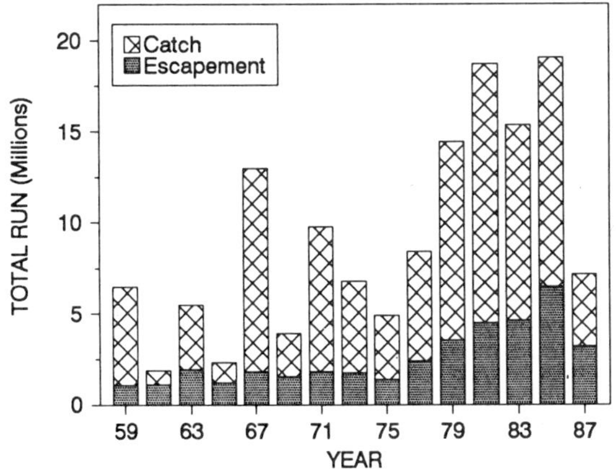

Figure 55. Total Fraser River pink salmon runs, 1959-1987. 1987 data from Pacific Salmon Commission.

Two important points are deduced from these data. First, it appears that total returns in the 1959-1975 era were suppressed due to Commission policy of managing for escapements at only the one- to two-million level. Secondly, the data for recent-year returns, from 1979-1985, suggest that the Fraser River system is capable of sustaining higher production and larger escapements if the spawners are properly distributed. While the Commission was gratified by the increased returns during the last four cycles (1979-1985), the average return of 16.82 million fish was only about one-half of the estimated runs prior to Hells Gate problems. Nevertheless, this was a 53 percent increase from the estimated 11 million run level in the 1947-1955 period. When compared to the low run status from 1961 to 1965 (annual average of 3.2 million), the recent runs have increased significantly. One of the beneficial results following the failure of funding for the $14 million Development Program was that the Commission was, in a sense, forced to increase escapements quickly. The stocks responded well.

Total catches of Fraser River pink salmon during the four cycle years from 1979-1985 averaged 11.9 million (range from 10.7 to 14.2 million). The catch in the 1935-1955 period, before Commission control, averaged 6.0 million fish (without Johnstone Strait catch) and ranged from 0.9 to 11.2 million fish each year.

The Commission was able to increase the commercial catch significantly from the low levels experienced from 1961-1965 of 1.8 million fish each year to the recent year average of 11.9 million pinks—a 6.6-fold increase. Nevertheless, there is room for improvement and catches well in excess of 20 million should be sought and attained in future years.

In terms of total catch resulting from Commission management policy from 1959-1985, Canadian fishermen caught 63,789,382 (68.2 percent) Fraser pink salmon in all fisheries while the United States catch was 29,716,829 (31.8 percent) pink salmon. The landed value (based on year of landing) to Canadian fishermen was $115.1 million and for the United States $49.5 million. Individual pink salmon stocks that have shown improvement are the Seton Creek, Thompson River and main Fraser River populations. Little or no improvement has occurred in the late-run races of the Harrison and Chilliwack-Vedder rivers. Production data (total run, catch and escapement) for separate races of pink salmon have not been compiled as they have for sockeye salmon; therefore, escapement trends are the best data available.

The pink salmon spawning populations in the Seton Creek/Portage Creek area increased to an average of 467,000 for the years 1981-1985 from an average of about 46,000 fish each year from 1957-1961. The peak escapement for the area was reached in 1979, when 713,000 pink salmon spawned in the various creeks and rivers. The increase of the Seton Creek population has been at least tenfold, due partly to the more restrictive fishing regulations that were in place during the 1957-1961 period. Also, large numbers of Seton Creek area pinks did not reach their native spawning sites in the 1981-1985 period due to the migration problems. The two Seton Creek spawning channels also played a major role in increasing production.

Escapements to the Thompson River area from 1957-1961 averaged 142,000 fish annually, compared to an average of 624,000 pink salmon for the years 1981-1985. A comparison of the rate of increase of the two upriver populations indicate that Seton Creek stocks increased more rapidly (10-fold) than the 4.4-fold increase at Thompson River. The increase in the escapement of early-run pink salmon above Hells Gate starting in 1945 was one of the major accomplishments of the Commission. Although management of the fisheries and the Seton Creek spawning channels played a role in this rehabilitation, the primary initial contributing factor was the fishways. Escapement of 1.2 million to the Thompson River area in 1981 was a modern record; however, only 193,000 reached the area in 1985. Disappointing is the lack of pink salmon in the Nicola River tributary to the Thompson River, where there were large numbers of pinks reported spawning prior to 1913. From 1957 to 1983, the maximum yearly number of pink salmon spawning in the Nicola was only 7,000 fish. In 1985 only 265 fish were present. Human impacts such as irrigation have affected river productivity.

The main Fraser River spawning area has always contained the largest number and the highest percentage of each year's escapement. For example, from 1957-1961 an average of 850,000 pink salmon used the area each year, which accounted for 55 percent of the total escapement. A very sharp increase in main Fraser River spawners occurred in the 1981-1985 period to average 3.6 million pinks and peaked at 5.3 million fish in 1985. During these last three years, main Fraser River spawners constituted 69 percent of the total escapement. It is not known how much of this increase was attributable to environmentally-related migration problems, or to an absolute increase in the productivity of the fish using those spawning grounds—the former reason is the more likely.

For the late runs, the Harrison River escapements averaged 294,000 from 1957-1961. This race has not improved; the escapement level from 1981 to 1985 was virtually the same at 300,000 each year. The status of the Chilliwack/Vedder rivers pink salmon spawners is worse, as escapements have declined from 164,000 each year in the early period to only 88,000 during the past three cycles (1981-1985). The relative strength of late-run pink salmon stocks is now at a level lower than when the Commission assumed control 30 years ago. Environmental problems such as flooding, gravel removal and dredging have all created fish-production problems.

In its $14 million development program, the Commission had proposed spawning channel construction for the Harrison, Chilliwack and Chehalis rivers. Land clearing for the Chilliwack River channel had been completed. The channel was large; it would have been 6.3 miles in length and would have accommodated 73,500 spawners. The channel on the Harrison River would have had spawning area for 117,350 pink salmon; the equivalent (in terms of fry produced) of 587,000 wild spawners. Based on the results of the Seton Creek spawning channels, it is clear that if these two channels had been built, the status and potential of late-run pink salmon would have been greatly enhanced. The need in these areas is very apparent.

FUTURE POTENTIAL AND COMMISSION ACHIEVEMENTS

Potential for further increases in total sockeye production are great, especially in the Stuart, Francois, Shuswap and Adams Lake systems. The Seton-Anderson system also has a large potential. Increased spawning ground area in the Quesnel system could be provided by fishways to utilize the large potential spawning area above the Horsefly River falls. When production of sockeye juveniles is brought on line in these areas to utilize the highly productive and abundant lake-rearing areas, a very substantial increase in adult sockeye returns can be expected. With regard to pink salmon, the late runs need assistance, as was proposed by the Commission. The primary factor presently limiting adult production of both sockeye and pink salmon appears to be the numbers of fry produced.

There are other unproven concepts (at least for the Fraser River system) that have been proposed for increasing Fraser sockeye production. One is lake enrichment through introduction of nutrients to increase phytoplankton and zooplankton for young sockeye. Lake-fertilization efforts have had some success elsewhere in British Columbia and Alaska, but must be administered carefully. Exhaustive investigations would be desirable before this procedure is applied and then initiated first only (if found suitable) to some of the smaller, less-productive lakes. Another concept mentioned from time to time was changing cyclic abundance. Changing the long-term, well-established numerical cyclic characteristics and interrelationships of the various races—especially with the major races—is an unproven concept. There is reason to believe that the current size of the subdominant runs of races such as the lower Adams, lower Shuswap, Stuart and Horsefly can and should be increased. However, any attempt to create dominant-size runs for each of the four years within a quadrennium presently cannot be justified based on current scientific evidence.

The total landed value of the commercial catch of Fraser River sockeye and pink salmon in all fisheries (using the year of landings value) during Commission management (sockeye 1950-1985, pinks 1959-1985) was $579.5 million for Canada and $304.4 million for the United States, a total of $884 million. Total contributions for each country from 1937-1985 were $21,390,736 by Canada and $21,328,362 by the United States. Income derived from the market value of the catch and supporting industries must be added to the total value of the resource. This value would be offset by costs incurred by Canada associated with protection of the resource, enforce-

ment of regulations, and many other administrative requirements involving the two governments and Washington State. In the end, both countries received excellent returns on their investments.

What would these values have been without the Commission and the expenditures? Undoubtedly much less, but it is impossible to know what might have happened. However, an estimate can be obtained representing the contribution of the Commission's efforts from 1950-1985. The year 1950 was selected as the first year of combined sockeye returns from the Commission's 1946 regulations and the fishways. For pink salmon, the evaluation period is from 1959-1985, since the Commission was first given regulatory control of the fishery in 1957.

The average annual catch of Fraser sockeye salmon in Convention Waters from 1917-1949 was 2,567,000 fish. If the Commission had not existed from 1946-1985, a catch of 92,412,000 sockeye could have been expected during the 36 years (1950-1985) assuming the factors prevailing from 1917-1949 continued. The actual commercial catch from 1950-1985 was 160,787,000 sockeye. Therefore, because of the Commission's efforts, 68,375,000 additional sockeye were available with a landed value of $586,658,000 (at 1985 prices).

For pink salmon a similar analysis can be made. The average odd-year Fraser River pink salmon catch from 1935-1955 calculated from Vernon's analysis indicates a catch of 6.04 million fish (not including Johnstone Strait). Assuming a catch of about 6.5 million fish each odd-year, the total expected Fraser River pink salmon catch from 1959-1985 would have been 91,000,000 fish. During the period of Commission tenure affecting the pink salmon returns (1959-1985), the total number of Fraser River pink salmon taken in all fisheries was 93,506,000 fish (Appendix H_3). Thus, the Commission's efforts for pink salmon produced a catch valued (1985 prices) at about $4.386 million more than what would have been received had the

Commission not been in place. In total, for both species, the Commission's work through 1985 increased the value (1985 prices) of the sockeye and pink salmon fisheries by $591,044,000. This can be compared with the total cost of the Commission's programs from 1937-1985 of $42.7 million ($110.2 million at 1985 value). This evaluation of benefits from Commission programs does not include commercial catches taken from 1986 to 1989 following the dissolution of the IPSFC in 1985, with benefits attributable to Commission management in brood years 1982-1985. For Fraser sockeye, a total commercial catch of 32,951,000 was taken from 1986 to 1989. In 1987, 3,666,000 Fraser pink salmon were landed. These fisheries were under the control of the Pacific Salmon Commission and catch data was obtained from their reports as amended (Doug Stelter, personal communication) (Pacific Salmon Commission 1988a, 1988b, 1989 and 1990). The additional net value (1985 prices) of those landings would be $194,363,000.

In total, the Commission's management efforts affecting returns from 1950 through 1989 increased the landed value to both countries by $785 million from the long-term status quo existent before Commission involvement. There will be additional value in subsequent years due, in part, to Commission rehabilitation efforts from 1946 through 1985. In addition, because a higher proportion of the total runs under IPSFC control were allowed to escape the fisheries (thus reducing the catch), the actual benefits accrued under the Commission's management were minimized. It is likely that the depleted status of the stocks prior to 1950 would have shown substantial further decline during the 1950-1985 period if the Commission had not taken steps to protect, manage and enhance the resource. Therefore, the benefits presented earlier are likely minimum estimates.

To a great extent much of the increase in returns has taken place in the recent years from the late 1970s to the 1980s. However, the foundation for these increases was established years earlier in the 1940s and 1950s.

Fraser River sockeye runs from 1982 to 1985 averaged 9.8 million fish. The increased escapements obtained by the IPSFC during those four years provided the basis for increased potential for the following four years and the benefits were realized. Even though the average yearly returns of 9.8 million fish (1982-1985) were approaching historical recorded levels, the runs during the next four years (1986-1989) continued increasing to average 11.4 million sockeye (Table 9).

Table 9. Average annual Fraser River sockeye salmon total runs, 1978-1989.

Years	Average (million)	Percent Increase from Previous 4 Year Period
1978-81	6.684	58.9
1982-85	9.758	46.0
1986-89	11.441	17.2

During the twelve years from 1978 through 1989, the average annual run during the four year periods covering cycle year 1979-1981 to 1986-1989 increased to 11.4 million fish from 6.7 million sockeye in the respective time periods. The increase in average run of 4.8 million fish was a 71 percent increase. Therefore, the Commission's last four years of management produced sockeye runs equal to those on record during the immediate pre-Hells Gate slide period (1893-1913) of 11.4 million fish.

The increased production which took place from 1982-1989 was not the result of fortuitous survival rates but rather it is the reflection of absolute increase in stock abundance and represents real future potential. Marine sea survival rates for Chilko sockeye averaged 10.2 percent from 1982-1989 compared with the long-term (1953-1989) average of 9.5 percent (Appendix K). In some years, marine survival reached 22 percent, the lowest being one percent.

While the status of the Fraser sockeye run is now excellent, lake rearing potential for further increases is available and for some races increased production can be realized rather quickly. For example, the Quesnel District populations, particularly Horsefly River sockeye, could be expected to produce returns on the dominant cycle of 15 to 20 million fish. With the inclusion of the potential of several other populations such as lower Shuswap River, upper Adams River, Nadina River and Stuart stocks, it should be possible to increase the average annual sockeye run to the Fraser River to a level considerably above the 1986 to 1989 level of 11.4 million fish annually, perhaps to 20 million fish average annual run.

Similarly, for pink salmon, the potential exists to provide total Fraser River pink salmon runs far in excess of those from 1979 to 1985 which averaged 17 million fish each return year. This will require escapements of at least six to seven million fish, and returns of more than 30 million pink salmon could be expected. Similar to sockeye, the large runs of pink salmon from 1979-1985 were not the result of unusually high survival rates. Both freshwater and marine survival were near the long-term average (Appendix O).

With more and better knowledge about salmon runs and their timing, routes and behavior, and in technology to develop and apply that information to resource management, the salmon resources of the Fraser River can continue to be managed efficiently and increase. However, it will be years before improved production can make up for losses related to the earlier environmental problems. It is hoped this will serve as an example to land and water-use managers, politicians and other resource related decision-makers of the long-term impacts of change in the environment.

The Commission's management program for Fraser River sockeye and pink salmon also contributed to the well-being of other Fraser River salmonids. The extensive early-season closures from 1946 to 1949 increased the escapements of several chinook salmon stocks. These no doubt also benefited from the fishways. Coho and steelhead (*Oncorhynchus mykiss*) migrating upriver in September and October would have been assisted for almost 40 years in their migration. These side benefits have gone almost unrecognized since the focus has been on sockeye and pink salmon. In addition, undetermined benefit has been provided the various Indian Bands of the Fraser system from increased food fishery catches.

The Commission's rehabilitation and scientific management program was very successful and this was recognized. Thor C. Tollefson, former Chairman of the Commission who also served 18 years as a U.S. Congressman and was the Director of the Washington State Department of Fisheries, stated at the Commission's 1973 Annual Meeting about international fishery commissions:

> … in my opinion none has been so unique, progressive and productive as has the International Pacific Salmon Fisheries Commission. It will continue to be productive so long as it exists.

Thor served many years on the Merchant Marine and Fisheries Committee of the U.S. House of Representatives and attended many fisheries meetings here and abroad as a participant, advisor and observer.

A recent statement, removed from Fraser River salmon management, about the Commission's achievements is found in a University of Oregon Ocean Law Memo:

> It is widely perceived that the ailing Fraser River Treaty has been phenomenally successful in revitalizing the Fraser River sockeye salmon runs. In fact, one will search the west coast of North America in vain for a comparable success story in the physical

rehabilitation of a moribund fishery. The Commission itself and its professional staff have been lauded by interests on both sides as competent and impartial, producing studies and data that are at once objective and easily obtainable. The Fraser River Treaty has worked, and has worked well. (Conner 1983)

On many occasions the federal governments were complimentary of the Commission's work. Minister of Fisheries for Canada, James Sinclair said that the IPSFC had a "very excellent record of accomplishment" (Sinclair 1956). In 1972, Ambassador D.L. McKernan of the U.S. State Department described the success of the Commission as "uniquely satisfactory."*

The achievements of the Commission were recognized many years earlier (Crutchfield and Pontecorvo 1969) and more recently by Petty (1979) as he stated: "The joint management scheme of the Sockeye Convention has been highly effective and the resulting achievements of the Commission are unprecedented in the field of fishery management."

The Commission's accomplishments can be summarized briefly as follows:

1. Fishway construction at Hells Gate has provided overall fish passage conditions that are probably better than those which existed prior to the Hells Gate slide.

2. The improvements at Bridge River Rapids and at Yale brought about by the fishway construction has resulted in fish migration conditions superior to those which existed under natural conditions.

3. The development of a racial management program based on cycle and spawning ground requirements led toward optimized production of major races.

4. Extensive scientific research programs provided the Commission with the necessary information to manage and understand the requirements of each race.

5. Spawning channels constructed by the Commission preserved some races from possible extinction and for some populations provided fry production potential far in excess of that existing in the natural environment.

6. Extensive restoration of depleted sockeye and pink salmon stocks as well as reestablishing exterminated runs such as upper Adams River.

---

* IPSFC Minutes of December 7, 1972.

Reasons for IPSFC Success

The Sockeye Convention of 1930 (Appendix A), as amended by the Pink Salmon Protocol of 1956 (Appendix A), contained eleven Articles and two Provisions. There were also two Understandings in the Protocol of Exchange of Ratifications. These four pages, as presented in the Commission's 1957 Annual Report, were sufficient to permit the Commission to carry out its duties in a deliberate and precise manner. An amendment (Appendix A) to the treaty was approved in 1980.

Article I clearly identified the area encompassing Convention Waters. There was seldom confusion on this matter except that, from time to time, it was not realized by everyone that "the Fraser River and the streams and lakes tributary thereto" were part of Convention Waters. On occasion, the question was posed, "could the Commission have exercised regulatory control of the Fraser River Indian fishery since the waters were part of Convention Waters?" This issue was not fully pursued.

Article II described the make-up of the Commission and procedures of appointment of the Commissioners and other aspects covering the Commissioners' terms of office, salaries and expenses.

Article III outlined the basic tasks the Commission would undertake. These were quite specific. It also covered the requirement for an annual report to the governments as well as equal appropriations from each government.

Articles IV, V, VI, VII, VIII and IX covered areas involving regulatory authority of the Commission, division of catch, and the governments' responsibilities for enforcement of the regulations. Voting procedures—requirements, government approval of regulatory proposals and the acquisition of land by the Parties were also included.

Article X stipulated that the Parties (governments) agreed to enact and enforce such legislation as was necessary to make effective the provisions of the Convention.

The two Understandings in the Ratification exchanges covered limitations on the Commission with respect to legal gear and the number and composition of the Advisory Committee as later amended in 1980. The original Convention stipulated five Advisors from each country, the Pink Salmon Protocol raised the number to six and the amendment in 1980 increased it to seven.

Provision I involved a coordinated study of pink salmon stocks and directed that the Parties would meet after the seventh year of the Protocol to examine further the status of the pink salmon stocks of common concern.

Provision II gave the Commission authority to gather such information on other stocks of salmon that it might require incidental to its activities directed at sockeye and pink salmon.

Article III in more detail, clearly identified the Commission's responsibilities under the Convention. It was to make a thorough investigation into the natural history of both species, hatchery methods, spawning ground conditions and any other matters affecting the status and productivity of the sockeye and pink salmon stocks of the Fraser River. The Commission was given the authority to conduct fish-cultural operations, improvement of spawning grounds and construction of such facilities as it determined necessary for the propagation and enhancement of the two species. It also was granted the right to overcome obstructions affecting the migration of sockeye and pink salmon. The cost of the work of the Commission was to be borne equally by both Parties (Appendix P). The Parties agreed to appropriate annually such funds as each deemed desirable for the Commission's work in light of the reports of the Commission.

At the request of the Commission in 1938 and with the concurrence of the Canadian government, the Canadian Fisheries Department was directed to advise and consult with the Commission on all proposed development projects that would possibly affect Fraser River sockeye and pink salmon. This was a very important supplement to the Convention.

It was clear that the Parties' intentions were that the Commission would have the authority to preserve, protect and extend the resource. There was no indication that this would be a responsibility shared with any other entity in any respect. The necessary biological and engineering investigations were the Commission's responsibility. The sole authority of the Commission in regard to management of the fisheries was clearly identified, with the exception that all (pre-season) regulations proposed by the Commission would be subject to approval of the governments. In-season adjustment in regulations did not require government approval. This provision facilitated rapid response by the Commission to in-season changes in salmon timing, abundance and availability, and to short-term environmental fluctuations affecting those factors.

There were three important aspects of the Convention worth noting. First, the Commission was relieved of the obligation of enforcing regulations. Secondly, it did not have controlling authority over development activities in the watershed and thirdly, it had no input as to types or numbers of fishing gear. These omissions permitted the Commission to focus its investigations directly on situations having an impact on the resource. This assured that Commission investigations were singular in intent: to identify the problem, to determine the factor or factors that required correction, and to make recommendations for actions on the basis of the facts.

The most serious inadequacy of the 1930 Convention and 1957 Pink Salmon Protocol was the restricted area of management jurisdiction which did not include the Canadian waters of the central and northern coast of Vancouver Island and the waters of Johnstone Strait. Since many salmon stocks originating in Washington State migrate through Juan de Fuca Strait (Convention Waters with fishing under Commission control), it would have been reasonable for the Commission to have had regulatory control in some form in the Canadian areas of Johnstone Strait, as was the case in the southern approach of Juan de Fuca Strait.

In the final analysis, the Commission's success was attributable primarily to three principles: the continuing scientific investigation which established the foundation for making sound decisions, the construction of the many mainstem fishways which resolved the fish migration problems in the Fraser River, and the concept and implementation of fishery management by distinctive races of salmon. In addition, success could be achieved only as long as water quality and other environmental factors were maintained at levels suitable for production of the species. The Commission was at the forefront of pollution control efforts on the Fraser River. The Commission's ability to get the job done was primarily related to the simplicity of its mandate and the efficient manner in which it was permitted to implement the decision-making process.

Relationship with Governments

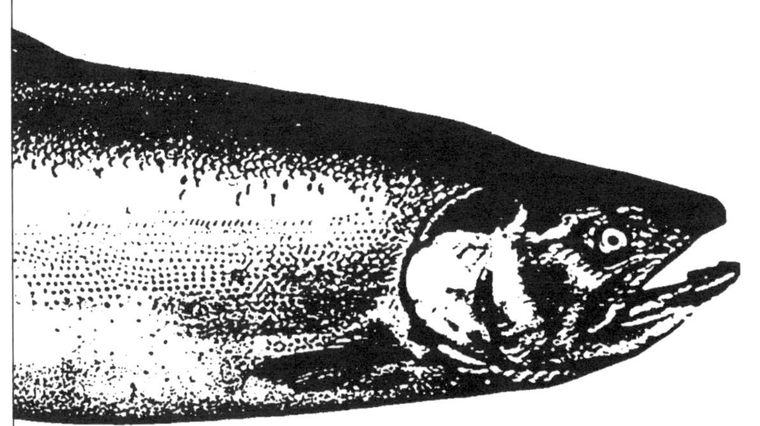

The general emotion at the Commission's beginnings can best be described as "final relief" by both Parties—relief that years of bickering and seemingly endless negotiations had finally been consummated and ratified in the treaty in 1937. As would be expected, neither country was completely satisfied with all of the terms of the new treaty but there was strong resolve by the governments to solve the problems of the resource and move forward on the aims of the Convention. The desperate condition of the resource was the common problem facing both countries, and this was a uniting force. In a real sense, crisis was the forerunner of the birth of the Commission. The governments were very supportive of the Commission's programs and achievements during its first thirty-five years of existence.

During the course of Commission investigations in the mid-1940s, strong and persistent differences of opinion on technical matters existed between Dr. W.E. Ricker of the Fisheries Research Board of Canada and former Director of IPSFC, Dr. W.F. Thompson and the Commission. This dispute (*Commercial Fishermen's Weekly* 1947) centered on the analysis and interpretation of data involving Hells Gate and the effects of commercial fishing as factors contributing to the decline of Fraser River sockeye stocks. Dr. Ricker was seen as the critic.

For several years there were exchanges between the Commission and Dr. Thompson with Dr. Ricker (Ricker 1947a, 1947b, 1947c, Thompson 1947a, 1947b, IPSFC 1947). The public statements by Dr. Ricker irritated the Commission. The technical position of Dr. Ricker in 1987 appeared to be much the same as forty years earlier when he considered the fishways "strictly a gamble—but not necessarily unwise" (Ricker 1987). Dr. W.A. Clemens of the University of British Columbia was also drawn into the conflict. He wrote a rebuttal to Dr. Thompson's comments which had referenced the Fisheries Research Board of Canada when Dr. Clemens was Director of the Biologi-

cal Station at Nanaimo (Clemens 1947).

There was little abatement in this controversy through the years. Minutes of one of the earlier Commission meetings (IPSFC 1948a) noted that Commissioner A.J. Whitmore expressed to the other Commissioners that the foreword of the Commission's 1946 Annual Report "had caused considerable consternation" to Canadian scientists of the Fisheries Research Board. Some of the comments in that report were evidently considered inflammatory and did nothing to improve the rapport and communication between the Commission, its scientists and Canadian government scientists. These feelings persisted for many years. The Hells Gate fishways were indeed required, as were more stringent controls of the harvest of the commercial fisheries.

In the 1960s, funding and budgeting problems were frequent, mostly involving untimely, delayed support from the United States. There were also personnel discussions concerning the staff, who were not civil servants of either country, and status with regard to taxation, salaries, exchange rates, pensions, border-crossing privileges and other matters resulting from the unique international aspect of their employment. In general, the relationship between governments and the Commission, until the early 1970s, was one of cooperation and understanding. It was a period of advancement: runs were increasing, fishways were being installed, threats of dam construction and water diversion projects were being abated, spawning channels were being put in place, and progress in pollution control measures was going ahead. Achievements and anticipation for improvements in management and enhancement of the resource was the mood of the times.

The 1970s and 1980s, however, became an era of increasing confrontation between the Commission and governments. This started in 1970 regarding government to government talks on common problems pertaining to salmonid stocks of both countries. Those Commissioners who were government employees tended to have a different view of the negotiations as compared to the other more independent Commissioners. As might be expected, this caused some friction within the Commission and staff.

The failure of the Canadian government to finance its share of the 1972 proposed development program was a severe blow to Canadian Commissioner Haig-Brown:

> The Commission was confounded by this reception of their plan. For Haig-Brown, who had diligently, hopefully and proudly given it all his energies as a new commissioner, such a dog-in-the-manger national policy was inconceivable madness in high places. (Metcalfe 1985)

Haig-Brown is quoted as stating he: "… digested the realization that politics ruled the Commission, which he had joined with so much pride and hope …" *(ibid.)*.

In the formative years of the Commission, there were many meetings of the Commission and the governments in Ottawa and in Washington, D.C. These courtesies even extended to meeting the prime minister and president. By the early 1970s, communication between the Commission and the two governments was strained and infrequent. Most of the meetings (mainly on the Pacific Coast) thereafter with senior government officials were over regulatory matters, most of which were related to the United States treaty Indian fishery.

The U.S. Federal Court 1974 Boldt Decision was the most contentious and unresolved issue the Commission faced. Staff were unanimous in opposition to government interference and involvement in the regulation of the fishery. Staff held firmly that the Convention was the final and authoritative direction from the governments.

Other significant departures from normal regulatory processes took place in Canada. The Canadian government took unilateral action in 1979 and 1980 when it opened waters that were closed by the Commission or altered Commission regulations in various ways. On July 25, 1979, the government authorized a chinook net fishery in the Fraser River area when the area was closed under Commission regulations. The Commission sent a telegram to the Minister of Fisheries protesting the action by the Department of Fisheries and Oceans. Subsequently, letters were sent to both governments concerning the incident. The United States industry sent several messages to both governments vehemently protesting unilateral government intervention.

In 1980, there were several occasions during the season where the Canadian government :

> … altered Commission regulations ranging from reducing fishing area to cancelling fishing days, thereby limiting the catch of sockeye below the desired exploitation rate. These actions affected division of catch between the United States and Canada and also resulted in excessive escapements at Weaver Creek and more than the desired escapement at Chilko River. (IPSFC 1981)

The motives behind the sudden unprecedented action by the Canadian government were not disclosed. Unfortunately, these events resulted in further tension between the staff of the Commission and with some Commissioners in their relationship with the governments.

In retrospect, it is not surprising that following the United States unilateral action of opening the U.S. treaty Indian fishery, the Canadian government initiated sockeye salmon research in the early 1980s in the Convention Waters area of the Strait of Georgia and at Chilko Lake in the Fraser River watershed. Thus, governments' encroachment into the Commission's historical authorities continued.

Both governments were petitioned in 1983 by the Commission for their views on the propriety of Canada's actions. It was not unexpected that in view of the ongoing treaty negotiations and government desires, both governments' answers were to the effect that there was nothing in the treaty preventing such investigations. The Convention was clear as to who should conduct the research and Canada had voluntarily ceased its Fraser sockeye investigations in 1938 when the Sockeye Convention was ratified. Thus, the respect held by governments, industry, and scientific peers as to the role and achievements of the IPSFC during most of its years, was being tarnished by forces outside the control or influence of the Commission. Perhaps some of this action was the result of over-anxious government employees anticipating that treaty changes were imminent.

There were three areas in which the Canadian government was particularly sensitive: the coastal troll fishery, the Johnstone Strait net fisheries, and the Indian food fishery in the Fraser River watershed. The Commission's attempts to limit the impact of the increasing troll catch of both sockeye and pink salmon in Convention and non-Convention Waters were for the most part unproductive, for a large proportion of the catch was outside the area of jurisdiction of the Commission. Also, within Convention Waters, other salmon species were of importance. Area 20 coho catches were of great concern to Washington State authorities. Government control of the troll fishery was essentially lacking, except in recent years, when other Canadian user-groups demanded that meaningful regulatory control be exercised. Although it was never stated in precise or direct terms, it was implied that the governments did not totally agree with the Commission's comments on the coastal troll fishery and its impacts. From the Commission's point of view, there was little tangible evidence that the government intended to restrict the fishery comparable to restrictions applied to Convention Waters net fisheries. In fact, the trollers were not restricted by Canada in their harvest and the interception of Fraser fish by the troll fleet escalated, taking more and more salmon before the fish reached Convention Waters. A conflict between the government

and Commission was inevitable. Thus, to some, the intent of the 1930 Convention, which implied that the fish were to be divided equally, continued to be eroded.

Johnstone Strait net fishermen were encouraged and supported to play a major role in the interception and harvest of large numbers of Fraser sockeye and pink salmon before the fish reached Convention Waters. Catches of sockeye not a part of the international division were enhanced in the late 1970s and 1980s by a series of exceptionally high northerly diversions through Johnstone Strait. The intensity of this fishery was openly flaunted to the United States fishing industry as a show of strength, hopefully leading to quick conclusion of treaty negotiations. The United States could not reciprocate in Washington State waters on Canadian stocks because of migration patterns. Thus, Canada held strong bargaining power.

The Indian food fishery in the Fraser River also presented challenges for the Commission. In the early years, the Commission was involved in gathering the catch statistics for this fishery. By the mid-1940s this task was taken over by the Canadian government. The Commission did not have a significant role in the fishery until the 1970s and 1980s, when staff members were requested to accompany government officers at meetings with the Indian fishing bands. The needs for the specific time and area closures requested were explained by the Commission. These requests for closures were generally specific to the early Stuart race.

The Commission expressed its concern from time to time during the years of increased catches in the Indian food fisheries and illegal fishing activities, but no official request was made to establish catch limits. The fishery was permitted to grow and the government did not appear to be willing to establish a ceiling on the catch. The Commission was unofficially given the feeling that responsibility for this fishery was a Canadian government matter and not that of the Commission.

On several occasions the Commission

reported to the governments on the issues. In the 1980 Annual Report the Commission clearly identified the problem:

> In effect, there now exists major fisheries in Canadian coastal waters by trollers, and net fisheries in Johnstone Strait and in the Fraser River area by Indians. These fisheries are essentially under the control of the Canadian Government. The United States Government is managing a separate treaty Indian fishery out of the United States share of the Convention Waters catch. The Commission manages the remainder of the allowable catch in Convention Waters for fishermen of the two countries. It is apparent that the Commission's control of the fishery has been reduced greatly and considering all fisheries where Fraser sockeye are caught, there are now three agencies regulating six general fishing areas. Clearly, this was not the intent of the Sockeye Salmon Convention when it was promulgated in 1937. This multi-agency regulation increases the potential for significant management errors.

On several occasions, the Commission staff, out of concern and sometimes desperation, gave advice to ensure there was no exploitation and damage to the stocks. Single-agency management of the resource and the firm and sure control of harvests, for the most part, had been greatly diminished.

Perhaps the governments' attitudes toward the Commission in the latter years were indicated more clearly by recent statements. In 1979 the Commission wrote the Minister of Fisheries following unilateral action by the Canadian Department of Fisheries opening the Fraser River for a chinook salmon fishery with large-mesh gillnets on June 25, 1979. The Commission had retained the services of its own legal counsel for examination of the action as to whether or not it was in conformance with the Convention. The Minister of Fisheries, James A. McGrath, responded to the Chairman on August 21, 1979. He said:

> I am frankly surprised that the Commission would retain its own lawyer to provide it with a legal interpretation of the International Convention. In my view, the Governments of Canada and the United States of America are the responsible authorities for the interpretation of the Convention in the absence of an agreement between the two governments to refer to a third party for this purpose. From time to time differences may arise in the interpretation of a Convention which is, after all, almost fifty years old. (McGrath 1979)

The Commission's legal counsel and the Commission and staff had their own opinion on the legality of the government action.

No doubt the United States was relieved that a new treaty was soon to be in effect as it was anticipated that many of the problems associated with other species of salmon would be resolved. In addition, another major problem would finally have an acceptable solution to both governments. The Chief Negotiator for the U.S. made this clear in his testimony of February 25, 1985 as he stated:

> As you know, following the Boldt decision, some difficulty was encountered in providing for U.S. treaty Indian participation in the regulatory process of the International Pacific Salmon Fisheries Commission. At long last, this matter will be put behind us. (Kronmiller 1985)

The Vice-Chairman of the newly-established Pacific Salmon Commission, Dr. A. W. May, in his opening statement on behalf of the Canadian government at the first meeting on September 25, 1985, said:

> This Treaty replaces the old Fraser River Treaty. We are optimistic that the arrangements to be put in place under the new Treaty will resolve some of the difficulties, misunderstanding and lack of coordination experienced in the past with respect to Fraser River sockeye and pink salmon management. (Pacific Salmon Commission 1986)

Relationships between the Commission and the Provincial Department of Fish and Game and the Washington State Department of Fisheries were in general congenial over the years. The Commission was from time to time involved in discussions concerning game fish policies at Cultus, Chilko, Shuswap and in the Seton-Anderson Lake system. For the latter mentioned lakes the Commission in 1942 voiced its opposition to the introduction of game fish. Extensive coordination and cooperation between Commission staff and both the federal and provincial fisheries authorities was necessary throughout the Commission's tenure.

NON-POLITICAL CONTROL

The Commission from 1937 to the early 1970s focused on the needs of the resource. Additional problems also arose during these years. However, the Commission enjoyed a free rein to pursue these without undue interference from the governments or the fishing industries. While the Commission was not always impervious to political forces, those forces were mostly supportive. Actions or statements by political entities outside the Commission did not deter the Commission from achieving its goals and continuing its commitment to the Convention.

THE IMPLEMENTING PROCESS

The Convention gave the Commission the freedom and ability to make and implement decisions rapidly. This was especially evident during the fishing season. For example, the Commission would meet weekly, usually on Fridays, or sometimes two or three times a week. Most meetings were in person and some by telephone conference calls. Extensive review of the status of the runs was given by senior staff. In most instances, decisions were made and fishing schedules were announced by noon of the meeting day. At infrequent times, emergency meetings were called to respond to quickly-developing crises. Telephone conference meetings were occasionally held within an hour of events on the fishing grounds that needed immediate attention. Commission staff also had ready access to other fishery agencies and the fishing industries of both countries for timely catch statistics and other information vital to the management of the resource. The catch statistics provided by the Washington State Department of Fisheries and Canadian Department of Fisheries both in and out of Convention Waters were invaluable to the Commission. The ability to open and close the commercial fishing periods on short notice was an advantage that even exceeded that given state and federal authorities to manage salmon stocks and fisheries under their respective jurisdictions.

The governments were cooperative in providing a funding process that ensured the Commission could fulfill its financial obligations readily without unnecessary delay or impediments once appropriations were received. In fact, in the most recent years the Commission was able to do its own banking and disbursement of funds.

In areas where large construction projects were proposed, such as fishway or spawning channel construction, the Commission would submit requests for funds, along with justification reports for government review and approval. When a project was approved, authorization of funding was usually provided quickly. Up to the early 1970s, the Commission's programs were truly a cooperative effort.

The Commission was cognizant of the need to communicate with the fishing industries of both countries. This was done by annual public and industry meetings to report on the current year's activities and predictions and regulations for the following year's run, annual reports, publications (bulletins and progress reports), and press releases during the fishing season to explain the status of the runs and reasons for regulatory adjustments in fishing times. In addition, the Commission released weekly updates on the catch totals for each country. Further close ties with the industry were maintained through contact by the staff and Commissioners with the Advisory Com-

mittee members. These avenues of communication were important to the Commission and contributed greatly to the close relationship the Commission and industry enjoyed throughout its history. Commission executive sessions and regulatory meetings however, were not open to the public or press.

The Commission

The Commissioners were the Commission. The first Commissioners, Dr. William A. Found (replaced early by A. J. Whitmore), A. L. Hager and Senator Thomas Reid for Canada; and Edward W. Allen, B. M. Brennan, and Charles E. Jackson for the United States; were men of considerable stature in the fishing industry or in their government. They set policies for the Commission and established standards of action that provided a firm and effective foundation for the future. They were quick to recognize the need for measures to protect the Fraser stocks from the adverse effects of dams, pollution and water diversions and they established the necessary liaison with the Department of Fisheries of Canada to deal with such matters. The Commissioners showed wisdom in their selection of the Commission's first Director, Dr. W. F. Thompson, who was one of the world's foremost fishery scientists. The concepts and policies of these gentlemen were carried through into every aspect of the Commission's work.

In addition to the first Commissioners, the governments were privileged to have other men of character, dignity and objectivity with high degrees of professionalism serve on the Commission. A. J. Whitmore served 30 years. Senator Tom Reid served 31 years, the longest of any member. W.R. Hourston was a Commissioner for 22 years. For the United States, DeWitt Gilbert had 18 years of service, followed by William G. Saletic with 10 years. It was no easy task for appointed Commissioners to spend one or more days each week during the summer traveling long distances to attend numerous Commission regulatory meetings. Many Commissioners were also busy at other tasks. Yet, throughout the history of the Commission, the excitement, sense of purpose, and dedication to the resource could be sensed from these dedicated individuals. Before the 1970s there was no national identity. There was no Canada or United States—only the Commission—one purpose and one body united to carry out the provisions of the Convention. Hutchison, ap-

propriately captured the significance of this commissioner/country relationship by stating: "For them the international boundary had ceased to exist" (Hutchison 1950). There were no U.S. or Canadian caucuses held at these meetings. Seldom did any Commissioner perform his tasks more specifically in this respect than A.J. (Joe) Whitmore, who had an important government role for Ottawa and for the federal fishery agency in the Pacific Region for many years, as well as his Commission position. During the Commission deliberations he wore only his Commission "hat," a hat he often spoke about. There were no compromises on principle; only adherence to the terms of reference as defined by the Convention. In this regard, the United States treaty Indian issue tested the Commission's resolve to the utmost. It was a blight on the Commission/government relationship.

Senator Thomas Reid was a strong and frequently vocal advocate for protection of the Fraser resource from the destructive effect of dams, water diversions and pollution. Although Senator Reid was known for his dim view on salary increases for the staff, this was somehow overlooked because of his strong and enthusiastic support of the Commission's programs. He was the Commission spokesman on sensitive issues. Commissioner Clarence F. Pautzke (Whitmore's counterpart in the United States) was effective on many occasions in presenting the Commission's needs before the U.S. government in Washington, D.C. DeWitt Gilbert often gave the Commission the benefit of his expertise in editorial assistance and was insistent that the Convention principles succeed. W.G. Saletic always exhibited a keen interest in every facet of the fishery and Commission programs. He was a strong advocate for concise and effective regulatory meetings. R. Haig-Brown was passionately in support of enhancement (spawning channels) but he recognized the serious adverse effects of hatcheries and was adamantly against any project which would have adverse impact on the resource.

There was a strong feeling of camaraderie towards a singular purpose in the formative years. Throughout the history of the Commission, each Commissioner played a distinctive and important role in Commission affairs.

It was more difficult for this relationship to continue into the 1970s and 1980s because of the turbulent events of those years. The government representatives particularly had their allegiance to the Commission weakened by frequent influences and admonitions from their respective governments and others to pursue, promote or adopt policies different from those prescribed by the Convention. State Commissioners were also at times urged to pursue state policies not always in concert with Commission objectives.

The high degree of cooperation and enthusiasm did not continue from 1974 to 1985. The Commission was irritated with the governments over the seemingly endless recommendations, many of which were viewed as artifices—on the U.S. side—to increase the treaty Indian catch. Canada was consistently opposed to having the Commission involved. There also was growing impatience with the Commission. Starting in 1977, word would filter back to the staff that U.S. Indians believed that Commission in-season changes in fishing schedules were made deliberately to thwart the Indians' opportunities to increase their catch. There were no winners; the Convention was not used to satisfy the court's decisions. The ill will between the industry and governments, Commission and governments, within the Commission, Indians and Commission and staff; and Indians and government continued for many years. Finally, by the early 1980s the Commission, staff and industry seemed to recognize, and perhaps the Canadian government accepted, that the U.S. government was going to prevail in the matter. Continued industry court battles seemed fruitless because of earlier judgments and industry funds for further legal efforts (which seemed hopeless) were reportedly exhausted.

The Commission, for most of its jurisdiction, was made up of administrators who had extensive background in fishery-related matters and other professional experience. This format was followed until the late 1970s and early 1980s at which time the Canadian government established a policy of bringing a fisherman or person with recent fishing background onto the Commission. Some of the Commissioners had prior experience with the industry as fishermen, processors or in other industry positions and, at times, preference was evident. For the most part however, biases were not extended into the decision-making process and did not harm the resource.

When the Canadian government established a precedent within the IPSFC and the Minister of Fisheries appointed an active fisherman (gillnetter) to the Commission in 1981, it created an additional challenge for the Commission. It would be difficult for any Commissioner in this circumstance to separate himself totally from his own fishing interests. It was an awkward position to be placed in as a Commissioner; only one of seven gear groups in both countries sitting on the Commission. Moreover, the other Commissioners were then required to adjust to a setting lacking complete objectivity. The Commissioner was, in effect, officially given permission to participate in and vote on the deliberations directly involving his livelihood. While it could be argued that some of the other Commissioners were intimately and directly involved in the daily affairs of the industry, their appointments would not be as visibly recognized or questioned compared with an active seiner, gillnetter or troller. This unprecedented action was perceived by many as preferential treatment toward one group of fishermen, namely Fraser River gillnetters. The Canadian government's decision could be viewed as akin to putting the fox in charge of the chicken house. The actions and decision-making process of the Commission were encumbered because of the sudden entrance of an active participating fisherman.

The Advisory Committee members' advice, and it was here where active fishermen had always participated, had been adequate for many years. And a fisherman as a Commissioner usurped the already well-established role of the Advisory Committee member. As a policy matter, user self-regulation over the stewardship of public resources is not desirable because of financial conflicts and possible conflict with good resource management. "Public resource trustees are needed in place of user group representatives" (American Fisheries Society 1989).

The placement of an active fisherman on the Commission was not within the Commission's purview—this was a government responsibility. This subject had previously been dealt with by the Honorable James Sinclair, Minister of Fisheries earlier in 1954, who "felt very strongly that our Commissions would be improved if we had on them a practicing commercial fisherman, who was in daily contact with the problems of the fishermen" (Sinclair 1954).

Sinclair was proud of the fact that he had appointed a fisherman to the International North Pacific Fisheries Commission and to the International Pacific Halibut Commission. There was very little similarity however, between those two organizations and the Salmon Commission with respect to seasonal management deliberations.

Sinclair emphasized that the fisherman did not "sit on these Commissions as special representative of the fishermen. They sit on the Committee as representatives of Canada."

The challenging aspect of each season was the long and sometimes arduous management of harvests and escapements. It was very rewarding and fulfilling for the Commissioners to experience the fruits of their labors when they received the weekly in-season reports that escapement and division of catch goals were met, especially when they and the fishermen recognized improvements in the returns, and when the additional benefits received by the industry were defined. Each season was different and there was always excitement, challenge, and anticipation. The responsibility was great and the stakes were high. Each management maneuver was contemplated and re-evaluated in response to daily and weekly staff reports. On many occasions, regulatory decisions were made at the last possible moment to gain additional information about the status of each stock in the returning run. The time spent waiting for predicted catches or escapements to materialize could be agonizing because of the serious biological and economic ramifications of the results.

The Commissioners held to a strong sense of purpose and dedicated commitment to fulfill the mandate of the Convention. The highlight and perhaps one reward at the end of the season each year was the Commissioners' visit to view the escapement at that year's major spawning grounds, spawning channels, or in the spring, to witness the smolt and fry migrations. Another benefit was received when in excess of 100,000 visitors, including thousands of school children, visited the Adams River spawning grounds to view and record the stimulating sight of millions of spawners. The Weaver Creek spawning channel was also a showcase display available for thousands of lower mainland visitors and school children. Hundreds of thousands of persons visited the Hells Gate fishways as well as other spawning channels and spawning grounds. The Commission was very much aware of the value of these public opportunities and events in protecting (through public awareness) and enhancing the resource. A professionally produced film depicting the Commission's work was also used by schools and many other organizations. Haig-Brown (1980) placed great significance in "simply watching fish". He said: "Any opportunity to watch fish is an educational experience".

The dedication and devotion of the Commissioners made the historical achievements of the Commission possible. The governments of both countries were fortunate to have had their service.

CHAPTER 22

The Search for a New Agreement

CANADA-UNITED STATES DISCUSSIONS

Several Commissioners, in dual roles, became involved in negotiations that started in 1970 for a new treaty. The Commission staff was not directly involved, but on occasion were asked to define Commission staff organization and responsibilities. Communication by the governments to the Commission on the treaty negotiations for the most part was non-existent. However, some Commissioners and staff were kept apprised of the status from contacts with industry representatives. It was a difficult time for many of the staff not knowing how much longer the Commission would be in operation. Some staff resigned and prospective employees took positions elsewhere because of the uncertainty of the Commission's future.

It was clear for some time that the governments wanted the Convention modified. Even the Canadian fishing industry, which had previously been very supportive of the IPSFC, changed its position to one favoring a new treaty—presumably because of the promise of a greater share of the Fraser salmon resources. This was, indeed, a strong motivation for a new agreement. At the end, the dissidents were the North Puget Sound net fishermen, United States processors, the Commission staff and some Commissioners who were opposed to many elements proposed in the new treaty. Strong opposition was given by some of these groups.

The intense and mostly deliberate fishing pressure exerted on Fraser River stocks outside Convention Waters played a significant role in bringing about the new treaty and exacerbated the poor status of the chinook stocks. Thus, it was partially the large migrations of Fraser River sockeye through Johnstone Strait from 1978-1983 and the large Johnstone Strait and Canadian troll catches in non-Convention Waters on the west coast of Vancouver Island that were important factors in terminating the IPSFC and bringing the new U.S./Canada Salmon Treaty into existence. There was also Canada's strong desire to establish greater management control over the Fraser River salmon resource and the desire to harvest a greater share of Fraser fish. Additionally, there was concern that greater international control be given to harvests of coho and chinook salmon. Chinook salmon stocks to many areas had been severely depressed due to excessive coastal troll fishery exploitation and this problem heightened in the early 1980's. Environmental degradation also affected Columbia River chinook stocks. Other factors involved were Canada's desire to conduct research in the Fraser system and the strong motivation from both countries to resolve the U.S. treaty Indian issue.

By the early 1980s criticisms from some were directed at the Commission's regulatory actions and its role with respect to other salmon species, and fishery agencies in relation to interference and cooperation. The final report of "The Commission on Pacific Fisheries Policy" stated:

> Arrangements for integrating the Commission's activities with the Department's efforts to conserve and manage other species, especially chinook, do not appear to have been effective.... . The isolation of these two agencies working side by side on the same Canadian river cannot be conducive to the most effective resource management. (Pearse 1982)

There is no evidence of any official correspondence from the Canadian government to the IPSFC stating that IPSFC management policies were responsible for the decline of any salmon species affected by its management. Moreover, Commission actions and policies were most likely beneficial to Fraser River Chinook stocks. Factors causing much of the decline of coho and chinook salmon stocks had their origin in non-Convention Waters.

By 1982 the two governments, had completed a draft version of the proposed new

Pacific Salmon Treaty (Anonymous 1982). The Chief Negotiators for the Canada/United States Salmon Treaty, Dr. M. Shepard for Canada and Dr. D. L. Alverson for the United States, solicited comments of the Commissioners and staff on the Exchange of Notes section of the proposed treaty since that portion of the document dealt with Fraser River sockeye and pink salmon. The Commissioners and senior staff met with the negotiators on January 17, 1983. Comments were submitted on the Exchange of Notes section; they contained numerous suggestions for changes, including many questions of an operational nature (Roos 1983). A few of the recommendations given to the negotiators were included in the final Pacific Salmon Treaty.

While the Commission and staff were not officially asked to comment on other sections of the proposed treaty it was noted that the Fraser River Panel and the Commission would be given the responsibility to "seek consistency with existing aboriginal rights … and domestic allocation objectives." This provision was viewed as positive, since it would provide to the Fraser River Panel management authority over the treaty Indian fishery to schedule fishing times and to target catch allocations for the Indians. Many other aspects of the proposed treaty, however, were viewed with trepidation. The Commission and its staff recognized the reasons for the historical successes under the Sockeye Convention. The changes proposed in the Pacific Salmon Treaty, which increased the scope of the new Commission, also increased the bureaucracy and costs of operation. This was conceived as increasing management and implementation difficulties. Staff was also opposed to limiting the Fraser River Panel management area to the existing Convention boundaries and not including Johnstone Strait and the West Coast - Vancouver Island troll fisheries. Many believed the governments should have retained the IPSFC structure and made the agreed-upon changes within the existing frame-work of the Commission.

It appeared that the new treaty would come into being in 1983 but there was disagreement between U.S. treaty Indians and the State of Alaska. Once that issue was resolved Alaska was willing to agree with the provisions of the proposed treaty.

One of the basic foundations of the IPSFC was that of oneness. This was mentioned by Senator Reid in a speech at Ottawa on May 10, 1963, in which he said: "We are appearing here today and have always acted in the past as one Commission, not two national sections." Most of those associated with IPSFC believe this unique and vital relationship ended on December 31, 1985. The concept of the new organization (PSC) has been changed from unity to duality. By returning significant resource and habitat management responsibilities to the Canadian government and increasing the share of the catch for Canadians, the majority utilization and control of the Fraser sockeye and pink salmon resource was assumed by Canada. This should assist Canada in controlling possible degradation of the environment (hydro development, etc.) in the watershed as the majority of the benefits of the fishery resource will accrue to Canadians.

The role of the new Pacific Salmon Commission was broadened in some respects compared to IPSFC but limited to primarily one of carrying out the allocation and monitoring enhancement objectives of the governments and in performing in-season management functions for Fraser River sockeye and pink salmon runs.

THE BASIC PROBLEMS

Strong nationalistic and user-group representations were present throughout the negotiations leading to the new treaty. One of the basic understandings of the new agreement was that both countries wished to reap the benefits from their own resource enhancement efforts. This philosophy was a drastic departure from the beginnings of the Sockeye Convention; each country needed the other's cooperation to save and rehabilitate the stocks, and control fishing. Therefore, they agreed to share costs and benefits equally. An argument heard quite often was that the U.S. had already received adequate payback for its investment. However, the United States had lost the opportunity to harvest millions of salmon at great economic loss after 1915 through no fault of its own.

Copes (1977) described Canada as perturbed over U.S. perceived rights for continuing access to Fraser River salmon as payment for its investment in habitat rehabilitation. The United States contribution to the success of the Fraser Convention was far more than shared contributions to construct fishways and eventually spawning channels as means of enhancing the resource. U.S. staff (as well as Canadian) members also played significant roles in the scientific management of the fisheries, research, engineering, pollution control investigation, and restoration of endangered and exterminated stocks. These contributions covered a period of almost 50 years and are additive to the more commonly referenced "one time" U.S. investment in the Hells Gate fishways.

Nevertheless, with strong national views regarding the resource and who should manage and harvest it, associated with trends in other international agreements based upon country of origin policy, it was inevitable that broad changes would be included in the new treaty. It is likely, however, that neither the financial contributions nor income lost by the United States played a significant role in the final outcome of treaty negotiations. National priorities had changed in Canada and coastal states, especially Washington State, wanted a better arrangement for protection of coho and chinook salmon.

The current excellent status of the Fraser sockeye and pink salmon stocks, coupled with the need for better management of other salmon stocks in many areas, were important factors leading to the formation of a new treaty. Sports fishing interests in Washington State strongly supported a new treaty, hoping for greater protection and increased access to U.S. coho and chinook salmon. Canada no longer needed the United States to help rebuild Canadian stocks. Also, each country wanted to harvest and have greater management authority over its own fish. This position is consistent with the United States policy with respect to the Japanese high-seas fishery on Bristol Bay sockeye salmon stocks. It is also consistent with policies of both Parties in adoption of a 200-mile coastal fishery management zone. It is hoped that the management of the international fisheries will continue to be effective and that the resource will flourish under the new agreement. There should be strong incentive for each country to increase substantially the enhancement of its own stocks and there was eager anticipation that this would be done once a new treaty was in place. Success can be achieved providing that all participants are committed first and foremost to the resource. Constraints on immediate decision-making, the inevitable result of a larger organization, must be kept to a minimum if effective management of the fisheries is to be achieved.

The apparent feelings of the United States government concerning the new treaty were expressed in the following manner: "The treaty also resolves an issue that has long been an irritant in relations with a valued trading partner and ally."* The treaty Indian issue would no longer be a major source of confrontation between the two governments.

Upon finalizing the new treaty there was the clear understanding in the U.S. that significant funds from the U.S. government would be forthcoming for enhancement projects.

---

* Letter to President Reagan from George P. Shultz, Secretary of State, Department of State. Februrary 4, 1985.

THE END OF AN ERA

The IPSFC conducted research and implemented rehabilitation efforts on Fraser River sockeye salmon from 1937 to 1985 and for pink salmon from 1957 to 1985. The Commission was responsible for the management of the sockeye fishery from 1946 to 1985 and the pink salmon fishery from 1957 to 1985.

When the Pacific Salmon Treaty was officially signed on March 18, 1985, it stipulated that the International Pacific Salmon Fisheries Commission would be terminated on December 31, 1985. All staff members lost their positions at that time. The Exchange of Notes section of the new treaty explicitly provided for the future employment of IPSFC staff within the new Commission or with the U.S. or Canadian governments. All but one of the former IPSFC staff (including U.S. staff) were subsequently offered (and accepted) employment with the new Commission or with the Canadian government. The United States government did not offer employment to any IPSFC staff, thereby excluding the Director. Thus, both governments did not completely fulfill their their mutually agreed upon commitment. This gave credence to presumptive evidence that this decision/action was an aftermath to the treaty Indian issue. The small staff of the Management Division was transferred to the new Commission to carry out in-season management functions assigned to the Fraser River Panel. In total, twelve former IPSFC staff members—professional, technical, clerical and support employees—were given jobs with the new Commission. Thus, only a small portion of the former IPSFC permanent staff (42) remained intact with the new Commission. The management division of IPSFC was the only division transferred intact to the PSC. Fraidenburg (1990 MS) described this as: "This is a significant legacy handed down from the predecessor Fraser River Treaty."

In view of the many distractions and difficulties encountered by the Commission from 1970 to 1985, it is remarkable that continued and significant improvement in stock abundance was accomplished. The Commission was able to operate adequately through many years of treaty negotiations, the cessation of government support in rehabilitation, and the protracted discussions and difficulties posed by the U.S. treaty Indian fishing issue.

The plaudits given the Commission and staff through the years by the governments, scientists, engineers and representatives from the commercial fishing industries were deeply appreciated and individually and collectively rewarding. Ironically, in the final analysis, the Commission's success at rehabilitation—by increasing current abundance and future potentials of a very valuable resource—was probably an important contributing factor leading to its demise. Perhaps its assigned task to re-build the resource had ended. The greatest satisfaction for the many Commissioners and staff members may be the knowledge of how the job was done and what was achieved. The legacies of the Commission's mandate, its methods and its achievements, will hopefully serve as a model and motivation for others to provide even greater benefits to the fishermen, processors and public of the two countries in future years and even for applied fishery management in other areas. At the termination of the Commission, one senior Canadian government official said: "… they can never remove from the record what the Commission has accomplished" (personal communication). The Commission's accomplishments are very noticeable at the present time and its contributions will continue to be evident. This occurred as recently as 1990 when the total Fraser sockeye run* of 22 million fish (PSC personal communication) was the largest since 1913.

Canada and the United States have given the world an example of the benefits that can be mutually derived through international diplomacy, respect, and cooperation. The con-

---

* Under PSC management.

servation of fisheries in each nation's waters has been an important topic for both countries for about 100 years. Joint management of the currently highly productive Fraser River sockeye and pink salmon runs and the Pacific halibut fishery are widely acclaimed as glowing examples of the benefits both nations received from their cooperative efforts. It is hoped that this spirit of mutual cooperation will continue.

The Fraser resource has recovered to excellent production levels because of the Commission's efforts. The founders of the Convention, as well as more recent Commissioners, undoubtedly had this intention. Unfortunately, many of them did not live to see the Commission's successes; however, they would have taken great pride in what was achieved for the resource and for the fishing industries of both countries. Those who participated in later years knew what was accomplished. The assignment of the Commission to be the trusted steward and re-builder of the valuable resource ended on December 31, 1985. The vision of a long-lasting and fruitful relationship was realized. It is hoped that this record of achievement will be expanded even further. The responsibility for future resource protection, extension and management control was repatriated to Canada.

American Fisheries Society. 1989. Testimony for Congressional Hearings on re-authorization of the Magnuson Fishery Conservation and Management Act. p. 10-13.

Andrew, F.J. and G.H. Geen. 1958. Sockeye and pink salmon investigations at the Seton Creek hydroelectric installation. Internat. Pacific Salmon Fish. Comm. Prog. Rep. No. 4, 74 p.

Andrew, F.J. and G.H. Geen. 1960. Sockeye and pink salmon production in relation to proposed dams in the Fraser River system. Internat. Pacific Salmon Fish. Comm. Bull. XI, 259 p.

Andrew, F.J., P.C. Johnson and L.R. Kersey. 1956a. Further experiments with an electric screen for downstream-migrant salmon at Baker Dam. Internat. Pacific Salmon Fish. Comm. Prog. Rept. No. 1, 29 p.

Andrew, F.J., P.C. Johnson and L.R. Kersey. 1956b. Electric screens for adult salmon. Internat. Pacific Salmon Fish. Comm. Prog. Rept. No. 2, 43 p.

Andrew, F.J., L.R. Kersey and P.C. Johnson. 1955. An investigation of the problem of guiding downstream-migrant salmon at dams. Internat. Pacific Salmon Fish. Comm. Bull. VIII, 65 p.

Anonymous. 1951. Report on the fisheries problems created by the development of power in the Nechako-Kemano-Nanika River Systems. Supplement No. 1 - Temperature changes in the Nechako River and their effects on the salmon populations. Department of Fisheries of Canada, Fish. Res. Bd. of Canada and International Pacific Salmon Fisheries Commission. 55 p. + 1 app.

Anonymous. 1955. A report on the fisheries problems related to the Fraser and Thompson River dam site investigations. Department of Fisheries of Canada, International Pacific Salmon Fisheries Commission, B.C. Fish. Dept. and B.C. Game Comm. Mimeo. 101 p. + 1 app. + 19 pl.

Anonymous. 1957. Conference on Coordination of Fisheries Regulations between Canada and the United States. Summary of Proceedings. Seattle, Washington Feb. 27-28, 1957. 15p.

Anonymous. 1969 Report on the fisheries problems associated with the proposed diversion of water from Shuswap River to Okanagan Lake. Department of Fisheries and Forestry, International Pacific Salmon Fisheries Commission and Fish and Wildlife Branch, British Columbia Department of Recreation and Conservation. Adm. Rep. 97 p.

Anonymous. 1971. Fisheries problems related to Moran Dam on the Fraser River. Canada Department of the Environment, Fisheries Service and International Pacific Salmon Fisheries Commission. 206 p.

Anonymous. 1982. Draft treaty between the Government of Canada and the Government of the United States of America concerning Pacific salmon. 61 p.

Aro, K.V. 1979. Transfer of eggs and young of Pacific salmon within British Columbia. Fish. Mar. Serv. Tech. Rep. 86, 147 p.

Babcock, J.P. 1902-1932. Spawning-beds of the Fraser River. Annual Report of the Commissioner of Fisheries (British Columbia) for the years 1901-09, 1911-31. Victoria, B.C. (Title varies and 1901-1908 do not have specific titles).

Babcock, J.P. 1904a. Investigations in British Columbia - From the Report of John Pease Babcock, Provincial Fish Commissioner. In: *Pacific Fisherman.* II(8). August 1904. p. 21.

Babcock, J.P. 1912a. Letter of September 28 to F.H. Cunningham, Inspector of Fisheries, New Westminster, B.C.

Babcock, J.P. 1914a. Report of the British Columbia Commissioner of Fisheries for 1913. The spawning beds of the Fraser. pp. 17-38.

Babcock, J.P. 1917a. Another blunder that should be investigated. *Canadian Fisherman* 4(12):497-499.

Babcock, J.P. 1918a. Salmon fishery of the Fraser River District. Ninth Annual Report of the Commission of Conservation, Ottawa. 16 p.

Babcock, J.P. 1919a. Letter of October 29 from British Columbia Commissioner of Fisheries to Samuel Hill. 7 p.

Babcock, J.P. 1920a. Fraser River salmon situation a reclamation project. Appendix V. Report of the Commissioner of fisheries for 1919. Department of Fisheries, Province of British Columbia. 11 p.

Babcock, J.P. 1920b. The Fraser River salmon situation-Canada's position. *Canadian Fisherman.* VII (6):101.

Babcock, J.P. 1924a. The spawning-beds of the Fraser River. Report of the Commissioner of Fisheries Province of British Columbia for 1923. 58 p.

Bell, F.H. 1981. The Pacific Halibut-The Resource and the Fishery. 267 p. Anchorage: Alaska Northwest Publishing Company.

Bell, M.C. 1945. Report on the engineering investigations of Hell's Gate, Fraser River. In Annual Report of IPSFC for 1944. p. 15-22.

Bell-Irving, H.B. 1930. Conditions in the Fraser River Canyon - A Canadian view. *Pacific Fisherman.* Vol. 28.

Bellingham Herald. 1978. Salmon fishing regulations yet to be set. April 12.

Blackbourn, D.J. 1984 MS. Variable migration in Fraser River sockeye and pink salmon. Internat. Pacific Salmon Fish. Comm. 50 p.

Blackbourn, D.J. 1987. Sea surface temperature and preseason prediction of return timing in Fraser River sockeye salmon (*Oncorhynchus nerka*). pp. 296-306 In: Smith, H.D., L. Margolis, and C.C. Wood (Ed.) 1987. Sockeye salmon (*Oncorhynchus nerka*) population biology and future management. Can. Spec. Publ. Fish. Aquat. Sci. 96, 486 p.

Blondin, C.J. 1974. Memo of October 2 from Assistant Director for International Fisheries, Nat. Mar. Fish. Serv., on meeting of September 28, 1974 of IPSFC Commissioners and interested parties on implementation of *U.S. Vs. Washington.*

Board of Engineers. 1928. Final report of the engineers enquiring into Fraser River conditions at Hell's Gate, 1926-28. Prepared for W.A. Found, Deputy Minister of Fisheries, Ottawa. 12 p. + 2 app.

Brannon, E.L. 1972. Mechanisms controlling migration of sockeye salmon fry. Internat. Pacific Salmon Fish. Comm. Bull. XXI, 86 p.

Brewster, R.C. 1977. Letter of June 17 from U.S. State Dept. to Donald R. Johnson, Chairman, International Pacific Salmon Fisheries Commission.

Brief for the United States in Opposition. 1975. On petitions for a writ of certiorari to the United States Court of Appeals for the Ninth Circuit. 24 p.

British Columbia Canners. 1904. Letter of December 29 to Honorable Raymond Prefontaine, Minister of Fisheries.

B.C. Fishermen. 1930. The Sockeye Treaty. 1(2):3.

B.C. Fishermen. 1931. Changing migration route ineffective. 1(9):4-5.

B.C. Fishermen's Protective Association. 1934. Minutes of November 19 meeting called by B.C. Fishermen's Protective Association for discussion of conditions affecting Fraser River fisheries. 44 p.

Burdis, W.D. 1912. Letter of June 25 to H.H. Stevens, Esq. M.P. Vancouver, B.C.

Canada Department of Naval Service, Fisheries Branch. 1915. Forty-eight Annual Report. Photographs. 9p.

Canadian Fisherman. 1921. Fisheries Conference held in Vancouver Dec. 12 & 13. Vol. VIII No. 12.

Canadian Fisherman. 1929. Sockeye salmon will be restored. XVI (4):18-19.

Clemens, W.A. 1947. A statement on the memorandum by W.F. Thompson entitled "The Hell's Gate Blockade and the Salmon". The University of B.C., Vancouver, B.C. April 9th, 1947.

Clingan, T.A., Jr. 1975a. Letter of July 21 from U.S. State Dept. to W.R. Hourston, Chairman, International Pacific Salmon Fisheries Commission.

Clingan, T.A., Jr. 1975b. Letter of August 8 from U.S. State Dept. to William G. Saletic, Manager, Seiners Association, Seattle, Wash.

Clutter, R.I. and L.E. Whitesel, 1956. Collection and interpretation of sockeye salmon scales. Internat. Pacific Salmon Fish. Comm. Bull. IX, 159 p.

Colgrove, D.J. and J.W. Wood. 1966. Occurrence and control of *Chondrococcus Columnaris* as related to Fraser River sockeye salmon. Internat. Pacific Salmon Fish. Comm. Prog. Rep. No. 15, 51 p.

Commercial Fishermen's Weekly. 1947. Salmon Commission hits back at critic. 13(10): 110, 112.

Committee on Foreign Relations. 1957. Pink Salmon Protocol. 85th Congress, 1st Session, Senate. Executive Report No. 2, 4 p.

325

Congressional Record. 1930a. Radio address by Governor Hartley of the State of Washington, December 4, 1930. 6 p.

Congressional Record. 1930b. Address by Miller Freeman in reply to the address of Governor Hartley delivered before the Seattle Bar Association, December 10, 1930. p. 7-15.

Conner, D. 1983. The troubled Pacific Salmon Treaty: why it must be ratified. Ocean Law Memo. Issue 24. Ocean and Coastal Law Center. School of Law, University of Oregon, Eugene, Oregon.

Cook, R.C. and I. Guthrie. 1987. In-season stock identification of sockeye salmon (*Oncorhynchus nerka*) using scale pattern recognition, p. 327-334 In: H.D. Smith, L. Margolis, and C.C. Wood (ed.) Sockeye salmon (*Oncorhynchus nerka*) population biology and future management. Can. Spec. Publ. Fish. Aquat. Sci. 96. 486 p.

Cooper, A.C. 1956. A study of the Horsefly River and the effect of placer mining operations on sockeye spawning grounds. Internat. Pacific Salmon Fish. Comm. Unpubl. 58 p.

Cooper, A.C. 1965. The effect of transported stream sediments on the survival of sockeye and pink salmon eggs and alevin. Internat. Pacific Salmon Fish. Comm. Bull. XVIII, 71 p.

Cooper, A.C. 1973. Temperature control during sockeye spawning period in McKinley Creek in 1969. Internat. Pacific Salmon Fish. Comm. Prog. Rep. No. 27, Part I. 34 p.

Cooper, A.C. 1977. Evaluation of the production of sockeye and pink salmon at spawning and incubation channels in the Fraser River system. Internat. Pacific Salmon Fish. Comm. Prog. Rep. No. 36. 80 p.

Cooper, A.C. and K.A. Henry. 1962. The history of the Early Stuart sockeye run. Internat. Pacific Salmon Fish. Comm. Prog. Rep. No. 10, 48 p.

Copes, P. 1977. The law of the sea and management of anadromous fish stocks. Ocean Development and International Law Journal (4)3:233-255.

Craddock, D.R. 1958. Spawning escapement of Okanagan River blueback salmon (*Oncorhynchus nerka*). 1957. U.S. Fish and Wildlife Serv., SP. Sci. Rept. 15:1-9.

Crutchfield, J. and G. Pontecorvo. 1969. The Pacific salmon fisheries. A study of international conservation. The John Hopkins Press. 220 p.

DeLacy, A.C. and F. Neave. 1948. Migration of pink salmon in southern British Columbia and Washington in 1945. Bull. Fish. Res. Bd. Canada, No. 74, 11 p.

Department of Fisheries State of Washington, 1989. 1987 Fisheries Statistical Report. 88 p.

Doyle, H. 1903. Letter of March 24 to Hon. R. Prefontaine, Minister of Marine and Fisheries, Ottawa, Ont. 7 p.

Doyle, H. 1920. History of the Pacific coast salmon industry. *Canadian Fishermen*. VII (6):72-74.

Edwards, J.T. 1911. Personal correspondence from John T. Edwards to F.H. Cunningham, Inspector of Fisheries, New Westminster, B.C. Source: Pacific Salmon Commission library, Vancouver, B.C.

Einarsen, A.S. 1929a. A report on an inspection trip to the Fraser River watershed. Wash. State Dept. Fish. 3 p.

Einarsen, A.S. 1929b. A further report on the Fraser River watershed. Wash. State Dept. Fish. 2 p.

Einarsen, A.S. 1930. A further report on the Fraser River. Wash. State Dept. Fish. 3 p.

Einarsen, A.S. 1931. Notes in regard to the sockeye treaty. Wash. State Dept. Fish. 2 p.

Einarsen, A.S. 1932. Periodic report on the Fraser River sockeye run. Wash State Dept. Fish. 3 p.

Einarsen, A.S. and L.E. Mayhall, 1928. Wash. State Dept. Fish. November 12, 1928, Memorandum.

Ellis, D.V. 1989. Environment at Risk - Case Histories of Impact Assessment. Chapter 2, Construction — Hell's Gate (Canada). p. 23-51. In Press. Springer-Verlag Berlin Heidelberg, New York.

Environment Canada. 1975. Canada expresses concern at United States salmon action. August 15 News Release.

Environment Canada. 1977. Fisheries Minister Le Blanc expresses concern on Fraser River salmon fishery. July 7 News Release.

Fish, F.F. 1948. The return of blueback salmon to the Columbia River. Sci. Monthly. 66 (4):283-92.

Foerster, R.E. 1938. An investigation of the relative efficiencies of natural and artificial propagation of sockeye salmon (*Oncorhynchus nerka*) at Cultus Lake, British Columbia. J. Fish. Res. Bd. Canada, 4(3): 151-161.

Foerster, R.E. 1968. The sockeye salmon, *Oncorhynchus nerka*. Fish. Res. Bd. Can. Bull. 162, 422 p.

Fraidenburg, M.E. 1990. The game's afoot understanding international conflict over salmon. Draft Master of Environmental Studies. Evergreen State College, Olympia, Washington. 275 p.

Fraser River Board, 1958. Preliminary report on flood control and hydro-electric power in the Fraser River basin. Victoria, B.C. 171 p.

Freeman, M. 1956. Memoirs of Miller Freeman, 1875-1955.

Fretwell, M.R. 1989. Homing behavior of adult sockeye salmon in response to a hydroelectric diversion of homestream waters at Seton Creek. Internat. Pacific Salmon Fish. Comm. Bull. XXV, 38 p.

Geen, G.H. and F.J. Andrew. 1961. Limnological changes in Seton Lake resulting from hydroelectric diversion. Internat. Pacific Salmon Fish. Comm. Prog. Rep. No. 8, 76 p.

Gilbert, C.H. 1913-1924. Contributions to the life history of the sockeye salmon. (1-10). Reports of the British Columbia Commissioner of Fisheries for the years 1913-15, 1917-19, 1921-24. Victoria, B.C.

Gilbert, DeWitt. 1988. Fish for tomorrow. University of Washington School of Fisheries. 162 p.

Gilhousen, P. 1960. Migratory behavior of adult Fraser River sockeye. Internat. Pacific Salmon Fish. Comm. Prog. Rept. No. 7, 78 p.

Gilhousen, P. 1980. Energy sources and expenditures in Fraser River sockeye salmon during their spawning migration. Internat. Pacific Salmon Fish. Comm. Bull. XXII, 51 p.

Gilhousen, P. 1989. Wounds, scars and marks on Fraser River sockeye salmon with some relationships to predation losses. Internat. Pacific Salmon Fish. Comm. Prog. Rept. No. 42, 64 p.

Gilhousen, P. 1990 MS. Historical efficiency of the Fraser River gill net sockeye fishery and calculation of annual escapements, 1893 to 1944. Internat. Pacific Salmon Fish. Comm. In preparation.

Gilhousen, P. 1990. Prespawning mortalities of sockeye salmon in the Fraser River system and possible causal factors. Internat. Pacific Salmon Fish. Comm. Bull. XXVI, 58 p.

Goodlad, J.C., T.W. Gjernes and E.L. Brannon. 1974. Factors affecting sockeye salmon (*Oncorhynchus nerka*) growth in four lakes of the Fraser River system. J. Fish. Res. Board Can. 31: 871-892.

Goodman, M.L. 1990. COMMENTS. Preserving the genetic diversity of salmonid stocks: A call for federal regulation of hatchery programs. Environmental Law. Vol. 80. p. 111-166.

Gordon, R.W. and J.A. Servizi. 1974. Acute toxicity and detoxification of kraft pulp mill effluent. Internat. Pacific Salmon Fish. Comm. Prog. Rep. No. 31, 26 p.

Government of Canada. 1981. Letter of September 14 from A. Campbell, Director General of International Directorate, to William G. Saletic, Chairman, International Pacific Salmon Fisheries Commission.

Groot, C. 1965. On the orientation of young sockeye salmon (*Oncorhynchus nerka*) during their seaward migration out of lakes. Behavior, Suppl. 14, 198 p.

Groot, C. and K. Cooke, 1987. Are the migrations of juvenile and adult Fraser River sockeye salmon (*Oncorhynchus nerka*) in near-shore waters related? p. 53-60. In: H.D. Smith, L. Margolis, and C.C. Wood. Sockeye salmon (*Oncorhynchus nerka*) population biology and future management. Can. Spec. Publ. Fish. Aquat. Sci. 96. 486 p.

Haig-Brown, R. 1972. The Fraser watershed and the Moran proposal. Nature Canada. 1(2):10.

Haig-Brown, R. 1980. Bright waters, bright fish. Douglas and McIntyre, Vancouver, British Columbia and Timber Press, Forest Grove, Oregon. 143 p.

Hamilton, J.A.R. and F.J. Andrew. 1954a. An investigation of the effect of Baker Dam on downstream-migrant salmon. Internat. Pacific Salmon Fish. Comm. Bull. VI, 73 p.

Hamilton, J.A.R. and F.J. Andrew. 1954b. A study of mortality rates of sockeye salmon fingerlings in a turbine at Ruskin Dam. Internat. Pacific Salmon Fish. Comm. Unpubl. 13 p.

Hamilton, K. 1985. A study of the variability of the return migration route of Fraser River sockeye salmon (*Oncorhynchus nerka*). Can. J. Zool. 63: 1930-1943.

Haring, D. 1981. Memorandum to Director Schmitten and Bill Wilkerson. 1981 IPSFC FISHERIES. April 23, 1981. State of Washington Department of Fisheries, Room 115, General Administration Building, Olympia, Washington.

Heckman, J.L. 1977. Northwest Indian Fisheries Commission letter of December 6 to John D. Negroponte, Department of State.

Henry, K.A. 1961. Racial identification of Fraser River sockeye salmon by means of scales and its applications to salmon management. Internat. Pacific Salmon Fish. Comm. Bull. XII. 97 p.

Hertslet's Commercial Treaties. 1895. Vol. xix, p. 907-8.

Hewes, G.W. 1973. Indian fisheries productivity in pre-contact times in the Pacific salmon area. North West Anthropol. Res. Notes 7: 133-155.

Hill, N. 1929. Fraser River salmon study - 1929. Bellingham Canning Co. 6 p.

Hourston, W.R. 1975. Letter of July 25 from IPSFC Chairman to Dr. Henry Kissinger, Secretary of State.

Hourston, W.R. and D. MacKinnon. 1956. Use of an artificial spawning channel by salmon. Trans. Amer. Fish. Soc. 86: 220-230.

Hourston, A.S., E.H. Vernon, and G.A. Holland. 1965. The migration, composition, exploitation and abundance of odd-year pink salmon runs in and adjacent to the Fraser River Convention area. Internat. Pacific Salmon Fish. Comm. Bull. XVII, 151 p.

Howard, G.V. 1948. Problems in enumeration of populations of spawning sockeye salmon. Internat. Pacific Salmon Fish. Comm. Bull. II, Part I. p. 1-66.

Hudson Bay Company Archives. 1822-1824. H.B.C. Arch. B.74/a/1. July 9 - December 7, July 12 - December 20 to January 23.

Hudson Bay Company Archives. 1851-1856. H.B.C. Arch. B.188/a/21. July 11 - November 11.

Hudson Bay Company Archives. 1898-1902. H.B.C. Arch. B.74/a/5. November 1 - December 31.

Hull, C. 1937. Statement by U.S. Secretary of State in July 28 Protocol of Exchange.

Hunter, M. 1978. Letter of February 27 from Canada Dept. Fish. to C.J. Blondin, Assistant Director for International Fisheries, National Marine Fisheries Service, Department of State.

Hutchison, B. 1950. Rivers of America : the Fraser. Clarke Irwin & Co. Limited, Toronto. 368 p.

Idler, D.R. and W.A. Clemens. 1959. The energy expenditures of Fraser River sockeye salmon during the spawning migration to Chilko and Stuart Lakes. Internat. Pacific Salmon Fish. Comm. Prog. Rept. No. 6, 80 p.

International Pacific Halibut Commission, 1987. The Pacific Halibut: Biology, Fishery, and Management Technical Rept. No. 22 (Revision of Nos. 6 and 16) 59 p.

IPSFC. 1939-1986. Annual Reports for 1937-1985.

IPSFC. 1941a. Minutes of the IPSFC meeting held in Seattle, WA. on June 13.

IPSFC. 1948a. Minutes of the Commission meeting of June 22 in Ottawa, Ontario.

IPSFC. 1949a. Interim Report on the Chilko River Watershed. Adm. Rept. 69 p.

IPSFC. 1953a. A review of the sockeye salmon problems created by the Alcan project in the Nechako River watershed. 28 p.

IPSFC. 1961a. The Quesnel story. An historical record of the decimation and rehabilitation of the Quesnel sockeye salmon run. 5 p.

IPSFC. 1965a. An examination of the factors affecting the migration of sockeye and pink salmon in the Fraser and Thompson Rivers at low river levels. Internat. Pacific Salmon Fish. Comm. 57 p.

IPSFC. 1966a. Problems in rehabilitating the Quesnel sockeye run and their possible solution. 85 p.

IPSFC. 1972a. Proposed program for restoration and extension of the sockeye and pink salmon stocks of the Fraser River. 91 p.

IPSFC. 1978a. Preliminary research and development of spawning and incubation channels for sockeye and pink salmon in the Fraser River system. A.C. Cooper, Ed. Unpubl. 108 p.

IPSFC. 1979a. Sockeye salmon studies on the Nechako River. Vol. 2 in: Salmon studies associated with the potential Kemano II hydroelectric development. B.C. Hydro and Power Authority. 95 p. + App.

IPSFC. 1979b. Statement of the International Pacific Salmon Fisheries Commission to the Senate Commerce Committee. Senator Warren Magnuson, U.S.S., Chairman, at hearings in Seattle, Washington. September 4, 1979.

IPSFC. 1983a. Potential effects of the Kemano Completion Project on Fraser River sockeye and pink salmon. 82 p. + App.

Jackson, R.I. 1950. Variations in flow patterns at Hell's Gate and their relationships to the migration of sockeye salmon. Internat. Pacific Salmon Fish. Comm. Bull. III, Part 2: 85-129.

Jackson, R.I. and W.F. Royce. 1986. Ocean Forum, an interpretative history of the International North Pacific Fisheries Commission. 240 p.

Johnson, D.R. 1977. Letter of June 27 from IPSFC Chairman to Robert C. Brewster, Department of State.

Johnson. R.L. and J.J. Burczynski. 1985. Application of dual-beam acoustic procedures to estimate limnetic juvenile sockeye salmon. Internat. Pacific Salmon Fish. Comm. Prog. Rep. No. 41, 33 p.

Kemerick, J. 1945. A review of the artificial propagation and transplantation of the sockeye salmon of the Puget Sound area in the State of Washington conducted by the Federal Government from 1896-1945. U.S. Fish. Wildl. Serv. Rep. 125 p.

Kew, J.E. 1937. Re Fraser River salmon runs. Canada Dept. Fish. Fisheries Inspectors Report of November 15. Quesnel, B.C. 2 p.

Killick, S.R. 1955. The chronological order of Fraser River sockeye salmon during migration, spawning and death. Internat. Pacific Salmon Fish. Comm. Bull. VII, 95 p.

Killick, S.R. and W.A. Clemens. 1963. The age, sex ratio and size of Fraser River sockeye salmon 1915 to 1960. Internat. Pacific Salmon Fish. Comm. Bull. XIV, 114 p. + App.

Kronmiller, T.G. Memorandum for James S.W. Drewry, Acting Assistant General Counsel. March 16, 1978. 3 p.

Kronmiller, T.G. 1985. Testimony to the Subcommittee on Fisheries and Wildlife - Conservation and the Environment on Merchant Marine and Fisheries. House of Representatives Ninety Ninth Congress. Washington D.C. February 25, 1985. p. 96.

Lay, D. 1936. North-eastern mineral survey district (No. 2). British Columbia Dept. of Mines. Annual Report for 1935, Pt. C, p. C17. Victoria.

Lucas, K.C. 1975. Letter of July 4 from Canada Dept. Fish. to W.R. Hourston, Chairman, International Pacific Salmon Fisheries Commission. 2 p.

Lyons, C. 1969. Salmon: our heritage. The story of a province and an industry. 768 p. Vancouver: Mitchell Press Limited.

McCloskey, R.J. 1975. Letter of March 25 from Assistant Secretary, U.S. State Dept. to Senator Henry M. Jackson.

McGrath, J.A. 1979. Letter of August 21 from the Minister of Fisheries of Canada to Gordon Sandison, Chairman of the International Pacific Salmon Fisheries Commission.

McHugh, J. 1915. Report on the work of removal of obstructions to the ascent of salmon on the Fraser River at Hell's Gate, Scuzzy Rapids, China Bar and White's Creek during the year 1914 and the early portion of the year 1915. V 48 1914/1915. Fisheries Branch, Dept. Naval Service, Canada. p. 263-75.

McIntosh, G.B. 1976. Letter of March 1 from IPSFC legal counsel to A.C. Cooper, Director, International Pacific Salmon Fisheries Commission.

Mackenzie, A. 1801. Voyages from Montreal through the continent of North America to the Fraser and Pacific Oceans in 1789 and 1793 with an account of the rise and state of the fur trade. Vol II. Republished in 1911 by the Courier Press, Ltd. Toronto. 360 p.

McKernan, D.L. 1971. Letter of May 10 from U.S. State Dept. to Rogers C.P. Morton, Secretary of Interior.

McKernan, D.L. 1973. Letter of April 23 from Department of State, Coordinator of Ocean Affairs and Special Assistant to the Secretary to Thor C. Tollefson, Chairman, International Pacific Salmon Fisheries Commission.

Manzer, J.I. 1958. Pink and chum salmon tagging experiments in Johnstone Strait and Discovery Passage, 1953. Fish. Res. Bd. Canada. Nanaimo. 34 p.

Martens, D.W. and J.A. Servizi. 1974. Acute toxicity of municipal sewage to fingerling sockeye salmon. Internat. Pacific Salmon Fish. Comm. Prog. Rep. No. 29, 18 p.

Martens, D.W. and J.A. Servizi. 1975. Dechlorination of municipal sewage using sulfur dioxide. Internat. Pacific Salmon Fish. Comm. Prog. Rep. No. 32, 24 p.

Maxwell, R. 1977. Telegram of July 6 from Fisheries Association of British Columbia to Fisheries Minister Romeo Le Blanc.

Mead, R.W. and W.L. Woodall. 1968. Comparison of sockeye salmon fry produced by hatcheries, artificial channels and natural spawning areas. Internat. Pacific Salmon Fish. Comm. Prog. Rep. No. 20, 41 p.

Metcalfe, E.B. 1985. A man of some importance; the life of Roderick Langmere Haig-Brown. 238 p. Seattle: James W. Wood.

Mikkelborg, J.A. and J.T. Mijich. 1976. Letter of March 3 from legal counsel for U.S. purse seine and gillnet associations to U.S. Commissioners Donald R. Johnson, William G. Saletic and Donald W. Moos.

Milne, D.J., E.A.R. Bell, H.M. Jensen and E. Jewell. 1959. Prog. Rep. on the 1957 coho salmon study in the Strait of Juan de Fuca by Canada and the United States. Fish. Res. Bd. Canada, Rept. Biol. No. 677, 68 p.

Mitchell, D.S. 1925. A story of the Fraser River's great sockeye runs, and their loss. Public Archives of British Columbia. 43 p.

Morice, D.M.I., Rev. A.G. 1904. The history of the northern interior of British Columbia, formerly new Caledonia (1660 to 1880). Toronto: William Briggs.

Morisset, M.D. 1976. Letter of April 23 from legal counsel for the Lummi and Makah tribes to Rozanne Ridgeway, Department of State.

Motherwell, J.A. 1927/28. Report of Western Fisheries Division (British Columbia) for 1927. Canada. Dept. of Marine and Fisheries. Fisheries Branch. Annual Report 61: 77.

Motherwell, J.A. 1931. Great run in British Columbia results in heavy spawning. *Pacific Fisherman* Statistical Number V.29, No.2, p. 111-12. Seattle.

Motherwell, J.A. 1934. Condition of British Columbia salmon-spawning grounds. Rept. of the Commissioner of Fisheries, 1934. Province of British Columbia. p. K59-K67. Victoria, B.C.

Napier, G.P. 1914. Report on the obstructed condition of the Fraser River at Scuzzy Rapids, China Bar, Hell's Gate and White's Creek. Rept. of the Commissioner of Fisheries, 1913. Province of British Columbia. p. R39-R42. Victoria, B.C.

Negroponte, J.D. 1977. Letter of August 1 from U.S. State Dept. to Donald R. Johnson, Chairman, International Pacific Salmon Fisheries Commission.

Northwest Indian Fisheries Commission. 1977. IPSFC Fishery - NWIFC calls for replacement of IPSFC Commissioners. Newsletter Vol. III. No. 5.

O'Malley, H. and W.H. Rich. 1919. Migration of adult sockeye salmon in Puget Sound and Fraser River. U.S. Bureau of Fisheries Document 873. Appendix VIII to the report of the United States Commission of Fisheries for 1918. p. 1-38.

Pacific Coast Fisheries. 1903. State Fish Hatcheries. 1(1):6.

Pacific Coast News. 1936a. Fishermen Appoint Committee to investiage salmon treaty. Vancouver, B.C. April 9. 1(36):4.

Pacific Coast News. 1936b. Meet emphasize powers of Commission as main point of disagreement. Vancouver, B.C. April 30. 1(39):8.

Pacific Coast News. 1936c. Delegates agree to support amended 1930 treaty. Vancouver, B.C. May 14. 1(41):6.

Pacific Coast News. 1938. Perpetuation of resources Commission's aim. Statement by Chairman of IPSFC, A. Hager. May 12. 3(41):8.

Pacific Fisherman. 1904. Will Fraser River sockeye go up the Skagit River? II(9):13.

Pacific Fisherman. 1905. The propagation of sockeye salmon on Puget Sound. III(9):16.

Pacific Fisherman. 1929. Treaty Meets Opposition. June 1929. 27(7):13.

Pacific Fisherman. 1931. Ratification of sockeye treaty blocked by Governor Hartley. 29(1):14-16.

Pacific Fisherman. 1937a. Puget Sound-Fraser River sockeye treaty is ratified. 35(5):33.

Pacific Fisherman. 1937b. Fraser River sockeye : discussed by Commission Chairman (A.L. Hager). 35(13):26.

Pacific Fisherman. 1943. International enterprise -restoring the sockeye salmon of the Fraser. 41(10):64

Pacific Salmon Commission. 1986. Annual Report for 1985/86. 41 p.

Pacific Salmon Commission. 1988a. Report of the Fraser River Panel to the Pacific Salmon Commission on the 1986 Fraser River sockcye salmon fishing season. 36 p.

Pacific Salmon Commission. 1988b. Report of the Fraser River Panel to the Pacific Salmon Commission on the 1987 Fraser River sockeye and pink salmon fishing season. 48 p.

Pacific Salmon Commission. 1989. Report of the Fraser River Panel to the Pacific Salmon Commission on the 1988 Fraser River sockeye salmon fishing season. 38 p.

Pacific Salmon Commission. 1990. Report of the Fraser River Panel to the Pacific Salmon Commission on the 1989 Fraser River sockeye and pink salmon fishing season. 51 p.

Parsons, T.A. 1972. The influence of the Fraser River in the productivity of the Strait of Georgia, B.C. Energy Board Provincial Power Study, V. 3, App. 1, Annex (X).

Patterson Jr., B.H. 1974. Memorandum of August 30 from the office of U.S. President to Robert Schoning, Director of National Marine Fisheries Service.

Pearse, P.H. 1982. Turning the tide. A new policy for Canada's Pacific fisheries. The Commission on Pacific Fisheries Policy final report. 292 p.

Peterson, A.E. 1954. The selective action of gillnets on Fraser River sockeye salmon. Internat. Pacific Salmon Fish. Comm. Bull. V, 101 p.

Petty, K.E. 1979. Accommodation of Indian Treaty rights in an international fishery: An international problem begging for an international solution. Washington Law Review 54:403-458.

Powell, S.J. Memorandum of June 2, 1977 of Understanding with the Department of the Interior on regulations of Indians in IPSFC fishery. 2 p.

Pritchard, A.L. 1932. Pacific salmon migration. The tagging of the pink salmon and the chum salmon in British Columbia in 1929 and 1930. Bull. Biol. Bd. Canada, No. 31, 16 p.

Pritchard, J., B. Adams and F.V. Hicks. 1975. Letter of February 6 from U.S. Congressman to Gerald R. Ford, U.S. President.

Pritchard, A.L. and A.C. DeLacy. 1944. Migration of pink salmon (*Oncorhynchus gorbuscha*) in southern British Columbia and Washington in 1943. Bull. Fish. Res. Bd. Canada, No. 66. 23 p.

333

Pugsley, W.H. 1924. Report on observations on the Fraser River August 22, 1922. Washington Department of Fisheries and Game. Fisheries Division. V.32-33, 1921/23, p. 14-15.

Pyper, J. and E.H. Vernon. 1958 MS. Physical and biological features of natural and artificial spawning grounds of Fraser River sockeye. Internat. Pacific Salmon Fish. Comm. Unpubl.

Rathbun, R. 1899. A review of the fisheries in the contiguous waters of the State of Washington and British Columbia. Report of the U.S. Fish Commission for 1899. Washington. p. 251-350

Report Joint Fisheries Commission. 1897. Canadian Sessional Papers. IID. P. vi.

Report of the Commissioners of Fisheries. 1933. Province of British Columbia for the year ending December 31, 1932. p. T27-T31.

Report of the Provincial Fisheries Department, 1937.

Ricker, W.E. 1947a. Hell's Gate and the sockeye. J. Wildlife Management. 11(1): 10-20.

Ricker, W.E. 1947b. Commission urged to guard early run. Commercial Fishermen's Weekly. 13(8):87-89.

Ricker, W.E. 1947c. Review of evidence suggested by Ricker. *Commercial Fishermen's Weekly.* 13(12):135-137.

Ricker, W.E. 1950. Cyclic dominance among the Fraser sockeye. Ecology 31(1):6-26.

Ricker, W.E. 1987. Effects of the fishery and of obstacles to migration on the abundance of Fraser River sockeye salmon (*Oncorhynchus nerka*). Can. Tech. Rep. Fish. Aquat. Sci. 1522, 75 p.

Ridgeway, R.L. 1976a. Letter of March 17 from U.S. State Dept. to U.S. Commissioners Donald R. Johnson, Donald W. Moos and William G. Saletic.

Ridgeway, R.L. 1976b. Letter of August 25 from U.S. State Dept. to Donald R. Johnson, Chairman, International Pacific Salmon Fisheries Commission.

Ridgeway, R.L. 1977. Letter of March 16, 1977 to Commissioner Don Moos.

Roos, J.F. 1961. Observation of possible winter smolt migration in the Fraser River. Memorandum of February 1 to Director of IPSFC. Internat. Pacific Salmon Fish. Comm.

Roos, J.F. 1969. MS Application of commercial and test fishing catches to management of Fraser River sockeye and pink salmon runs. Internat. Pacific Salmon Fish. Comm. 188 p.

Roos, J.F. 1983. IPSFC suggestions for improvement of "Exchange of Notes" section of proposed new salmon treaty. Internat. Pacific Salmon Fish. Comm. 7 p.

Roos, J.F. 1984. Memorandum of July 10 from IPSFC Director to all Commissioners. Fraser River Indian Fishery. Internat. Pacific Salmon Fish. Commission.

Roos, J.F., P. Gilhousen, S.R. Killick and E.R. Zyblut. 1973. Parasitism on juvenile Pacific salmon (*Oncorhynchus*) and Pacific herring (*Clupea harengus pallasi*) in the Strait of Georgia by the river lamprey (*Lampetra ayresi*). J. Fish. Res. Bd. Can. 30(4):565-568.

Rounsefell, G.A. 1949. Methods of estimating total runs and escapements of salmon. Biometrics 5(2):115-126.

Rounsefell, G.A. and G.B. Kelez. 1938. The salmon and salmon fisheries of Swiftsure Bank, Puget Sound, and the Fraser River. Bulletin of the Bureau of Fisheries, Vol. XLIX, Bull. No. 27:693-823.

Royal, L.A. 1953. The effects of regulatory selectivity on the productivity of Fraser River sockeye. Can. Fish Cult. 14:1-12.

Royal, L.A. and A. Seymour. 1940. Building new salmon runs : Puget Sound sockeye salmon plantings show varying degrees of success. Prog. Fish. Cult. 52:1-7.

Royce, F., D.E. Bevan, J.A. Crutchfield, G.J. Paulik and R.L. Fletcher. 1963. Salmon gear limitation in Northern Washington waters. An economic, biological and legal survey of the salmon resource of North Puget Sound and Strait of Juan de Fuca. Univ. of Washington Publication in Fisheries, New Series 2(1), 123 p.

Russell, F.J. 1971. Letter of January 28 from Undersecretary of U.S. Interior Dept. to U.S. Secretary of State William P. Rogers.

Schaefer, M.B. 1951. A study of the spawning populations of sockeye salmon in the Harrison River system, with special reference to the problem of enumeration by means of marked members. Internat. Pacific Salmon Fish. Comm. Bull. IV, 207 p.

Schoning, R. 1975. Speech given to Marine Fisheries Advisory Committee by Director, National Marine Fisheries Service.

Servizi, J.A. and D.W. Martens. 1974. Preliminary survey of toxicity of chlorinated sewage to sockeye and pink salmon. Internat. Pacific Salmon Fish. Comm. Prog. Rep. No. 30, 42 p.

Servizi, J.A. and D.W. Martens. 1978. Effects of selected heavy metals on early life of sockeye and pink salmon. Internat. Pacific Salmon Fish. Comm. Prog. Rep. No. 39, 26 p.

Servizi, J.A., D.W. Martens and R.W. Gordon. 1970. Effects of decaying bark on incubating salmon eggs. Internat. Pacific Salmon Fish. Comm. Prog. Rep. No. 24, 28 p.

Servizi, J.A., E.T. Stone and R.W. Gordon. 1966. Toxicity and treatment of kraft pulp bleach plant waste. Internat. Pacific Salmon Fish. Comm. Prog. Rep. No. 13, 34 p.

Sinclair, J. 1954. Address of March 22 to the annual convention of the United Fishermen and Allied Workers' Union at Vancouver, B.C. by the Minister of Fisheries of Canada.

Sinclair, J. 1955. Address of April 25 to the National Institute of Fisheries at New Orleans, Louisiana by the Minister of Fisheries of Canada.

Sinclair, J. 1956. Letter of March 1 from Minister of Fisheries of Canada to Robert J. Schoettler, Chairman, IPSFC.

Smith, R., G. Riley, A. Morrison and E.E. Prince. 1903. Report of the British Columbia Salmon Commission to the Honorable Raymond Prefontaine, Minister of Marine and Fisheries, Ottawa. 4 p.

State of Washington Department of Fisheries, 1989. 1987 Fisheries Statistical Report. 88 p.

335

State of Washington Department of Fisheries and Game. 1902. 13th Annual Report of the State Fish Commissioner. Fraser River Hatchery. Report by J. Crawford. p. 11.

State of Washington Department of Fisheries and Game. 1903-04. 14th and 15th Annual Reports of the State Fish Commissioner. Skagit River Experimental Work. p. 32-33.

Summary Record. 1974. July 18 meeting of IPSFC and United States and Canadian representatives on problems posed by special treaty fishing rights of certain U.S. Indian tribes. Washington, D.C. July 18.

Sullivan, Jr., W.L. 1974. Letter of October 1 from Acting Coordinator of Ocean Affairs, U.S. State Dept. to Donald R. Johnson, Regional Director, Department of Commerce, National Marine Fisheries Service.

Supreme Court of the United States. 1979. *Washington et al. v Washington State Commercial Passenger Fishing Vessel Assn. et al.* 37 p.

Talbot, G.B. 1950. A biological study of the effectiveness of the Hell's Gate fishways. Internat. Pacific Salmon Fish. Comm. Bull. III, Part I: 1-80.

Taylor, S.G., J.H. Landingham, D.G. Mortenson and A.C. Wertheimer. 1986. Pink salmon early life history in Auke Bay: residence, growth, diet, and survival. Northwest Alaska Fisheries Center, National Oceans & Atmospheric Administration, National Marine Fisheries Service. Auke Bay Lab., Auke Bay, AK.

The Fisherman. 1977. U.S. stand hits at Commission. Vancouver, B.C. April 22.

Thompson, M. 1974. Letter of November 22 from Commissioner of Indian Affairs, U.S. Interior Dept. to A.C. Cooper, Director, IPSFC.

Thompson, W.F. 1939. Report on the investigations of International Pacific Salmon Fisheries Commission on the Fraser River sockeye for the year 1938. IPSFC Annual Report for 1937 and 1938. Appendix B. 21 p.

Thompson, W.F. 1945a. Effect of the obstruction at Hell's Gate on the sockeye salmon of the Fraser River. Internat. Pacific Salmon Fish. Comm. Bull. I, 175 p.

Thompson, W.F. 1945b. Scientific management of the Fraser River sockeye. p. 23-28 In: Annual Report of the IPSFC for the year 1944. 74 p.

Thompson, W.F. 1947a. The Hell's Gate blockade and the salmon. Internat. Pacific Salmon Fish. Comm. Section 1, 19 p., Section 2, 20 p.

Thompson, W.F. 1947b. Dr. Thompson replies to biologists' criticism of Hell's Gate fishways. *Commercial Fisherman.* 14(1):6-12.

Thompson, W.F., L.P. Schultz and L.R. Donaldson. 1935. Letter of July 3 from Univ. of Wash. School of Fisheries to Hon. Clarence D. Martin, Governor, State of Washington.

Tomasovich, J. 1943. International agreements on conservation of marine resources with special reference to the North Pacific. 297 p.

United Fishermen and Allied Workers' Union. 1977. Telegram of July 18 to Romeo Le Blanc, Minister of Fisheries of Canada.

United States Court of Appeals for the Ninth Circuit. 1975. On appeal from the United States District Court for the Western District of Washington. 31 p.

United States Department of State. 1981. Letter of September 23 from Theodore G. Kronmiller to William G. Saletic, Chairman, International Pacific Salmon Fisheries Commission.

United States District Court for the Western District of Washington at Tacoma. 1974a. *United States of America v State of Washington.* Civil No. 9213. 149 p.

United States District Court for the Western District of Washington at Tacoma. 1974b. *United States of America v State of Washington.* Civil No. 9213 - Fisheries' motion for reconsideration. Clarification of March 22, 1974.

United States District Court for the Western District of Washington at Tacoma. 1975. Affidavit of William L. Sullivan, Jr. Civil No. 9213.

United States District Court for the Western District of Washington. 1977. *Purse Seine Vessel Owners Association, et al. v United States Department of State, et al.* No. C77-471M.

Verhoeven, L.A. and E.B. Davidoff. 1962. Marine tagging of Fraser River sockeye salmon. Internat. Pacific Salmon Fish. Comm. Bull. XIII, 132 p.

Vernon, E.H. 1958. An examination of the factors affecting the abundance of pink salmon in the Fraser River. Internat. Pacific Salmon Fish. Comm. Prog. Rep. No. 5, 49 p.

Vernon, E.H. 1966. Enumeration of migrant pink salmon fry in the Fraser River estuary. Internat. Pacific Salmon Fish. Comm. Bull. XIX, 83 p.

Vernon, E.H., A.S. Hourston and G.A. Holland. 1964. The migration and exploitation of pink salmon runs in and adjacent to the Fraser River Convention area in 1959. Internat. Pacific Salmon Fish. Comm. Bull. XV, 296 p.

Ward, F.J. 1959. Character of the migration of pink salmon to Fraser River spawning grounds in 1957. Internat. Pacific Salmon Fish. Comm. Bull. X, 70 p.

Ward, F.J. and P.A. Larkin. 1964. Cyclic dominance in Adams River sockeye salmon. Internat. Pacific Salmon Fish. Comm. Prog. Rep. 11, 116 p.

Wickett, P.W. 1979. Changes in inshore migration route of adult Fraser River sockeye salmon (*Oncorhynchus nerka*) as a response to oceanographic variables. Resource Services Branch, Department of Fisheries and Oceans, Pacific Biological Station, Nanaimo, B.C. 17 p.

Williams I.V. 1973. Investigation of the prespawning mortality of sockeye in Horsefly River and McKinley Creek in 1969. Internat. Pacific Salmon Fish. Comm. Prog. Rep. 27, Part II, 42 p.

Williams, I.V., J.R. Brett, G.R. Bell, G.S. Troxler, J. Bagshaw, J.R. McBride, U.H.M. Fagerlund, H.M. Dye, J.P. Sumpter, E.M. Donaldson, E. Bilinski, H. Tsuyuki, M.D. Peters, E.M. Choromanski, J.H.Y. Cheng and W.L. Coleridge. 1989a. The 1983 early run Fraser and Thompson River pink salmon: morphology, energetics, and fish health. Internat. Pacific Salmon Fish. Comm. Bull. XXIII, 55 p.

Williams, I.V. and U.H.M. Fagerlund, J.R. McBride, G.A. Strasdine, H. Tsuyuki and E.J. Ordal. 1977. Investigation of prespawning mortality of 1973 Horsefly River sockeye salmon. Internat. Pacific Salmon Fish. Comm. Prog. Rep. No. 37, 37 p.

Williams, I.V. and P. Gilhousen. 1968. Lamprey parasitism on Fraser River sockeye and pink salmon during 1967. Internat. Pacific Salmon Fish. Comm. Prog. Rept. No. 18, 22 p.

Williams, I.V., P. Gilhousen, W. Saito, T. Gjernes, K. Morton, R. Johnson and D. Brock. 1989b. Studies of the lacustrine biology of the sockeye salmon (*Oncorhynchus nerka*) in the Shuswap system. Internat. Pacific Salmon Fish. Comm. Bull. XXIV, 108 p.

Wood, W. 1965. A report on fish disease as a possible cause of pre-spawning mortalities of Fraser River sockeye. Internat. Pacific Salmon Fish. Comm. 24 p.

Woodey, J.C. 1987. In-season management of Fraser River sockeye salmon (*Oncorhynchus nerka*): meeting multiple objectives p. 367-374. In: H.D. Smith, L. Margolis, and C.C. Wood (ed.) Sockeye salmon (*Oncorhynchus nerka*) population biology and future management. Can. Spec. Publ. Fish. Aquat. Sci. 96. 486 p.

Yanagida, J.A. 1987. The Pacific Salmon Treaty. The American Journal of International Law (81)3:577-592.

Ziontz, A.J. 1975. Letter of May 15 from attorney for Indian tribes to Commissioners Thor Tollefson, Donald R. Johnson, William G. Saletic and State Fisheries Director Donald W. Moos.

A P P E N D I X   A

The 1930 Convention, 1956 Protocol, Sockeye Salmon Fishery Act of 1947 and 1980 Amendment

By the President of the United States of America.

A PROCLAMATION.

WHEREAS a Convention between the United States of America and Canada for the protection, preservation and extension of the sockeye salmon fishery of the Fraser River system was concluded and signed by their respective Plenipotentiaries at Washington, on the twenty-sixth day of May, one thousand nine hundred and thirty, the original of which Convention is word for word as follows:

The President of the United States of America and His Majesty the King of Great Britain, Ireland and the British dominions beyond the Seas, Emperor of India, in respect of the Dominion of Canada, recognizing that the protection, preservation and extension of the sockeye salmon fisheries in the Fraser River system are of common concern to the United States of America and the Dominion of Canada; that the supply of this fish in recent years has been greatly depleted and that it is of importance in the mutual interest of both countries that this source of wealth should be restored and maintained, have resolved to conclude a Convention and to that end have named as their respective plenipotentiaries:

The President of the United States of America: Mr. Henry L. Stimson, Secretary of State of the United States of America; and

His Majesty, for the Dominion of Canada: The Honorable Vincent Massey, a member of His Majesty's Privy Council for Canada and His Envoy Extraordinary and Minister Plenipotentiary for Canada at Washington;

Who, after having communicated to each other their full powers, found in good and due form, have agreed upon the following Articles:

ARTICLE I

The provisions of this Convention and the orders and regulations issued under the authority thereof shall apply, in the manner and to the extent hereinafter provided in this Convention, to the following waters:

1. The territorial waters and the high seas westward from the western coast of the United States of America and the Dominion of Canada and from a direct line drawn from Bonilla Point, Vancouver Island, to the lighthouse on Tatoosh Island, Washington,—which line marks the entrance to Juan de Fuca Strait,—and embraced between 48 and 49 degrees north latitude, excepting therefrom, however, all the waters of Barklay Sound, eastward of a straight line drawn from Amphitrite Point to Cape Beale and all the waters of Nitinat Lake and the entrance thereto.

4334—37   (1)

2. The waters included within the following boundaries:

Beginning at Bonilla Point, Vancouver Island, thence along the aforesaid direct line drawn from Bonilla Point to Tatoosh Lighthouse, Washington, described in paragraph numbered 1 of this Article, thence to the nearest point of Cape Flattery, thence following the southerly shore of Juan de Fuca Strait to Point Wilson, on Quimper Peninsula, thence in a straight line to Point Partridge on Whidbey Island, thence following the western shore of the said Whidbey Island, to the entrance to Deception Pass, thence across said entrance to the southern side of Reservation Bay, on Fidalgo Island, thence following the western and northern shore line of the said Fidalgo Island to Swinomish Slough, crossing the said Swinomish Slough, in line with the track of the Great Northern Railway, thence northerly following the shore line of the mainland to Atkinson Point at the northerly entrance to Burrard Inlet, British Columbia, thence in a straight line to the southern end of Bowen Island, thence westerly following the southern shore of Bowen Island to Cape Roger Curtis, thence in a straight line to Gower Point, thence westerly following the shore line to Welcome Point on Seechelt Peninsula, thence in a straight line to Point Young on Lasqueti Island, thence in a straight line to Dorcas Point on Vancouver Island, thence following the eastern and southern shores of the said Vancouver Island to the starting point at Bonilla Point, as shown on the United States Coast and Geodetic Survey Chart Number 6300, as corrected to March 14, 1930, and on the British Admiralty Chart Number 579, copies of which are annexed to this Convention and made a part thereof.

3. The Fraser River and the streams and lakes tributary thereto.

The High Contracting Parties engage to have prepared as soon as practicable charts of the waters described in this Article, with the above described boundaries thereof and the international boundary indicated thereon. Such charts, when approved by the appropriate authorities of the Governments of the United States of America and the Dominion of Canada, shall be considered to have been substituted for the charts annexed to this Convention and shall be authentic for the purposes of the Convention.

The High Contracting Parties further agree to establish within the territory of the United States of America and the territory of the Dominion of Canada such buoys and marks for the purposes of this Convention as may be recommended by the Commission hereinafter authorized to be established, and to refer such recommendations as the Commission may make as relate to the establishment of buoys or marks at points on the international boundary to the International Boundary Commission, United States-Alaska and Canada, for action pursuant to the provisions of the Treaty between the United States of America and His Majesty, in respect of Canada, respecting the boundary between the United States of America and the Dominion of Canada, signed February 24, 1925.

Article II

The High Contracting Parties agree to establish and maintain a Commission to be known as the International Pacific Salmon Fisheries Commission, hereinafter called the Commission, consisting of six members, three on the part of the United States of America and three on the part of the Dominion of Canada.

3

The Commissioners on the part of the United States of America shall be appointed by the President of the United States of America. The Commissioners on the part of the Dominion of Canada shall be appointed by His Majesty on the recommendation of the Governor General in Council.

The Commissioners appointed by each of the High Contracting Parties shall hold office during the pleasure of the High Contracting Party by which they were appointed.

The Commission shall continue in existence so long as this Convention shall continue in force, and each High Contracting Party shall have power to fill and shall fill from time to time vacancies which may occur in its representation on the Commission in the same manner as the original appointments are made. Each High Contracting Party shall pay the salaries and expenses of its own Commissioners, and joint expenses incurred by the Commission shall be paid by the two High Contracting Parties in equal moieties.

341

Article III

The Commission shall make a thorough investigation into the natural history of the Fraser River sockeye salmon, into hatchery methods, spawning ground conditions and other related matters. It shall conduct the sockeye salmon fish cultural operations in the waters described in paragraphs numbered 2 and 3 of Article I of this Convention, and to that end it shall have power to improve spawning grounds, construct, and maintain hatcheries, rearing ponds and other such facilities as it may determine to be necessary for the propagation of sockeye salmon in any of the waters covered by this Convention, and to stock any such waters with sockeye salmon by such methods as it may determine to be most advisable. The Commission shall also have authority to recommend to the Governments of the High Contracting Parties removing or otherwise overcoming obstructions to the ascent of sockeye salmon, that may now exist or may from time to time occur, in any of the waters covered by this Convention, where investigation may show such removal of or other action to overcome obstructions to be desirable. The Commission shall make an annual report to the two Governments as to the investigations which it has made and other action which it has taken in execution of the provisions of this Article, or of other Articles of this Convention.

The cost of all work done pursuant to the provisions of this Article, or of other Articles of this Convention, including removing or otherwise overcoming obstructions that may be approved, shall be borne equally by the two Governments, and the said Governments agree to appropriate annually such money as each may deem desirable for such work in the light of the reports of the Commission.

Article IV

The Commission is hereby empowered to limit or prohibit taking sockeye salmon in respect of all or any of the waters described in Article I of this Convention, provided that when any order is adopted by the Commission limiting or prohibiting taking sockeye salmon in any of the territorial waters or on the High Seas described in paragraph numbered 1 of Article I, such order shall extend to all such territorial waters and High Seas, and, similarly, when in any of the waters of the United States of America embraced in paragraph numbered 2 of Article I, such order shall extend to all such waters of the United States of America, and when in any of the Canadian waters embraced in paragraphs numbered 2 and 3 of Article I, such order shall extend to all such Canadian waters, and provided further, that no order limiting or prohibiting taking sockeye salmon adopted by the Commission shall be construed to suspend or otherwise affect the requirements of the laws of the State of Washington or of the Dominion of Canada as to the procuring of a license to fish, in the waters on their respective sides of the boundary, or in their respective territorial waters embraced in paragraph numbered 1 of Article I of this Convention, and provided further that any order adopted by the Commission limiting or prohibiting taking sockeye salmon on the High Seas embraced in paragraph numbered 1 of Article I of this Convention shall apply only to nationals and inhabitants and vessels and boats of the United States of America and the Dominion of Canada.

Any order adopted by the Commission limiting or prohibiting taking sockeye salmon in the waters covered by this Convention, or any part thereof, shall remain in full force and effect unless and until the same be modified or set aside by the Commission. Taking sockeye salmon in said waters in violation of an order of the Commission shall be prohibited.

ARTICLE V

In order to secure a proper escapement of sockeye salmon during the spring or chinook salmon fishing season, the Commission may prescribe the size of the meshes in all fishing gear and appliances that may be operated during said season in the waters of the United States of America and/or the Canadian waters described in Article I of this Convention. At all seasons of the year the Commission may prescribe the size of the meshes in all salmon fishing gear and appliances that may be operated on the High Seas embraced in paragraph numbered 1 of Article I of this Convention, provided, however, that in so far as concerns the High Seas, requirements prescribed by the Commission under the authority of this paragraph shall apply only to nationals and inhabitants and vessels and boats of the United States of America and the Dominion of Canada.

Whenever, at any other time than the spring or chinook salmon fishing season, the taking of sockeye salmon in waters of the United States of America or in Canadian waters is not prohibited under an order adopted by the Commission, any fishing gear or appliance authorized by the State of Washington may be used in waters of the United States of America by any person thereunto authorized by the State of Washington, and any fishing gear or appliance authorized by the laws of the Dominion of Canada may be used in Canadian waters by any person thereunto duly authorized. Whenever the taking of sockeye salmon on the High Seas embraced in paragraph numbered 1 of Article I of this Convention is not prohibited, under an order adopted by the Commission, to the nationals or inhabitants or vessels or boats of the United States of America or the Dominion of Canada, only such salmon fishing gear and appliances as may have been approved by the Commission may be used on such High Seas by said nationals, inhabitants, vessels or boats.

ARTICLE VI

No action taken by the Commission under the authority of this Convention shall be effective unless it is affirmatively voted for by at least two of the Commissioners of each High Contracting Party.

ARTICLE VII

Inasmuch as the purpose of this Convention is to establish for the High Contracting Parties, by their joint effort and expense, a fishery that is now largely nonexistent, it is agreed by the High Contracting Parties that they should share equally in the fishery. The Commission shall, consequently, regulate the fishery with a view to allowing, as nearly as may be practicable, an equal portion of the fish that may be caught each year to be taken by the fishermen of each High Contracting Party.

ARTICLE VIII

Each High Contracting Party shall be responsible for the enforcement of the orders and regulations adopted by the Commission under the authority of this Convention, in the portion of its waters covered by the Convention.

Except as hereinafter provided in Article IX of this Convention, each High Contracting Party shall be responsible, in respect of its own nationals and inhabitants and vessels and boats, for the enforcement of the orders and regulations adopted by the Commission, under the authority of this Convention, on the High Seas embraced in paragraph numbered 1 of Article I of the Convention.

Each High Contracting Party shall acquire and place at the disposition of the Commission any land within its territory required for the construction and maintenance of hatcheries, rearing ponds, and other such facilities as set forth in Article III.

ARTICLE IX

Every national or inhabitant, vessel or boat of the United States of America or of the Dominion of Canada, that engages in sockeye salmon fishing on the High Seas embraced in paragraph numbered 1 of Article I of this Convention, in violation of an order or regulation adopted by the Commission, under the authority of this Convention, may be seized and detained by the duly authorized officers of either High Contracting Party, and when so seized and detained shall be delivered by the said officers, as soon as practicable, to an authorized official of the country to which such person, vessel or boat belongs, at the nearest point to the place of seizure, or elsewhere, as may be agreed upon with the competent authorities. The authorities of the country to which a person, vessel or boat belongs alone shall have jurisdiction to conduct prosecutions for the violation of any order or regulation, adopted by the Commission in respect of fishing for sockeye salmon on the High Seas embraced in paragraph numbered 1 of Article I of this Convention, or of any law or regulation which either High Contracting Party may have made to carry such order or regulation of the Commission into effect, and to impose penalties for such violations; and the witnesses and proofs necessary for such prosecutions, so far as such witnesses or proofs are under the control of the other High Contracting Party, shall be furnished with all reasonable promptitude to the authorities having jurisdiction to conduct the prosecutions.

ARTICLE X

The High Contracting Parties agree to enact and enforce such legislation as may be necessary to make effective the provisions of this Convention and the orders and regulations adopted by the Commission under the authority thereof, with appropriate penalties for violations.

ARTICLE XI

The present Convention shall be ratified by the President of the United States of America, by and with the advice and consent of the Senate thereof, and by His Majesty in accordance with constitutional practice, and it shall become effective upon the date of the exchange of ratifications which shall take place at Washington as soon as possible and shall continue in force for a period of sixteen years, and thereafter until one year from the day on which either of the High Contracting Parties shall give notice to the other of its desire to terminate it.

In witness whereof, the respective plenipotentiaries have signed the present Convention, and have affixed their seals thereto.

Done in duplicate at Washington on the twenty-sixth day of May, one thousand nine hundred and thirty.

[SEAL]   HENRY L STIMSON
[SEAL]   VINCENT MASSEY

AND WHEREAS the said Convention has been duly ratified on both parts, and the ratifications of the two Governments were exchanged in the city of Washington on the twenty-eighth day of July, one thousand nine hundred and thirty-seven;

AND WHEREAS the said Convention was ratified by the United States of America subject to three understandings, made a part of the ratification, as follows:

(1) That the International Pacific Salmon Fisheries Commission shall have no power to authorize any type of fishing gear contrary to the laws of the State of Washington or the Dominion of Canada;

(2) That the Commission shall not promulgate or enforce regulations until the scientific investigations provided for in the convention have been made, covering two cycles of Sockeye Salmon runs, or eight years; and

(3) That the Commission shall set up an Advisory Committee composed of five persons from each country who shall be representatives of the various branches of the industry (purse seine, gill net, troll, sport fishing, and one other), which Advisory Committee shall be invited to all non-executive meetings of the Commission and shall be given full opportunity to examine and to be heard on all proposed orders, regulations or recommendations.

AND WHEREAS the aforesaid three understandings have been accepted by the Government of Canada, as is recorded in the Protocol of Exchange of ratifications [1] of the said Convention;

Now, THEREFORE, be it known that I, Franklin D. Roosevelt, President of the United States of America, have caused the said Convention to be made public, to the end that the same and every article and clause thereof may be observed and fulfilled with good faith by the United States of America and the citizens thereof, subject to the three understandings herein recited.

IN TESTIMONY WHEREOF, I have hereunto set my hand and caused the Seal of the United States of America to be affixed.

DONE at the city of Washington this fourth day of August in the year of our Lord one thousand nine hundred and thirty-[SEAL] seven, and of the Independence of the United States of America the one hundred and sixty-second.

FRANKLIN D ROOSEVELT

By the President:
CORDELL HULL
Secretary of State.

---

[1] See p. 9.

345

PROTOCOL BETWEEN THE GOVERNMENT OF THE UNITED STATES OF AMERICA AND THE GOVERNMENT OF CANADA TO THE CONVENTION FOR THE PROTECTION, PRESERVATION AND EXTENSION OF THE SOCKEYE SALMON FISHERIES IN THE FRASER RIVER SYSTEM SIGNED AT WASHINGTON ON THE 26TH DAY OF MAY 1930.

The Government of the United States of America and the Government of Canada, desiring to coordinate the programs for the conservation of the sockeye and pink salmon stocks of common concern by amendment of the Convention between the United States of America and Canada for the Protection, Preservation and Extension of the Sockeye Salmon Fisheries in the Fraser River System, signed at Washington on the 26th day of May, 1930, hereinafter referred to as the Convention,

TS 918.
50 Stat. 1355.

Have agreed as follows:

Article I

The Convention as amended by the present Protocol shall apply to pink salmon with the following exception:

The understanding stipulated in the Protocol of Exchange of Ratifications signed at Washington on the 28th day of July, 1937, which provides that "the Commission shall not promulgate or enforce regulations until the scientific investigations provided for in the Convention have been made, covering two cycles of sockeye salmon runs, or eight years;" shall not apply to pink salmon.

TS 918.
50 Stat. 1361.

Article II

The following words shall be deleted from the first sentence of Article IV of the Convention:

". . . that when any order is adopted by the Commission limiting or prohibiting taking sockeye salmon in any of the territorial waters or on the High Seas described in paragraph numbered 1 of Article I, such order shall extend to all such territorial waters and High Seas, and, similarly, when in any of the waters of the United States of America embraced in paragraph numbered 2 of Article I, such order shall extend to all such waters of the United States of America, and when in any of the Canadian waters embraced in paragraphs numbered 2 and 3 of Article I, such order shall extend to all such Canadian waters, and provided further. . . . ."

4

Article III

The following paragraph shall be added to Article VI of the Convention:

"All regulations made by the Commission shall be subject to approval of the two Governments with the exception of orders for the adjustment of closing or opening of fishing periods and areas in any fishing season and of emergency orders required to carry out the provisions of the Convention."

Article IV

Article VII of the Convention shall be replaced by the following Article:

"The Commission shall regulate the fisheries for sockeye and for pink salmon with a view to allowing, as nearly as practicable, an equal portion of such sockeye salmon as may be caught each year and an equal portion of such pink salmon as may be caught each year to be taken by the fishermen of each Party."

Article V

Paragraph (3) of the understandings stipulated in the Protocol of Exchange of Ratifications signed at Washington on the 28th day of July, 1937, shall be amended to read as follows:

"That the Commission shall set up an Advisory Committee composed of six persons from each country who shall be representatives of the various branches of the industry including, but not limited to, purse seine, gill net, troll, sport fishing and processing, which Advisory Committee shall be invited to all non-executive meetings of the Commission and shall be given full opportunity to examine and to be heard on all proposed orders, regulations or recommendations."

Article VI

1. The Parties shall conduct a coordinated investigation of pink salmon stocks which enter the waters described in Article I of the Convention for the purpose of determining the migratory movements of such stocks. That part of the investigation to be carried out in the waters described in Article I of the Convention shall be carried out by the Commission.

2. Except with regard to that part of the investigation to be carried out by the Commission, the provisions of Article III of the Convention with respect to the sharing of cost shall not apply to the investigation referred to in this Article.

3. The Parties shall meet in the seventh year after the entry into force of this Protocol to examine the results of the investiga-

tion referred to in this Article and to determine what further arrangements for the conservation of pink salmon stocks of common concern may be desirable.

Article VII

Nothing in the Convention or this Protocol shall preclude the Commission from recording such information on stocks of salmon other than sockeye or pink salmon as it may acquire incidental to its activities with respect to sockeye and pink salmon.

Article VIII

The present Protocol shall be ratified and the exchange of the instruments of ratification shall take place in Ottawa as soon as possible. It shall come into force on the day of the exchange of the instruments of ratification.

IN WITNESS WHEREOF the undersigned, duly authorized by their respective Governments, have signed this Protocol and have affixed thereto their seals.

Done in duplicate at Ottawa this 28th day of December, 1956.

For the Government of the
United States of America:

[SEAL]   LIVINGSTON T. MERCHANT

WM C HERRINGTON

For the Government of Canada:

[SEAL]   JAMES SINCLAIR

[Public Law 255—80th Congress]
[Chapter 345—1st Session]

[H. R. 3767]

AN ACT

To provide for the protection, preservation, and extension of the sockeye salmon fishery of the Fraser River system, and for other purposes.

Be it enacted by the Senate and House of Representatives of the United States of America in Congress assembled, That this Act may be cited as the "Sockeye Salmon Fishery Act of 1947".

Sec. 2. When used in this Act—

(a) Convention: The word "convention" means the convention between the United States of America and the Dominion of Canada for the protection, preservation, and extension of the sockeye salmon fishery of the Fraser River system, signed at Washington on the 26th day of May 1930.

(b) Commission: The word "Commission" means the International Pacific Salmon Fisheries Commission provided for by article II of the convention.

(c) Person: The word "person" includes individuals, partnerships, associations, and corporations.

(d) Convention waters: The term "convention waters" means those waters described in article I of the convention.

(e) Sockeye salmon: The term "sockeye salmon" means that species of salmon known by the scientific name Oncorhynchus nerka.

(f) Vessel: The word "vessel" includes every type or description of water craft or other contrivance used, or capable of being used, as a means of transportation in water.

(g) Fishing: The word "fishing" means the fishing for, catching, or taking, or the attempted fishing for, catching, or taking, of any sockeye salmon in convention waters.

(h) Fishing gear: The term "fishing gear" means any net, trap, hook, or other device, appurtenance or equipment, of whatever kind or description, used or capable of being used, for the purpose of capturing fish or as an aid in capturing fish.

Sec. 3. (a) It shall be unlawful for any person to engage in fishing for sockeye salmon in convention waters in violation of the convention or of this Act or of any regulation of the Commission.

(b) It shall be unlawful for any person to ship, transport, purchase, sell, offer for sale, import, export, or have in possession any sockeye salmon taken in violation of the convention or of this Act or of any regulation of the Commission.

(c) It shall be unlawful for any person or vessel to use any port or harbor or other place subject to the jurisdiction of the United States for any purpose connected in any way with fishing in violation of the convention or of this Act or of any regulation made by the Commission.

(d) It shall be unlawful for any person or vessel to engage in fishing for sockeye salmon in convention waters without first having obtained

349

such license or licenses as may be used by or required by the Commission, or to fail to produce such license, upon demand, for inspection by an authorized enforcement officer.

(e) It shall be unlawful for any person to fail to make, keep, submit, or furnish any record or report required of him by the Commission or to refuse to permit any officer authorized to enforce the convention, this Act, and the regulations of the Commission, or any authorized representative of the Commission, to inspect any such record or report at any reasonable time.

(f) It shall be unlawful for any person to molest, interfere with, tamper with, damage, or destroy any boat, net, equipment, stores, provisions, fish-cultural stations, rearing pond, weir, fishway, or any other structure, installation, experiment, property, or facility acquired, constructed, or maintained by the Commission.

(g) It shall be unlawful for any person or vessel to do any act prohibited or to fail to do any act required by the convention or by this Act or by any regulation of the Commission.

SEC. 4. Any person who fails to make, keep, or furnish any catch return, statistical record, or any report that may be required by the Commission, or any person who furnishes a false return, record, or report, upon conviction shall be subject to such fine as may be imposed by the court not to exceed $1,000, and shall in addition be prohibited from fishing for and from shipping, transporting, purchasing, selling, offering for sale, importing, exporting, or possessing sockeye salmon from the date of conviction until such time as any delinquent return, record, or report shall have been submitted or any false return, record, or report shall have been replaced by a duly certified correct and true return, record, or report to the satisfaction of the court.   The penalties imposed by section 5 of this Act shall not be invoked for failure to comply with requirements respecting returns, records, and reports.

SEC. 5. (a) Except as provided in section 4, any person violating any provision of the convention or of this Act or the regulation of the Commission upon conviction shall be fined not more than $1,000 or be imprisoned not more than one year, or both, and the court may prohibit such person from fishing for, or from shipping, transporting, purchasing, selling, offering for sale, importing, exporting, or possessing sockeye salmon for such period of time as it may determine.

(b) The catch of fish of every vessel or of any fishing gear employed in any manner, or any fish caught, shipped, transported, purchased, sold, offered for sale, imported, exported, or possessed in violation of this Act or the regulations of the Commission shall be forfeited; and upon a second and subsequent violation the catch of fish shall be forfeited and every such vessel and any fishing gear and appurtenances involved in the violation may be forfeited.

(c) All procedures of law relating to the seizure, judicial forfeiture, and condemnation of a vessel for violation of the customs laws and the disposition of such vessel or the proceeds from the sale thereof shall apply to seizures, forfeitures, and condemnations incurred, or alleged to have been incurred, under the provisions of this Act insofar as such provisions of law are applicable and not inconsistent with this Act.

(d) In cases of minor violations of the provisions of the convention or of this Act or the regulations of the Commission, and in cases where immediate arrest of the person or seizure of fish, fishing gear, or of a vessel, together with its tackle, apparel, furniture, appurtenances, and cargo, would impose an unreasonable hardship, the person authorized to make such arrest or seizure or any court of competent jurisdiction may, in his or its discretion, issue a citation requiring such person to appear before the proper official of the court having jurisdiction thereof within a specified time, not exceeding fifteen days; or in the case of property, post such citation upon said property and require its delivery to such court within such specified time. Upon the issuance of such citation and the filing of a copy thereof with the clerk of the appropriate court the person so cited and the property so seized and posted shall thereupon be subject to the jurisdiction of the court to answer the order of the court in such cause. Any property so seized shall not be disposed of except pursuant to the order of such court or the provisions of subsection (e) of this section.

(e) When a warrant of arrest or other process in rem, including that specified in subsection (d) of this section, is issued in any cause of admiralty jurisdiction under this section, the marshal or other officer shall stay the execution of such process, or discharge any property seized if the process has been levied, on receiving from the claimant of the property a bond or stipulation with sufficient sureties or approved corporate surety in such sum as the court shall order, conditioned to deliver the property seized, if condemned, without impairment in value (or, in the case of sockeye salmon, to pay its equivalent in money) or otherwise to answer the decree of the court in such cause. Such bond or stipulation shall be returned to the court and judgment thereon against both the principal and sureties may be recovered in the event of any breach of the conditions thereof as determined by the court.

Sec. 6. (a) The President of the United States shall designate a Federal agency which shall be responsible for the enforcement of the provisions of the convention and this Act and the regulations of the Commission, except to the extent otherwise provided for in the convention and this Act. It shall be the duty of the Federal agency so designated to take appropriate measures for enforcement at such times and to such extent as it may deem necessary to insure effective enforcement and for this purpose to cooperate with other Federal agencies, State officers, the Commission, and with the authorized officers of the Dominion of Canada.

(b) The Federal agency designated by the President for enforcement purposes may authorize officers and employees of the State of Washington to enforce the provisions of the convention and of this Act and the regulations of the Commission. When so authorized such officers may function as Federal law-enforcement officers for the purposes of this Act.

(c) Enforcement of the convention and this Act and the regulations of the Commission shall be subject to and in accordance with the provisions of article IX of the convention.

(d) Any duly authorized officer or employee of the Federal agency designated by the President for enforcement purposes under the provisions of subsection (a) of this section 6; any officer or employee of the State of Washington who is authorized by the Federal agency so designated by the President; any enforcement officer of the Fish and Wildlife Service of the Department of the Interior, any Coast Guard officer, any United States marshal or deputy United States marshal, any collector or deputy collector of customs, and any other person authorized to enforce the provisions of the convention, this Act, and the regulations of the Commission, shall have power, without warrant or other process, but subject to the provisions of the convention, to arrest any person committing in his presence or view a violation of the convention or of this Act or of the regulations of the Commission and to take such person immediately for examination before an officer or trial before a court of competent jurisdiction; and shall have power, without warrant or other process, to search any vessel within convention waters when he has reasonable cause to believe that such vessel is subject to seizure under the provisions of the convention or this Act, or the regulations of the Commission, and to search any place of business or any commercial vehicle when he has reasonable cause to believe that such place or vehicle contains fish taken, possessed, transported, purchased, or sold in violation of any of the provisions of the convention, this Act, or the regulations of the Commission. Any person authorized to enforce the provisions of the convention and of this Act and the regulations of the Commission shall have power to execute any warrant or process issued by an officer or court of competent jurisdiction for the enforcement of this Act, and shall have power with a search warrant to search any person, vessel, or place, at any time. The judges of the United States courts and the United States commissioners may, within their respective jurisdictions, upon proper oath or affirmation showing probable cause, issue warrants in all such cases. Subject to the provisions of the convention, any person authorized to enforce the convention and this Act and the regulations of the Commission may seize, whenever and wherever lawfully found, all fish caught, shipped, transported, purchased, sold, offered for sale, imported, exported, or possessed contrary to the provisions of the convention or this Act or the regulations of the Commission and may seize any vessel, together with its tackle, apparel, furniture, appurtenances and cargo, and all fishing gear, used or employed contrary to the provisions of the convention or this Act or the regulations of the Commission, or which it reasonably appears has been used or employed contrary to the provisions of the convention or this Act or the regulations of the Commission.

(e) Evidence of any regulation made by the Commission may be given in any court proceedings by the production of a copy of such regulation certified by the Secretary of the Commission to be a true copy and no proof of the signature of the Secretary on such certification shall be required.

(f) Any authorized representative of the Commission, or any person authorized to enforce this Act and the regulations of the Commission may inspect any licenses issued to persons or vessels engaging in fishing for sockeye salmon in convention waters and for this purpose may at any reasonable time board any vessel or enter upon any premises where such fishing is or may be conducted.

Sec. 7. (a) All agencies of the Federal Government are authorized, upon request by the Commission, to furnish facilities and personnel for the purpose of assisting the Commission in carrying out its duties of scientific investigation and improvement of the fishery, as specified in the convention.

(b) None of the prohibitions contained in this Act, or in the laws and regulations of the States, shall prevent the Commission from conducting or authorizing the conduct of fishing operations and biological experiments at any time for purposes of scientific investigation, or shall prevent the Commission from discharging any other duties prescribed by the convention.

SEC. 8. There is authorized to be appropriated, out of any moneys in the Treasury not otherwise appropriated, such sums, from time to time, as may be necessary to enable the Commission and agencies of the Federal Government to carry out the provisions of the convention and of this Act, including purchase, operation, maintenance, and repair of aircraft, motor vehicles (including passenger-carrying vehicles), boats, research vessels, and other necessary facilities; and printing.

SEC. 9. If any provision of this Act is held invalid for any cause, such invalidity shall not affect the other provisions hereof.

SEC. 10. This Act shall be effective thirty days from the date of its approval.

Approved July 29, 1947.

353

PROTOCOL BETWEEN THE GOVERNMENT OF CANADA AND THE GOVERNMENT OF THE UNITED STATES OF AMERICA TO AMEND THE CONVENTION FOR THE PROTECTION, PRESERVATION AND EXTENSION OF THE SOCKEYE SALMON FISHERIES IN THE FRASER RIVER SYSTEM, AS AMENDED

The Government of Canada and the Government of the United States of America, parties to the Convention for the Protection, Preservation and Extension of the Sockeye Salmon Fisheries in the Fraser River System signed at Washington on May 26, 1930[1], and to the Protocol signed at Ottawa, December 28, 1956[2], amending the aforesaid Convention,

Have agreed as follows:

ARTICLE I

Paragraph 3 of the understandings stipulated in the Protocol of Exchange of Ratifications signed at Washington on July 28, 1937, and amended by the Protocol signed at Ottawa on December 28, 1956, amending the Convention for the Protection, Preservation and Extension of the Sockeye Salmon Fisheries in the Fraser River System, signed at Washington on May 26, 1930, shall be further amended to read as follows:

"That the Commission shall set up an Advisory Committee composed of seven persons from each country who shall be representatives of the various branches of the industry, including, but not limited to, purse seine, gill net, troll, sport fishing and processing, which Advisory Committee shall be invited to all non-executive meetings of the Commission and shall be given full opportunity to examine and to be heard on all proposed orders, regulations, or recommendations."

ARTICLE II

The present Protocol shall be subject to ratification and the exchange of the instruments of ratification shall take place in Ottawa as soon as possible. This Protocol shall come into force on the day of the exchange of instruments of ratification.

---

[1] Treaty Series 1937 No. 10

[2] Treaty series 1957 No. 21

1980   No. 20   4

IN WITNESS WHEREOF the undersigned, duly authorized by their respective Governments, have signed this Protocol.

DONE in duplicate, in the French and English languages, both equally authentic, at Washington this twenty-fourth day of February, 1977.

J. R. McKINNEY
For the Government of Canada

FREDERICK IRVING
For the Government of the United States of America

International Pacific Salmon Fisheries Commission

MEMBERS AND PERIOD OF SERVICE

CANADA

Dr. William A. Found	1937-1939
A.L. Hager	1937-1948
Senator Thomas Reid	1937-1967
A.J. Whitmore	1939-1966
	1968-1969
Olof Hanson	1948-1952
Dr. H.R. MacMillan, C.B.E.	1952-1956
F.D. Mathers	1956-1960
W.R. Hourston	1960-1981
Richard Nelson	1966-1976
Dr. Roderick Haig-Brown	1970-1976
Richard A. Simmonds	1976-1980
Alvin W. Dixon	1978-1984
C. Wayne Shinners	1981-1985
Michael W.C. Forrest	1981-1985
David C. Schutz	1984-1985

DIRECTOR OF INVESTIGATIONS

Dr. W.F. Thompson	1937-1942
B.M. Brennan	1943-1949
Dr. Loyd A. Royal	1951-1970
A.C. Cooper	1971-1981
John F. Roos	1982-1985

ASSOCIATE DIRECTOR

Milo C. Bell	1950-1951

ASSISTANT DIRECTOR

Dr. J.L. Kask	1941-1942
F. H. Bell	1943-1943
Dr. Donald C.G. MacKay	1944-1945
Roy I. Jackson	1951-1955
John F. Roos	1971-1981

UNITED STATES

Edward W. Allen	1937-1951
	1957-1957
B.M. Brennan	1937-1942
Charles E. Jackson	1937-1946
Fred J. Foster	1943-1947
Milo Moore	1946-1949
	1957-1961
Albert M. Day	1947-1954
Alvin Anderson	1949-1950
Robert J. Schoettler	1951-1957
Elton B. Jones	1951-1957
Arnie J. Suomela	1954-1961
DeWitt Gilbert	1957-1974
George C. Starlund	1961-1966
Clarence F. Pautzke	1961-1969
Thor C. Tollefson	1966-1975
Charles H. Meacham	1969-1970
Donald R. Johnson	1971-1980
William G. Saletic	1974-1983
Donald W. Moos	1975-1977
Gordon Sandison	1977-1980
Herbert A. Larkins	1980-1983
Rolland A. Schmitten	1981-1985
Ted A. Smits	1983-1985
Dr. Thomas E. Kruse	1984-1984
William R. Wilkerson	1985-1985
Edward P. Manary	1985-1985

International Pacific Salmon Fisheries Commission

CANADA

Salmon Processors

Richard Nelson	1938-1966
Ken Fraser	1966-1971
Lloyd Monk	1971-1977
J. O'Connor	1977-1980
Brian Fraser	1980-1985

Purse Seine Fishermen

M.E. Guest	1938-1939
W.T. Burgess	1941-1945
George Miller	1945-1949
H. Martinick	1949-1950
W.J. Petrie	1950-1956
George T. Brajcich	1956-1957
C.N. Clarke	1957-1967
F. Buble	1967-1972
John Lenic, Jr.	1972-1973
John Brajcich	1973-1984
Larry Wick	1985-1985

Gill Net Fishermen

F. Rolley	1938-1944
Homer Stevens	1944-1949
P. Jenewein	1949-1970
Frank Nishii	1970-1985

Troll Fishermen

W.A. Hawley	1938-1939
A.E. Carr	1944-1952
M. Berg	1952-1955
H. North	1955-1960
R.H. Stanton	1960-1969
M. Guns	1969-1971
	1975-1979
W. Edwards	1971-1973
M. Ellis	1973-1975
John Makowichuk	1979-1983
Brian Fahey	1983-1985

UNITED STATES

Salmon Processors

C.J. Collins	1938-1949
J. Plancich	1949-1972
D. Franett	1972-1980
J. Lind	1980-1983
	1985-1985
J. Theodore	1984-1984

Purse Seine Fishermen

L. Makovich	1938-1946
N. Mladinich	1946-1976
W. Green	1976-1984
V. Barcott	1984-1985

Gill Net Fishermen

C. Karlson	1938-1958
J.F. Jurich	1946-1946
J. Erisman	1958-1964
V. Blake	1964-1967
R. Christensen	1967-1982
R. Suggs	1982-1985

Troll Fishermen

S. Leite	1938-1945
E. Larum	1939-1943
C.J. Dando	1946-1948
A. Anderson	1948-1949
J.R. Brown	1949-1957
B.J. Johnson	1958-1962
F. Bullock	1962-1966
C. Mechals	1966-1972
F. Lowgren	1972-1973
G.D. Simmons	1973-1981
W. Kimzey	1981-1982
C. Finley	1982-1984
M. Davis	1984-1985

CANADA

Purse Seine Crew Members

H. Staveness	1958-1975
Nick Carr	1976-1985

Sport Fishermen

M.W. Black	1938-1961
J.C. Murray	1961-1965
R.H. Wright	1965-1972
H. English	1972-1980
A. Downs	1980-1984

Native Indians

D. Guerin	1981-1983
S. Douglas	1984-1985

UNITED STATES

Reef Net Fishermen

J.R. Brown	1958-1974
G.H. Schuler	1974-1978
T. Philpott	1978-1985

Sport Fishermen

K. McLeod	1938-1953
H. Gray	1953-1972
E. Engman	1972-1985

Native Indians

C. Peterson	1981-1985

Biographies of Early Participants - Commissioners - Commission Directors/Staff

JOHN PEASE BABCOCK
1855-1936

John Pease Babcock, L.L.D. was born in St. Paul, Minnesota. He moved to Calfornia and was an Executive Officer of the California Fish and Game Commission. He gained a reputation as an expert in salmon propagation, although he was not a fisheries scientist.

He moved to British Columbia, Canada about the turn of the century and was appointed Assistant Commissioner of Fisheries in 1901 for the British Columbia government. He entered the provincial Fisheries Department at a time of serious conflict with the federal government.

For many years Dr. Babcock made extensive field surveys of the many spawning ground areas in the Fraser River watershed. His observations were published in the B.C. Commissioner of Fisheries' Annual Reports. The extensive and thorough information contained in those reports was very useful to the Commission. Babcock's reports were instrumental to the clear identification and understanding of the migratory problem relating to the decline of Fraser River sockeye. His close and detailed observations of the various spawning populations of sockeye established a standard that has been built upon and stands to this time.

He was one of the first individuals to sound the alarm for international control of the fisheries on sockeye salmon of the Fraser River system. He played a major role in the development of regulations and was a guiding force in seeing that a Convention was achieved. Dr. Babcock died one year before the treaty came into fruition.

LEIGH MILLER FREEMAN
1875-1955

Miller Freeman was born in Ogden, Utah. At the age of eight, his family moved to Yakima, Washington, and then on to Seattle in 1897. He published the first issue of the Pacific Coast Fisheries in 1903 which soon became Pacific Fisherman. He was a strong conservationist who used this publication very effectively to proclaim fishing issues. On many occasions he was the "silent" spokesman pushing towards a sockeye treaty. Later, as the official spokesman for Governor Martin of Washington State, he vigorously advocated a sockeye treaty.

Freeman was very active in local, state and national affairs. For thirty years from 1907 to 1937 he was a major U.S. influence in promoting and supporting a sockeye treaty. He was appointed as an original member of the International Fisheries (Halibut) Commission in 1924 and served there until 1932. At that time he resigned to work towards a sockeye treaty. He remained a strong spokesman for the halibut treaty.

It is said that he would not accept a position on the IPSFC as Commissioner because of possible suggestion that his advocacy for the sockeye treaty might be misinterpreted. With his Halibut Commission background, he strongly supported a sockeye treaty of simple, but effective, design. He became a close personal friend of Dr. Babcock and these two individuals were probably the most vocal proponents for the treaty.

The School of Fisheries at the University of Washington came into being as a result of his suggestion. He was also instrumental in the development of the North Pacific Fisheries Treaty in 1953 between Japan, Canada and the United States.

EDWARD WEBER ALLEN
1885-1976

E W. Allen was born in Oshkosh, Wisconsin. He was admitted to the American Bar Association in 1909 and had extensive experience in international law. He served for 23 years on the International Fisheries (Halibut) Commission. Allen served on the International Pacific Salmon Fisheries Commission from 1937 to 1951 and in 1957, a total of 16 years. He was also appointed to the International North Pacific Fisheries Commission and was a member from 1954 to 1973. His long and distinguished career with international fisheries Commissions is without equal.

As a successful practicing lawyer, Mr. Allen was involved in many aspects of the fishing industry. His foresight, knowledge and abilities were utilized in his work with the IPSFC during its formative years. Mr. Allen, because of his extensive legal background, was frequently asked to draft replies for the Commission on difficult matters. His contributions to the Commission were varied, numerous and longstanding.

In addition to his long career with three International Fisheries Commissions, Allen served in several other different capacities for the U.S. government. He was the author of many publications in the legal and fisheries field.

BERTRAM M. BRENNAN
1889-1949

B ertram M. Brennan was born in Grand Rapids, Wisconsin, and moved to Washington State in 1905. He was active in the lumber industry for 25 years and was Manager of the Clear Lake Lumber Company from 1915 to 1928. He was Chief Inspector of the Patrol Division of the Washington State Department of Fisheries from 1931 to 1932. Brennan became Director of the Washington State Department of Fisheries in 1932 and served until 1941. While he was Director of Fisheries, he was involved in the discussions which led to the Convention on Fraser River sockeye. He was later appointed as one of the original United States Commissioners to the IPSFC in 1937 and served in this capacity until 1942. After resigning as a Commissioner in 1942, Brennan then was appointed as the Commission's second Director in 1943 and retained this position until his death in 1949.

Brennan used his practical and administrative background to assist the Commission in formulating policies and decisions required by the Convention. He was Director of the Commission during the period when the fishway construction was initiated at Hells Gate and Bridge River Rapids. He was present at a time when the Commission needed a person of strong leadership and decision-making capability.

· · · · · · · · · · · · · · · · · · · ·

WILLIAM A. FOUND
1873-1940

William A. Found was born in New London, Prince Edward Island, Canada. He was a school teacher from 1891 to 1897 and in 1898 entered Dominion Public Service in the Department of Marine and Fisheries in 1898. Dr. Found was Superintendent of Dominion Fisheries and Director of Fisheries from 1911 to 1928. He was appointed to the Canadian-American Fisheries Commission in 1911. He also was assigned in 1918 to consider a settlement of outstanding fisheries issues between the United States and Canada—chief of which was the Fraser River problem. He was known as a hard worker, vitally interested in all fishery matters and a man of high personal integrity, fairness and free from political influences.

Dr. Found had very extensive background knowledge of the Fraser situation, having been involved in the discussions for many years. He also was one of the original Members of the International Halibut Commission. His appointment as an original Member of IPSFC in 1937, where he served for three years, was further recognition of his standing as a fishery expert.

Dr. Found was one of the principal Canadian proponents for the Sockeye Convention and his efforts as Deputy Minister of Fisheries over many years were in large part responsible for the treaty being completed.

· · · · · · · · · · · · · · · · · · · ·

ALVAH L. HAGER
1877-1948

Alvah L. Hager was born in Olewin, Iowa. As a young man he moved to Boston and entered the fish business where most of the East Coast fish market was controlled by the New England Fish Co. At the age of 24 he ventured to the West Coast and established a fish facility at Kalama, Washington. His business became well-known for his iced boxes of fresh Columbia River Chinook salmon that he shipped by railroad to the East Coast market.

He established the Northwestern Fisheries Co. in 1903 which was purchased in 1908 by the New England Fish Co. For several years the East Coast company attempted to purchase both the company and the services of Hager—reportedly doubling the offers each year (Doug Hager, personal communication). The purchase was contingent on his staying on the West Coast.

Hager relocated to Vancouver, B.C. in 1908 as general manager of New England Fish Co.'s Pacific operations. In the following year New England Fish Co. purchased the Canadian Fishing Co. In 1921 Hager became a Canadian citizen and in 1931 president of the company. The Canadian Fishing Co. went on to be one of the dominant fishing companies in Canada.

One of the traits of A.L. Hager was his deep conservation concern for the resource. This was evident in earlier years with regard to the Pacific Halibut where he was a Commissioner on the first Halibut Convention in 1923.

Hager was instrumental in recognizing the need for international control of harvests and the rebuilding of the Fraser River sockeye stocks. He was appointed in 1937 as one of the original Canadian Commissioners on the IPSFC and served until his death in 1948. He served as the Commission's first chairman and was chairman for several years. He met personally with President Roosevelt in the Oval Office to petition and secure funds for the Fraser River investigations and fishway construction. He was known as a stern but fair employer whose employees were loyal because of his outstanding character.

At his death, the Vancouver newspapers reported that "he put more fish back into the sea than he took out."

CHARLES E. JACKSON
1898-1977

Charles E. Jackson was born in Columbia, South Carolina. He was active in both political and fisheries fields. He had been Secretary to Senator E.D. Smith of South Carolina from 1922 to 1923. He later became Deputy Commissioner of the U.S. Bureau of Fisheries from 1933 to 1939 and Acting Commissioner from 1939 to 1940. From that position he moved on to become Assistant Director, U.S. Fish and Wildlife Service from 1940 to 1945. He became General Manager of the National Fisheries Institute, Inc. in Washington, D.C., in 1946. He was appointed to the IPSFC in 1937 and served until 1945.

Mr. Jackson served on several Committees at the Democratic National Conventions from 1928 to 1936. He was also active in several fisheries organizations as well as the American Bar Association. His legal, political and fisheries background were utilized by the Commission on many occasions during its formative years. His broad background and expertise provided an important East Coast link needed by the Commission in the 1930s and 1940s.

THOMAS REID
1886-1968

Thomas Reid was born in Cambuslang, Scotland, in 1886. He moved to Canada in 1907. While in Scotland he studied steel and iron metallurgy and later became Manager of Pacific Car and Foundry Co. He settled into a farming operation in Surrey, B.C. Reid was a councillor and reeve for the Municipality of Surrey.

After his election to the House of Commons as MP in 1930, he served as Parliamentary Assistant to the Minister of Fisheries; Minister of National Revenue; and Minister of National Health and Welfare. Tom Reid was appointed to the Senate in 1949.

Senator Reid was an excellent debater—strong in his retorts—and was characterized as speaking with the fervor of an Old Testament prophet. He applied this vitality and vigor very successfully to the Fraser River task.

He was a strong and sometimes contentious friend of the fishing industry. He was an ardent spokesman for protecting the salmon resource of the Fraser River system. His long time hope was that the Fraser runs would eventually be restored to equal those that returned prior to 1913.

Senator Reid served the resource and the Commission for 31 years, the longest term of any member. He made significant contributions on behalf of the Commission's work.

A. J. WHITMORE
1901-1984

Joe Whitmore was born in New Westminster, B.C. He joined the Canadian Department of Fisheries in 1917 and became Chief Inspector in 1920. In 1929, he was appointed Chief-Western Fisheries with headquarters in Ottawa. He returned to British Columbia in 1947 as Chief Supervisor of Fisheries and later as Director of Fisheries for the Pacific Area until retirement in 1960.

Whitmore was closely involved in the discussions leading to the formation of the IPSFC; he was appointed to the Commission in 1939, a position in which he served for 30 years. He also was a member of the International Pacific Halibut Commission.

Joe Whitmore was instrumental in guiding the Commission to adopting strong policies to fulfill the Commission's mandate and seeing that the Commission achieved its purpose under the Convention. He played a major role in the separation of government and Commission, yet at the same time he encouraged and supported approval by Canada of the Commission's programs. His broad administrative background and wisdom was utilized by other Commissioners and the staff on many occasions.

WILLIAM FRANCIS THOMPSON
1888-1965

W F. Thompson was born in St. Cloud, Minnesota. He moved to the State of Washington in 1903. He received his Ph.D. from Stanford University in 1930. From 1911 to 1915 he conducted shellfish studies in California and British Columbia and studied the Pacific Halibut for the Province.

Thompson was appointed as the first Director of the International Fisheries Commission (Halibut) in 1924. The application of his findings to halibut management led to the restoration of that fishery. Dr. Thompson was requested to be the first Director of the IPSFC in 1937 while at the same time retaining his position with IPHC. Thompson served the Commission from 1937 to 1942 as Director until he became Director of the School of Fisheries at the University of Washington in 1943, and later was consultant to IPSFC. In 1946 he became Director of the Fisheries Research Institute at the University, a position he held until he retired in 1958.

The investigative approaches established by Thompson at the IPSFC were solid and productive. Many of the research programs instigated under his leadership are still being used today in many areas. His findings with regard to Hells Gate published as the first Commission Bulletin set the course for the Commission's restoration of Fraser River sockeye and pink salmon stocks.

LOYD A. ROYAL
1908-

L oyd Royal was born in Olympia, Washington. He graduated from the School of Fisheries at the University of Washington. From 1932 to 1948 he was Chief Biologist and Assistant Director of the Washington Department of Fisheries. He served as one of the first scientific advisors to the Commission.

Royal was appointed Chief Biologist of the IPSFC in January 1949, Acting Director in August 1949, and Director in December 1950. He was Director until his retirement in March 1971. He then was retained by the Commission as Consultant for a period of two years.

The state fisheries background which Royal brought to the Commission enabled him to plan and implement far-reaching pollution control standards which were effective in protecting Fraser River salmon. He also established innovative principles that are still applicable in modern day fisheries management. The principle of individual racial management of particular races as advocated by Royal enabled the Commission to rehabilitate many Fraser sockeye and pink salmon runs. He also was instrumental in directing the Commission into other productive stock development approaches such as spawning channels.

Royal was awarded an honorary degree, L.L.D, from the University of British Columbia, Vancouver, B.C. in 1966.

ALEX C. COOPER
1921-

Al Cooper was born in Vancouver, B.C. He graduated from the University of British Columbia with a B.A. Science in Civil Engineering in 1944. He spent two summers employed by the International Pacific Salmon Fisheries Commission surveying at the fishway sites on the Fraser River at Hells Gate and at Bridge River Rapids.

He joined the engineering staff at the Commission in 1944 and initially was engaged in the preparation of the plans for the fishways at Hells Gate and Bridge River Rapids. In 1950 he was admitted as a member of the Association of Professional Engineers of British Columbia. After joining the Commission his duties required extensive cooperation with other fisheries agencies and negotiations with other government branches and with industries. These involved many aspects of water use such as hydraulic research, hydrology, flood control, water diversions, hydroelectric installations, pollution and construction activities. During this time he prepared 21 reports and collaborated in 17 reports on these subjects. In 1957 he was appointed Chief Engineer of the Commission and continued in that capacity until his appointment as Director of Investigations in 1971. During this time the Commission constructed two additional fishways at Hells Gate, four fishways at Yale Rapids, the Sweltzer Creek Research Laboratory, the Upper Pitt River Field Station, two spawning channels for pink salmon on Seton Creek, the sockeye spawning channels at Weaver Creek and Gates Creek and at Nadina Lake, and the temperature control dam and pipeline at McKinley Creek.

Cooper's achievements for the Commission were readily apparent to his professional peers. He made significant contributions to the rehabilitation of Fraser sockeye and pink salmon.

Cooper retired in March 1982 but continued as a consultant to the Commission for a time, primarily on the fisheries aspects of the Alcan Nechako development.

JOHN F. ROOS
1932-

John Roos was born in Seattle, Washington. He graduated from the School of Fisheries, University of Washington in 1955. In 1953, and from 1955 to 1959 he was with the Fisheries Research Institute (FRI), University of Washington conducting research on sockeye salmon at Bristol Bay and Chignik, Alaska.

In 1960, Roos joined the IPSFC as a fisheries scientist evaluating various Commission programs. By 1963, in addition to other research responsibilities, he became closely involved in management of the commercial fishery and was later appointed as Chief of the Management Division. He became Chief Biologist in 1968 and was appointed Assistant Director in 1971 and Director in 1982.

The extensive research background which Roos had at FRI and IPSFC enabled him to direct and instigate advances in test fishing/escapement estimation procedures, racial identification technique and pre-season and in-season predictions of stock abundance estimates. He was instrumental in ensuring that increased escapements were obtained from the fisheries in the rehabilitation efforts of the Commission.

Upon dissolution of the IPSFC, a contract between the two governments and Roos was let for writing the history of the Commission. In 1988 he accepted a position as Vice President with Pacific Seafood Processors Association, Seattle, Washington. He is a Fellow of the American Institute of Fishery Research Biologists and member of American Fisheries Society.

MILO C. BELL
1905-

Milo Bell was born in Marion, Iowa. He received his education at the University of Washington (B.S. in M.E. in 1930). As a professional engineer he was deeply involved in the academic field, starting as a special lecturer at the College of Fisheries, University of Washington in 1940 and then Research Associate Professor (1953), Associate Professor (1958), Professor (1963), and Professor Emeritus (1975). He also conducted courses at several other universities. He was the recipient of numerous professional honors.

Mr. Bell was Chief Engineer of the Washington Department of Fisheries from 1930 to 1933, and from 1935 to 1943. From 1933 to 1935 he was a consultant to the fishery agencies of the States of Washington and Oregon, assigned to the Bonneville Dam project, Columbia River; he was also a consultant for the U.S. Army Corps of Engineers, which was the lead agency constructing the dam. He is a co-holder of basic patents for the fishway systems developed for Bonneville Dam.

From 1943 to 1951 he was Chief Engineer and Associate Director on the IPSFC staff. His work with IPSFC began in 1941 as a consultant. He inaugurated the field work and experimentation for the fish passage facilities at Hells Gate. From 1941 to 1986 he was actively involved as a consultant on at least fifty-five fish-passage studies and constructions, both inside and outside the United States. Mr. Bell's expertise was used extensively by the Commission. He supervised the design and construction of the Hells Gate, Bridge River Rapids and Farwell Canyon fishways, the Horsefly Lake Field Station, and a suspension bridge across the Fraser River at Hells Gate. He played a significant role in advising and guiding the Commission at a crucial time in the history of the resource.

He represented the United States at international conferences on fisheries management, and participated in local, national and international conferences on fisheries problems, and served as a citizen-member on committees for the Legislature of the State of Washington. Milo Bell had a unique and widely varied career and the results of his engineering and teaching achievements are evident in many areas.

COMMISSION STAFF

This account of the Commission's activities would not be complete without reference to the staff that carried out the programs of the Commission so effectively. Some professional staff members devoted their entire career to the tasks of the Commission: F.J. Andrew was Chief Engineer and served 37 years, A.C. Cooper was Chief Engineer, Director and Consultant with 38 years full-time service. P. Gilhousen, Biologist, (32); S.R. Killick, Chief of Operations Division, (39); J. Pyper, Engineer, (39); W. Tomkinson, Administration, (39); and J. Weir, Biologist, (26).

Other staff members with two or more years service include the following:

ADMINISTRATION

M.C. Bell, Associate Director, Chief Engineer, Consultant	29
Dr. E.L. Brannon, Chief Biologist	9
B.M. Brennan, Director	6
Dr. K.A. Henry, Chief Biologist	7
R.I. Jackson Chief Engineer & Ass't. Director	17
Dr. J.L. Kask, Ass't Director	5
Dr. D.C.G. McKay, Ass't. Director	3
J.F. Roos, Chief Biologist, Chief of Management Div., Ass't. Director & Director	26
Dr. L.A. Royal, Chief Biologist Acting Director, Director & Consultant	24
P.B. Saxvik, Chief Engineer	31
Dr. J.A. Servizi, Chief of Environment Conservation Div.	22
Dr. W.F. Thompson, Director & Consultant	9
Dr. R. Van Cleve, Chief Biologist	3
I.V. Williams, Chief of Biology Div.	20
Dr. J.C. Woodey, Chief of Management Div.	14

BIOLOGICAL

C.E. Atkinson	10
Y. Bishop	2
Dr. D.J. Blackbourn	11
D. Brock	6
R. Burkhalter	2
J. Cave	5
A. Chapman	7
P. Cheng	11
R.I. Clutter	6
Dr. D. Colgrove (DVM)	2
Dr. G. Colgrove (DVM)	2
Dr. R. Cook	2
E. Davidoff	5
W. Davis	4
J. Demelker	2
Dr. R.E. Foerster	2
M.R. Fretwell	8
J. Gable	10
Dr. G.H. Geen	3
T. Gjernes	8
J.C. Goodlad	20
R. Goodlad	19
R. Gordon	21
Dr. J.A.R. Hamilton	15
Dr. H.H. Harvey	6
E.A.C. Haskell	6
G.V. Howard	8
Dr. D. Hurley	4
Dr. C.P. Idyll	7
R.L. Johnson	8
P.C. Johnson	5
E. Knight	23
B. Krug	3
R. MacLeod	4
J. Mason	9
D. Martens	19
Dr. R. Mead	4
K. Morton	7
A.E. Peterson	15
J. Remington	4
G. Robins	2
W. Saito	14
Dr. M.B. Schaefer	3
A. Sowerby	2
G.B. Talbot	9

S. Tremper	3
L.A. Verhoeven	6
E.H. Vernon	9
Dr. F.J. Ward	16
Dr. A.D. Welander	4
L.E. Whitesel	18

ENGINEERING

L. Bomberger	7
W. Boresky	2
C.H. Clay	5
E. Gibson	4
B. McAthey	2
C.R. Walters	8
J. Wild	2
A. Wilson	2

D. Johnson was the only staff member who later became a Commissioner. B. Brennan was the only Commissioner who later became a staff member.

Dr. Richard VanCleve was Chief Biologist from 1946 to 1948. Although his time with the Commission was brief, he contributed significantly to the planning and establishment of the Commission's harvest-management program. Dr. VanCleve later became Dean of the College of Fisheries at the University of Washington, where many Commission biologists were trained. Several Commission staff received their Ph.D's while working on IPSFC projects. Dr. Douglas Chapman, a consultant to the Commission while he was at the University of British Columbia, developed and refined the mathematical methodology of determining the numbers of salmon on the spawning grounds by the tag-and-recovery method. This basic method was used throughout the Commission's existence to enumerate the spawning escapement.

Dr. W.A. Clemens was appointed as a scientific advisor to IPSFC in 1938. Holding senior positions with the Fisheries Research Board of Canada and later with the University of British Columbia, he worked on a part-time basis for 14 years advising and assisting IPSFC biological staff in data analysis and report preparation.

Peter C. Johnson, a 31-year-old biologist, tragically lost his life in 1957 while engaged in fish-power research at Baker Dam in Washington State. In five years of employment with IPSFC he demonstrated a keen interest in the work and proved to be a very capable scientist. His death was the only fatal accident despite the extreme danger in many aspects of the Commission's work.

The technical staff constituted an important component of the Commission's operations. B. Tasaka worked for the IPSFC for 31 years and read hundreds of thousands of scales used in the racial identification program. R. Stewart (Chief of Operations Division) was with the Commission for 35 years and was involved in many of the Commission's programs. Long-term field studies were supervised by D. Loyst (27) at Chilko and similarly at the Shuswap Lake system (and Chilko) by F.G. Scott (30). Other employees such as F. Sato (Librarian, Catch Statistics) (26), H.S. Dunlop (32) and F.R. Johnston (Hells Gate fishways) (30), R.B. Kent (Statistician, Asst. Chief of Operations Division) (25), L.W. Johnston (Draftsman) (29), H.K. Hiltz (32) and D.F. Stelter (Statistician) (20) each served the Commission in specific areas of importance.

There were many other employees who worked for the Commission during its tenure. Those working ten or more years were: J. Bailey (15); D. Barnes (14); O. Brockwell (16); P.M. Buck (15); C. Carmichael (10); D.E. Chandler (11); M. Coventry (15); J. Elderkin (20); H. Enzenhofer (10); D. Hembrough (31); B. Van Horlick (17); Nina Hughes (10); J. Jensen (24); W.E. Keillor (16); P. Lenaghan (16); A. MacLean (18); G. McNaughton (20); E.B. Phillips (15); E. Pierce (10); B. Rannie (10); H. Smardon (14); A. Third (19); V.A. Tolvanen (15); S. Usher (14); W.E. Wells (10); R. Wien (28); W.L. Woodall (13); L.V. Woods (22); and W. Zingg (19).

Others with two or more years of service included:

Name	Years
P. Babb	2
J. Cameron	3
M. Cassidy	2
R.J.G. Cooper	6
G. Coupar	6
D.E. Courtney	7
R.A. Dick	7
S.J. Dykstra	5
T.R. Eburne	7
C.W. Ellis	5
V.E. Ewert	9
M.G. Ferguson	8
V.G. Gloster	8
C.L. Stewart Gordon	6
G.M. Grant	7
B.J. Greaves	4
E.M. Green	9
O.F.W. Hughes	4
G. Kirkpatrick	2
K.E. Kroeker	2
V.B. Law	3
A.H. Lesburg	5
S.G. MacLellan	5
C.J. Mack	6
I. Maeda	7
W. Mast	7
E. Maxwell	8
K.N. Medlock	2
S. Morelli Morton	7
D.C. Nelson	6
E.I. Nessel	3
K.L. Peters	9
E. Foye Peterson	8
M.N. Pond	6
W. Preston	6
G.F. Randall	5
C. Schoen	7
D.J. Short	6
E.M. Stevens	4
W.J. Stevenson	4
E.R. Stewart	8
D.M. Stillwell	9
A.R. Stobbart	2
J. Stobbart	8
E.T. Stone	9
D.B. Sundvick	8
G. Suther	3
B.J. Thompson	7
R.B. Van Horlick	5
K.E. Warkentin	6
M.E. Bowden Weir	4

There were comments from time to time that staff leadership favored U.S. citizens. Of the five directors and one associate director, five were Americans. However, to a great extent, the nationality of the director was dictated by the nationality of the Halibut Commission Director, who had generally been a Canadian. There seemed to be an unwritten understanding that these two Commissions would not have directors (at the same time) who were nationals of the same country. Since the office was in Canada, the vast majority of the staff were Canadians. In 1985 more than 85 per cent were Canadians. This did not create serious problems for the Commission in carrying out its programs. In most instances, the assistant director was of a nationality opposite to the director.

During the final 15 years of the IPSFC, the Commission structure consisted of six divisions: Administration, Management, Biology, Engineering, Field Operations and Environment-Conservation.

Even though the funds for the Commission's operations were provided by the governments, Commission staff were not government employees. Overall, benefits were not comparable to government employees. Salary schedules beginning in 1948 were to track those of the federal governments, whichever was highest. Pensions were administered through an International Fisheries Commissions Pension Society starting in 1957 although the Commission examined the possibility of a pension program in 1947. Staff were paid in the currency of their own country. This understandably caused considerable consternation to Canadian staff members in the early 1980s when U.S. staff living in Canada received a considerable bonus on exchange rates because of the low valued Canadian dollar. By the late 1970s, overall benefits were in most respects similar to government employees. Several Commission staff members obtained Ph.D.'s while working on IPSFC projects.

The Commission, throughout its existence, was recognized by government, industry and the scientific community for the ability to get the job done with minimum expenditures. At times there was slight derision intended in Commissioners' remarks; perhaps on occasion rightfully justified. Commissioner Joe Whitmore, in 1962, was "impressed" and at the same time perturbed because a frugal staff member retrieved discarded lumber as it floated downriver from bridge construction at Spences Bridge to construct fish-counting towers on the bank of the Thompson River. He remarked that "government simply did not operate in this manner."

Most recommendations by staff regarding fishery management and watershed problems were accepted by the Commission. However, by no means was this a "rubber-stamp" affair, but rather the meshing of scientific fact-finding with the expertise of a highly-qualified seasoned decision-making body free from external encumbrances. This relationship is rare now in most fishery management agencies because of the political structure and political sensitivities of the organizations.

Finally, special thanks are extended to the Commissioners, who were men of great character and understanding. They had, for the most part, already established themselves in their respective fields. They encouraged trust and permitted the staff to guide them through each of the many fishing seasons and proposed projects and, collectively, with their backgrounds, a highly successful team relationship existed. Seldom did their own personal desires or interests conflict with the management needs of the resource. There was a unique and close relationship between the Commission and the staff.

Leigh Miller Freeman

A.J. Whitmore

W.A. Found

John Pease Babcock

Active participants in promoting the establishment of IPSFC. In addition to these pioneers, E.W. Allen, A.L. Hager, C.E. Jackson, T. Reid and W.F. Thompson played important roles in establishing IPSFC.

1939-1942 Commissioners: Edward W. Allen, A.J. Whitmore, T. Reid, B.M. Brennan, Charles E. Jackson, A.L. Hager. This group formed the original panel of Commissioners in 1937 except that William A. Found, an original member, retired in 1939 and was replaced by A.J. Whitmore. In all photographs, the parties are named from left to right.

1943-1946 Commissioners: Fred J. Foster, A.J. Whitmore, Edward W. Allen, A.L. Hager, T. Reid, Charles E. Jackson. This group took the very bold step of approving the staff's recommendation for a multi-million dollar program of fishway construction and stream improvement at Hells Gate and many other sites in the Fraser watershed. They recommended appropriations of funds by the two governments and they succeeded in convincing the governments to approve the required funding despite the fact that the proposed fishway design was entirely new and despite the pressing demands for other expenditures of funds and materials for the war effort at that time.

1951-1952 Commissioners: Olof Hanson, Robert J. Schoettler, Edward W. Allen, A.J. Whitmore, T. Reid, Elton B. Jones, Albert M. Day on the occasion of Mr. Allen's retirement in 1951. Mr. Allen served again in 1957 to fill a short-term vacancy on the Commission.

1952-1954 Commissioners: Elton B. Jones, Robert J. Schoettler, A.J. Whitmore, H.R. MacMillan, T. Reid. Albert M. Day was absent.

1957-1960 Commissioners: F.D. Mathers, Milo Moore, T. Reid, DeWitt Gilbert, A.J. Whitmore. Arnie J. Suomela was absent.

1960-1961 Commissioners at a meeting in Vancouver, B.C. on January 19, 1961: T. Reid, Milo Moore, W.R Hourston, DeWitt Gilbert, A.J. Whitmore, Arnie J. Suomela.

1961-1966 Commissioners: W.R. Hourston, A.J. Whitmore, DeWitt Gilbert, Clarence F. Pautzke, T. Reid, George C. Starlund.

1970-1974 Commissioners: Richard Nelson, DeWitt Gilbert, W.R. Hourston, Donald R. Johnson, Thor C. Tollefson, Roderick Haig-Brown.

(Left to right) H.A. Larkins, R.A. Schmitten, M.W. Forrest, W.G. Saletic, Chairman, W.C. Shinners, A.W. Dixon, Vice Chairman & Secretary, 1982.

Commissioners W.R. Hourston, G. Starlund, T. Reid, A.J. Whitmore and C. Pautzke met with government officials Fisheries Minister Robichaud (third from left) and Deputy Minister A. Needler (extreme right) in Ottawa on May 10, 1963. Frequent meetings with United States and Canadian officials were held during the Commission's early years.

A meeting was held in Vancouver on March 8-10, 1946 to review progress of the Commission's work at Hells Gate and to consider fishing regulations for 1946. Among those attending were Commissioners T. Reid, F.J. Foster, A.L. Hager, A.J. Whitmore and M. Moore (seated left to right in front center). Advisory Committee members for Canada were R. Nelson (10), G. Miller, H. Stevens (11), A.E. Carr and M.W. Black. For the United States, the Advisors were C.J. Collins, L. Makovich (12), C. Karlson (2), C.J. Dando, K. McLeod (3) and J.F. Jurich (18). Guests included Major Motherwell (seated on left), C.L. Anderson (5), Tom Taylor (15), F.A. Warne (16), J.A. Craig (17) and J.N. Plancich (8). Staff shown are Director B.M. Brennan (7), Chief Biologist R. Van Cleve, (6) and scientific consultants W.F. Thompson, (seated on the right) and W.A. Clemens, (13). The eighteen persons standing are identified by numbers starting from the left.

Advisory Committee members: Howard Gray, John Brown, Chester Karlson, Nick Mladnich, John Plancich, Ted Black, Herbert North and Richard Nelson.

Advisory Committee members at a January 8, 1956 meeting: Standing: Howard Gray, Nick Mladnich, William Pitre and Peter Jenewein. Seated: John Plancich, Chester Karlson, Herbert North, Jerry Anderson (alternate for John Brown) and Richard Nelson.

Advisory Committee members: Howard Gray, John Brown, John Plancich, Chester Karlson and Nick Mladnich.

At a meeting in Washington, D.C. in 1944, Commissioners A.L. Hager, C.E. Jackson, B.M. Brennan, E.W. Allen and A.J. Whitmore explained to Secretary of State Ickes (seated) and two aides (standing behind Mr. Ickes) their reasons for requesting funds for the Hells Gate fishways and other projects.

W.F. Thompson

Loyd A. Royal

Milo C. Bell

A.C. Cooper

John F. Roos

W.F. Thompson, Loyd A. Royal, A.C.Cooper and John F. Roos were four of the five Directors appointed by the Commissioners. B.M. Brennan was the other Director and since he was a Commissioner before being appointed Director, his photograph is included with the other Commissioners. Milo C. Bell was appointed as Associate Director.

On the annual end-of-the-season field trip for 1985, Commissioners, Advisory Committee members, former employees, staff and other interested parties visited the Horsefly River spawning area to witness the return of the largest Quesnel run since 1913. The group included (from the left):

B. Suggs, J. Roos, B. Fraser, D. Johnson, F. Fraser, D. Schutz, A. Dixon, P. Saxvik, T. Philpott, E. Engman, R. Stewart, R. Kent, D. Kent, J. Woodey, R. Hourston, L. Loomis, E. Manary, R. Schmitten (kneeling), R. Jackson, P. Jackson, L. Wick, M. Davis. (Not pictured, B. Tasaka)

A P P E N D I X   D

Sockeye Escapement by Area, Stream and Year
1938 - 1985*

Location	1938	1939	1940	1941	1942	1943
Lower Fraser						
Cultus	13,342	73,189	74,121	18,164	37,305	11,875
Upper Pitt Area	8,457	16,446	12,386	8,763	19,068	39,257
Miscellaneous	463	2,301	3,031	18,226	1,268	431
Harrison-Lillooet						
Harrison	7,500	200	11,000	37,870	139	2,021
Weaver Creek	18,012	4,646	17,220	8,831	19,000	3,879
Birkenhead	11,100	15,435	29,580	46,490	93,099	50,668
Miscellaneous	2,707	544	8,033	3,514	1,959	214
Seton-Anderson						
Gates Creek	13	0	221	0	0	P
Portage	56	P	1,211	1,881	N.O.	N.O.
Miscellaneous	162	N.O.	1,346	1,055	0	67
Early South Thompson						
Seymour	N.O.	250	P	0	2,412	67
Scotch	N.O.	0	0	0	0	N.O.
Anstey	N.O.	0	0	0	0	N.O.
Eagle	N.O.	0	N.O.	N.O.	N.O.	N.O.
Momich	N.O.	N.O.	N.O.	0	N.O.	N.O.
Upper Adams	N.O.	N.O.	N.O.	0	N.O.	N.O.
Miscellaneous	N.O.	P	0	0	N.O.	N.O.
Late South Thompson						
Adams	620,000	16,200	8,194	61	1,967,553	94,325
Little River	175,000	15,687	2,430	0	400,000	a
South Thompson	N.O.	N.O.	100	0	P	P
Big Shuswap Lake	1,130	N.O.	N.O.	0	P	a
Lower Shuswap	N.O.	0	N.O.	0	2,000	0
Middle Shuswap	N.O.	0	N.O.	0	0	0
Miscellaneous	2,039	0	N.O.	N.O.	206,797[b]	0

a   Included in Adams
b   Adams Lake and tributaries (except those listed) 200,607
c   Includes Bowron tributaries
*   P - indicates fish present, but no estimate
    N.O. - indicates no observation made

Location	1938	1939	1940	1941	1942	1943
North Thompson						
Raft	587	2,687	16,021	412	713	5,267
Barriere	N.O.	0	0	0	0	N.O.
Fennell	N.O.	N.O.	0	N.O.	0	N.O.
North Thompson	N.O.	0	N.O.	0	0	N.O.
Miscellaneous	N.O.					
Chilcotin						
Chilko	10,870	2,000+	341,760	367,506	30,080	10,474
S. End Chilko Lake	N.O.	N.O.	15,000	N.O.	N.O.	N.O.
Taseko	N.O.	N.O.	N.O.	N.O.	N.O.	N.O.
Miscellaneous						
Quesnel						
Horsefly	0	7	74	918	P	0
Little Horsefly	N.O.	0	0	27	0	0
McKinley	N.O.	0	0	N.O.	0	0
Mitchell	0	0	N.O.	41	0	0
Miscellaneous	4	0	0	4	0	0
Nechako						
Early Nadina	54	P	0	133	112	0
Late Nadina River	N.O.	P	N.O.	45	N.O.	N.O.
Stellako	6,943	2,585	3,276	8,566	91,840	14,897
Miscellaneous	218	23	70	423	1,141	83
Early Stuart						
Ankwill	N.O.	N.O.	N.O.	25	N.O.	N.O.
Driftwood	0	P	N.O.	25	N.O.	N.O.
Forfar	4,694	160	112	1,776	3,244	363
Gluske	N.O.	0	0	597	1,728	0
Kynoch	2,835	806	219	2,477	2,224	1,976
Narrows	115	9	0	196	97	5
Rossette	18	0	0	1,193	1,145	394
Miscellaneous	9	4	4	246	36	N.O.
Late Stuart						
Kuzkwa	N.O.	N.O.	5	0	N.O.	N.O.
Middle	62	N.O.	261	9,200	P	N.O.
Tachie	N.O.	N.O.	N.O.	1,000	N.O.	N.O.
Miscellaneous	N.O.	N.O.	13	31	0	0
Northeast						
Bowron	2,205	4,584	6,853	1,912	1,634	6,461[c]
TOTALS	888,595	157,763	552,541	541,608	2,884,594	242,724

Location	1944	1945	1946	1947	1948	1949
Lower Fraser						
Cultus	14,200	9,227	33,284	8,898	13,086	9,301
Upper Pitt Area	N.O.	35,688	18,512	106,685	55,380	9,290
Miscellaneous	1,050	952	1,404	751	N.O.	490
Harrison-Lillooet						
Harrison	92	16,060	15,631	14,000	26,186	8,000
Weaver Creek	22,289	12,944	37,363	8,174	19,970	12,751
Birkenhead	69,111	96,664	93,243	123,627	122,424	74,085
Miscellaneous	2,619	1,543	1,723	500	12,481	2,534
Seton-Anderson						
Gates Creek	400	0	N.O.	N.O.	N.O.	N.O.
Portage	N.O.	227	200	50	N.O.	N.O.
Miscellaneous	18	22	95	0	N.O.	N.O.
Early South Thompson						
Seymour	200	P	3,778	19,795	4,099	10,772
Scotch	N.O.	P	N.O.	0	50	1,764
Anstey	N.O.	N.O.	N.O.	0	0	4
Eagle	N.O.	N.O.	N.O.	0	N.O.	11
Momich	N.O.	N.O.	N.O.	N.O.	N.O.	N.O.
Upper Adams	N.O.	N.O.	N.O.	N.O.	N.O.	N.O.
Miscellaneous	N.O.	N.O.	N.O.	0	N.O.	N.O.
Late South Thompson						
Adams	1,404	57,780	1,835,000	187,798	15,384	11,742
Little River	200	6,000	419,000	16,251	1,313	9,571
South Thompson	P	1,000	92,000	100	202	9
Big Shuswap Lake	0	0	29,814	0	0	0
Lower Shuswap	0	0	1,237	0	0	13
Middle Shuswap	N.O.	0	0	0	0	0
Miscellaneous	0	3,500	5,150	103	0	7

Location	1944	1945	1946	1947	1948	1949
North Thompson						
Raft	1,875	4,936	3,257	10,009	10,359	6,113
Barriere	N.O.	N.O.	N.O.	N.O.	N.O.	43
Fennell	N.O.	N.O.	N.O.	N.O.	N.O	N.O.
North Thompson	N.O.	N.O.	N.O.	N.O.	N.O.	N.O.
Miscellaneous	0	N.O.	N.O.	N.O.	N.O.	N.O.
Chilcotin						
Chilko	194,276	172,931	59,819	54,931	671,025	58,310
S. End Chilko Lake	N.O.	N.O.	N.O.	N.O.	P	N.O.
Taseko	N.O.	N.O.	N.O.	N.O.	P	100
Miscellaneous		4,005	305			
Quesnel						
Horsefly	5	4,441	104	11	100	30,000
Little Horsefly	4	N.O.	P	0	0	0
McKinley	N.O.	N.O.	0	0	0	0
Mitchell	N.O.	N.O.	4	0	0	664
Miscellaneous	N.O.	N.O.	N.O.	0	0	0
Nechako						
Early Nadina	N.O.	29	119	83	50	20,457
Late Nadina River	N.O.	205	N.O.	N.O.	N.O.	N.O.
Stellako	5,768	20,826	245,172	59,904	16,213	104,835
Miscellaneous	29	1,376	1,282	506	241	4,711
Early Stuart						
Ankwill	N.O.	0	0	N.O.	N.O.	715
Driftwood	N.O.	P	18	0	N.O.	407
Forfar	83	7,081	1,822	1,328	2,650	80,484
Gluske	N.O.	3,643	2,752	175	2,477	103,483
Kynoch	630	11,963	1,843	10,004	13,336	185,850
Narrows	0	140	248	0	0	19,577
Rossette	4	8,753	2,502	2,342	2,497	160,770
Miscellaneous	N.O.	333	58	N.O.	N.O.	30,942
Late Stuart						
Kuzkwa	N.O.	N.O.	N.O.	N.O.	N.O.	N.O.
Middle	41	61,000	2,000	52	820	86,373
Tachie	N.O.	5,740	60	0	80	20,000
Miscellaneous	5	1,224	56	0	149	1,379
Northeast						
Bowron	1,658	4,130[c]	7,411[c]	24,039[c]	25,218	22,330[c]
TOTALS	315,961	554,363	2,916,266	650,116	1,015,790	1,087,887

Location	1950	1951	1952	1953	1954	1955
Lower Fraser						
Cultus	30,595	13,143	18,910	13,000	24,150	26,000
Upper Pitt Area	40,061	37,837	48,899	18,693	17,624	17,950
Miscellaneous	595	745	1,433	1,844	1,000	100
Harrison-Lillooet						
Harrison	33,864	17,145	25,794	21,328	28,800	5,595
Weaver Creek	30,743	12,979	28,100	9,395	28,450	21,672
Birkenhead	72,567	42,063	77,386	55,823	40,453	24,450
Miscellaneous	438	422	7,416	511	279	191
Seton-Anderson						
Gates Creek	N.O.	N.O.	7,070	74	45	77
Portage	N.O.	29	N.O.	200	3,379	43
Miscellaneous	N.O.	N.O.	N.O.	N.O.	N.O.	N.O.
Early South Thompson						
Seymour	12,471	24,344	6,428	5,947	24,876	9,011
Scotch	50	0	338	1,364	0	0
Anstey	0	0	0	0	16	11
Eagle	0	0	0	0	0	0
Momich	0	0	0	0	0	0
Upper Adams	0	0	0	0	194	0
Miscellaneous	0	0	0	0	0	0
Late South Thompson						
Adams	1,100,081	134,964	8,692	165,678	1,740,067	60,233
Little River	137,939	9,690	1,861	37,659	199,004	4,328
South Thompson	40,171	450	200	14,533	57,675	0
Big Shuswap Lake	5,900	0	0	0	6,786	0
Lower Shuswap	12,000	0	0	0	17,462	23
Middle Shuswap	50	0	0	0	61	0
Miscellaneous	8,400	0	0	884	9,958	0

Location	1950	1951	1952	1953	1954	1955
North Thompson						
Raft	6,404	8,561	15,617	7,909	9,995	5,081
Barriere	63	108	0	0	0	97
Fennell	0	0	0	0	0	0
North Thompson	N.O.	N.O.	N.O.	N.O.	N.O.	N.O.
Miscellaneous	0	0	0	0	0	0
Chilcotin						
Chilko	26,447	118,110	490,065	201,245	37,743	132,146
S. End Chilko Lake	N.O.	N.O.	N.O.	N.O.	N.O.	N.O.
Taseko	500	500	3,647	4,422	3,500	4,400
Miscellaneous						
Quesnel						
Horsefly	385	49	7,013	105,440	274	63
Little Horsefly	13	0	2	0	7	0
McKinley	0	0	0	3,141	0	0
Mitchell	0	0	2	2,344	18	0
Miscellaneous	0	0	0	2	0	0
Nechako						
Early Nadina	1,908	119	1,649	24,136	1,449	94
Late Nadina River	774	175	38	13,617	770	108
Stellako	145,108	96,208	40,466	43,688	141,882	51,746
Miscellaneous	2,417	886	1,180	3,799	583	704
Early Stuart						
Ankwill	67	25	240	5,913	56	0
Driftwood	144	56	38	8,656	387	0
Forfar	10,372	13,600	6,975	18,185	5,702	68
Gluske	11,009	3,787	5,911	16,074	5,292	99
Kynoch	23,907	32,825	13,439	16,734	13,860	1,037
Narrows	2,265	401	1,453	20,604	2,756	27
Rossette	6,425	10,019	3,577	6,355	3,836	916
Miscellaneous	5,254	328	1,949	61,791	3,169	23
Late Stuart						
Kuzkwa	N.O.	N.O.	N.O.	3,686	0	0
Middle	5,200	4,000	952	244,405	3,853	3,598
Tachie	400	200	728	107,506	1,529	4,000
Miscellaneous	243	202	295	8,079	88	20
Northeast						
Bowron	16,266[c]	22,292[c]	18,672	13,517	10,774	9,355
TOTALS	1,791,496	606,262	846,435	1,288,181	2,447,802	383,266

Location	1956	1957	1958	1959	1960	1961
Lower Fraser						
Cultus	14,133	20,647	14,097	48,461	17,689	15,428
Upper Pitt Area	32,094	12,338	10,385	15,740	24,510	11,162
Miscellaneous	1,000	1,200	20,448	637	400	1,293
Harrison-Lillooet						
Harrison	2,586	3,812	14,701	27,885	17,280	42,778
Weaver Creek	8,772	20,866	36,199	8,379	7,042	4,383
Birkenhead	57,899	24,168	33,055	38,604	39,848	49,627
Miscellaneous	6,235	389	N.O.	64	4,711	466
Seton-Anderson						
Gates Creek	7,848	891	81	868	5,449	252
Portage	0	452	4,748	572	0	527
Miscellaneous	N.O.	N.O.	445	76	0	0
Early South Thompson						
Seymour	2,562	14,295	78,578	52,325	3,048	5,822
Scotch	155	2,230	0	0	11	598
Anstey	0	0	112	0	0	0
Eagle	0	0	31	0	0	0
Momich	0	0	0	0	1,000	0
Upper Adams	9	0	291	0	P	0
Miscellaneous	0	0	0	0	0	0
Late South Thompson						
Adams	3,245	253,573	1,730,609	113,257	1,848	56,988
Little River	1,255	34,964	409,480	21,080	66	8,253
South Thompson	0	14,645	123,864	472	0	254
Big Shuswap Lake	0	0	14,300	0	0	0
Lower Shuswap	5	490	9,387	289	0	342
Middle Shuswap	0	0	499	0	0	0
Miscellaneous	0	3,556	1,018,012[d]	0	0	20

[d] Miscellaneous and diverted sockeye into Little River, South Thompson and Big Shuswap Lake.

Location	1956	1957	1958	1959	1960	1961
North Thompson						
Raft	9,037	6,881	10,215	10,210	5,553	7,301
Barriere	2	38	0	203	23	335
Fennell	0	0	5	27	0	0
North Thompson	N.O.	N.O.	N.O.	N.O.	1,208	225
Miscellaneous	0	0	0	0	0	0
Chilcotin						
Chilko	647,479	140,765	137,081	471,162	426,607	40,315
S. End Chilko Lake	N.O.	N.O.	N.O.	N.O.	P	N.O.
Taseko	1,995	3,667	7,538	16,410	2,726	80
Miscellaneous						
Quesnel						
Horsefly	2,556	214,254	1,784	49	3,029	277,305
Little Horsefly	5	38	14	27	23	40
McKinley	94	6,698	0	0	0	18,400
Mitchell	14	2,677	65	0	5	6,601
Miscellaneous	0	0	0	0	0	228
Nechako						
Early Nadina	1,228	29,998	88	351	1,566	18,432
Late Nadina River	83	27,553	635	1,013	157	17,544
Stellako	38,459	38,921	112,273	79,355	38,884	47,241
Miscellaneous	385	3,339	849	1,755	189	146
Early Stuart						
Ankwill	117	8,285	617	0	54	18,468
Driftwood	50	45,567	1,897	5	34	79,087
Forfar	5,497	17,975	8,715	281	1,755	13,073
Gluske	4,619	21,899	2,111	97	2,138	6,377
Kynoch	9,535	13,473	9,477	1,123	4,154	13,171
Narrows	697	16,184	2,344	167	598	8,636
Rossette	3,863	7,087	4,802	911	4,558	5,146
Miscellaneous	779	104,563	8,849	86	1,281	55,178
Late Stuart						
Kuzkwa	0	50,006	0	0	0	39,245
Middle	500	336,301	8,388	4,773	1,269	177,516
Tachie	600	118,252	13,738	3,445	1,687	177,047
Miscellaneous	3,541	26,823	1,512	7	5	17,297
Northeast						
Bowron	6,996	12,069	14,871	29,247	7,620	7,460
TOTALS	872,742	1,661,829	3,867,190	949,413	628,025	1,250,087

Location	1962	1963	1964	1965	1966	1967
Lower Fraser						
Cultus	27,070	20,571	11,143	2,532	17,464	33,492
Upper Pitt Area	16,585	12,680	13,804	6,981	20,867	10,300
Miscellaneous	599	353	667	275	884	1,006
Harrison-Lillooet						
Harrison	8,162	22,287	2,202	15,034	32,672	20,577
Weaver Creek	15,962	14,469	1,370	13,539	13,875	19,730
Channel				4,436	6,541	2,887
Birkenhead	52,146	67,151	69,939	30,008	81,134	58,036
Miscellaneous	270	85	3,967	646	329	0
Seton-Anderson						
Gates Creek	1,046	4,858	19,971	1,679	592	1,665
Portage	12,034	2,011	9	2,108	31,844	4,048
Miscellaneous	0	0	0	0	0	256
Early South Thompson						
Seymour	58,104	71,690	2,784	6,954	28,754	13,361
Scotch	7	0	0	1,910	459	0
Anstey	77	2	0	0	0	0
Eagle	169	5	0	0	288	4
Momich	0	0	823	0	0	0
Upper Adams	79	5	162	0	63	0
Miscellaneous						
Late South Thompson						
Adams	991,728	154,086	716	55,041	1,197,336	755,238
Little River	67,398	2,436	0	3,274	55,952	74,490
South Thompson	14,441	45	0	192	4,313	270
Big Shuswap Lake	29,916	0	0	0	22,102	237
Lower Shuswap	31,205	7,379	0	583	24,629	5,951
Middle Shuswap	457	0	0	0	1,872	58
Miscellaneous	15,997	0	0	439	16,276	9,686

Location	1962	1963	1964	1965	1966	1967
North Thompson						
Raft	7,613	8,724	5,201	6,624	6,250	1,303
Barriere	14	92	85	104	4	16
Fennell	0	439	146	0	0	920
North Thompson Miscellaneous	90	70	38	P	46	N.O.
Chilcotin						
Chilko	92,467	1,002,252	238,601	39,902	226,702	176,337
S. End Chilko Lake	0	P	P	0		
Taseko Miscellaneous	657	31,667	433	P	353	5,700
Quesnel						
Horsefly	1,001	86	15,315	359,232	1,607	119
Little Horsefly	72	0	217	0	4	N.O.
McKinley	0	N.O.	b	b	0	0
Mitchell	5	N.O.	169	5,335	142	N.O.
Miscellaneous	0	0	0	149	N.O.	N.O.
Nechako						
Early Nadina	450	1,019	1,397	3,884	83	1,595
Late Nadina River	1,683	7,304	232	11,293	1,784	7,790
Stellako	124,495	138,805	31,047	39,418	101,684	91,525
Miscellancous	308	3,344	200	36	10	2,637
Early Stuart						
Ankwill	290	4	2	2,806	86	20
Driftwood	374	14	2			
Forfar	4,719	652	27	2,221	1,739	4,815
Gluske	1,841	0	218	2,200	1,876	1,368
Kynoch	8,672	2,147	1,147	2,885	3,591	6,694
Narrows	666	180	22	1,377	322	454
Rossette	4,887	1,600	952	1,165	1,645	6,566
Miscellaneous	5,267	31	51	10,391	1,922	1,152
Late Stuart						
Kuzkwa	14	0	0	10,000	295	2
Middle	11,706	1,838	743	139,186	4,917	972
Tachie	6,750	1,035	1,157	62,469	3,600	576
Miscellaneous	219	364	0	3,303	220	92
Northeast						
Bowron	6,292	25,144	1,500	2,660	2,480	31,695
TOTALS	1,624,004	1,606,924	426,459	852,271	1,919,608	1,353,640

Location	1968	1969	1970	1971	1972	1973
Lower Fraser						
Cultus	25,736	6,739	15,149	9,145	10,660	858
Upper Pitt Area	16,988	25,084	6,657	15,469	13,412	11,928
Miscellaneous	1,602	715	364	394	742	427
Harrison-Lillooet						
Harrison	5,391	15,006	12,675	3,790	1,399	3,060
Weaver Creek	2,606	41,857	6,373	2,887	15,505	27,807
Channel	1,910	17,089	4,723	2,736	11,043	22,366
Birkenhead	83,750	64,527	72,760	32,672	113,097	139,295
Miscellaneous	1,234	139	261	0	2,979	378
Seton-Anderson						
Gates Creek	4,005	205	68	797	1,762	231
Channel	6,284	676	735	1,502	6,807	668
Portage	173	1,040	3,901	281	1,460	4,272
Miscellaneous	142	0	0	0	0	32
Early South Thompson						
Seymour	3,957	7,327	11,991	19,028	2,889	2,856
Scotch	126	3,395	304	0	47	6,235
Anstey	0	0	196	23	0	N.O.
Eagle	0	18	23	9	4	11
Momich	617	0	0	0	1,003	0
Upper Adams	0	0	4	0	31	0
Miscellaneous	0	0	0	0	0	0
Late South Thompson						
Adams	3,983	45,908	1,297,990	280,176	4,325	33,312
Little River	0	6,842	168,881	2,821	81	6,689
South Thompson	0	630	5,931	10	0	545
Big Shuswap Lake	0	0	0	0	0	0
Lower Shuswap	P	1,703	29,074	6,117	290	7,452
Middle Shuswap	0	0	4,559	284	0	0
Miscellaneous	15	236	50,389	885	2	0

Location	1968	1969	1970	1971	1972	1973
North Thompson	8,121	5,593	4,474	840	11,151	2,729
Raft	275	40	2	5	94	22
Barriere	954	52	9	1,300	1,931	205
Fennell	N.O.	N.O.	270	888	465	0
North Thompson Miscellaneous	0	0	0	0	342	0
Chilcotin						
Chilko	414,446	76,518	145,049	161,943	562,333	61,707
S. End Chilko Lake	P	0	0	12,323	2,132	N.O.
Taseko Miscellaneous	N.O.	P	P	10,500	2,287	N.O.
Quesnel						
Horsefly	5,686	242,051	1,350	171	2,859	238,278
Little Horsefly	73	40	0	0	18	4
McKinley	e	27,936	0	0	526	15,106
Mitchell	4	8,939	23	N.O.	85	24,673
Miscellaneous	N.O.	0	N.O.	N.O.	N.O.	0
Nechako						
Early Nadina	902	8,541	78	1,222	827	2,705
Late Nadina River Channel	1,496	27,898	3,939	14,525	2,702	7,951 8,786
Stellako	30,420	49,341	45,876	39,726	36,771	30,755
Miscellaneous	119	140	0	2,080	139	54
Early Stuart						
Ankwill	0	15,795	220	220	5	21,790
Driftwood	N.O.	37,028	1,983	335	50	131,172
Forfar	149	9,922	6,476	25,178	835	18,924
Gluske	18	4,660	5,702	14,305	591	19,450
Kynoch	833	12,380	4,676	22,932	2,534	22,485
Narrows	41	5,746	144	3,467	104	5,726
Rossette	518	1,566	7,664	16,454	834	4,156
Miscellaneous	28	22,721	5,882	13,051	133	76,950
Late Stuart						
Kuzkwa	0	8,370	90	227	50	20,124
Middle	288	111,322	12,115	873	972	91,879
Tachie	149	86,431	2,776	360	7,527	97,445
Miscellaneous	33	934	74	75	155	4,895
Northeast						
Bowron	3,634	3,872	1,341	25,497	4,138	4,700
TOTALS	626,706	1,006,972	1,943,221	747,523	830,128	1,181,093

e Included in Horsefly

Location	1974	1975	1976	1977	1978	1979
Lower Fraser						
Cultus	9,814	11,478	4,450	353	7,265	32,045
Upper Pitt Area	20,792	39,942	36,530	13,887	24,835	37,558
Miscellaneous	2,063	4,105	4,428	1,097	4,688	3,247
Harrison-Lillooet						
Big Silver	837	31	1,642	349	1,253	119
Harrison	16,920	5,987	5,130	2,246	19,747	45,706
Weaver Creek	42,143	12,195	22,867	22,105	43,989	28,515
Channel	24,664	19,816	28,211	33,040	32,248	22,888
Birkenhead	173,463	92,928	108,121	43,139	99,857	78,088
Miscellaneous	1,094	76	160	0	378	36
Seton-Anderson						
Gates Creek	146	788	2,889	1,176	931	907
Channel	1,645	3,768	14,855	1,713	1,639	4,118
Portage	8,986	3,829	3,800	7,974	10,230	3,663
Miscellaneous	54		74	493		
Early South Thompson						
Seymour	45,189	37,024	8,489	5,911	62,929	49,321
Scotch	464	0	41	13,586	2,056	0
Anstey	666	14	0	0	886	0
Eagle	263	9	0	0	189	0
Momich	0	38	1,998	11	0	0
Upper Adams	13	23	40	0	0	0
Miscellaneous	67	13	0	105	355	416
Late South Thompson						
Adams	889,613	148,187	5,013	57,964	1,493,473	275,616
Little River	122,112	7,268	175	8,684	81,055	10,443
South Thompson	14,466	16	0	432	9,986	144
Big Shuswap Lake	0	0	0	0	100,493	2,140
Lower Shuswap	86,396	11,652	400	14,695	187,167	10,092
Middle Shuswap	3,064	227	0	0	10,890	578
Miscellaneous	41,882	1,215	0	0	17,339	700

Location	1974	1975	1976	1977	1978	1979
North Thompson						
Raft	2,396	2,664	8,684	648	2,500	1,780
Barriere	4	0	85	16	0	0
Fennell	243	4,127	4,090	355	675	15,590
North Thompson	343	123	500	1,372	P	1,009
Miscellaneous	0	0	0	0	0	18
Chilcotin						
Chilko	128,131	220,554	364,311	54,322	151,835	240,294
S. End Chilko Lake	14,464	55,144	23,156	2,460	7,339	32,400
Taseko	N.O.	4,394	634	N.O.	N.O.	N.O.
Miscellaneous	0	0	0	0	0	0
Quesnel						
Horsefly	4,459	101	1,277	431,920	7,287	511
Little Horsefly	0	0	32	106	0	
McKinley	0	100	0	0	0	0
Upper	0	N.O.	2	33,064	5	N.O.
Lower	0	100	783	8,024	85	N.O.
Mitchell	N.O.	0	101	42,396		
Miscellaneous	N.O.	0	2	713	1,237	0
Nechako						
Early Nadina	0	481	101	1,453	0	1,809
Late Nadina River	2,930	4,013	279	610	227	14,474
Channel	895	11,306	1,394	16,286	2,555	41,212
Stellako	41,473	176,079	150,741	23,452	60,421	290,116
Miscellaneous	36	1,336	70	150	0	2,594
Early Stuart						
Ankwill	544	101	32	6,287	1,363	192
Driftwood	1,894	20	N.O.	54,568	4,903	247
Forfar	5,495	6,818	1,249	3,628	9,579	15,805
Gluske	5,548	10,370	966	4,646	4,295	10,040
Kynoch	10,652	25,124	6,727	5,893	10,649	34,228
Narrows	486	1,704	244	2,844	709	1,575
Rossette	5,675	8,543	2,090	2,261	7,452	15,893
Miscellaneous	9,350	14,791	1,340	37,890	11,147	14,783
Late Stuart						
Kuzkwa	718	N.O.	0	9,031	742	2,090
Middle	8,990	6,704	330	80,420	4,061	18,111
Tachie	4,680	7,525	2,637	54,282	8,028	10,940
Miscellaneous	239	0	33	2,896	196	777
Northeast						
Bowron	1,850	29,700	2,250	2,500	3,150	35,000
TOTALS	1,758,311	992,551	823,453	1,113,453	2,514,318	1,407,828

Location	1980	1981	1982	1983	1984	1985
Lower Fraser						
Cultus	1,687	1,159	17,222	19,952	1,147	571
Upper Pitt Area	17,135	25,327	8,725	16,858	15,797	3,574
Miscellaneous	1,988	1,464	7,229	3,489	1,959	2,138
Harrison-Lillooet						
Big Silver	610	173	1,919	230	155	106
Harrison	5,092	3,193	9,189	4,239	1,267	5,097
Weaver Creek	33,244	24,138	237,542	21,024	14,511	17,621
Channel	41,595	19,655	57,932	19,243	45,859	21,839
Birkenhead	90,922	65,495	128,771	48,841	42,849	37,612
Miscellaneous	189	0	412	261	216	97
Seton-Anderson						
Gates Creek	4,354	821	232	927	2,678	1,140
Channel	21,140	3,988	1,977	7,498	26,394	4,664
Portage	1,998	6,086	23,965	7,945	1,768	2,083
Seton	0	1,200	0	0	0	50
Miscellaneous	125	2,500	0	14	0	502
Early South Thompson						
Seymour	8,390	11,529	63,306	29,838	17,172	6,435
Scotch	205	18,952	4,709	239	428	3,442
Eagle						
Momich	3,345	63	0	0	5,854	56
Upper Adams	560	P	124	0	3,502	83
Miscellaneous	0	0	7,477	800	0	0
Late South Thompson						
Adams	2,560	31,097	2,070,813	201,669	4,183	10,715
Little River	32	8,169	239,278	P	49	972
South Thompson	0	182	73,603	P	11	0
Big Shuswap Lake	0	0	72,974	1,814	0	0
Lower Shuswap	81	7,358	513,925	7,308	79	3,123
Middle Shuswap	0	N.O.	40,302	27	0	180
Miscellaneous	2	101	50,363	606	17	263

Location	1980	1981	1982	1983	1984	1985
North Thompson						
Raft	5,418	873	2,992	2,857	19,098	3,638
Barriere	133	0	0	0	86	0
Fennell	8,437	2,113	1,139	5,018	11,021	1,620
North Thompson	36	0		750	31	1,883
Miscellaneous	4	13	0	0	130	0
Chilcotin						
Chilko	468,658	35,909	242,263	331,510	452,968	86,120
S. End Chilko Lake	30,168	240	11,288	55,061	127,696	2,000
Taseko	679	0	0	1,630	2,771	0
Miscellaneous	0	0	0	0	0	0
Quesnel						
Horsefly	2,815	661,614[f]	30,317	1,998	5,606	1,020,222
Little Horsefly	0	2	0	0	45	17,030
McKinley, Upper	347	15,775	3,829	0	0	14,999
Lower	N.O.	—	5,657	38	472	82,553
Mitchell	14	66,106	0	119	63	204,579
Miscellaneous	0	5,155	38	0	0	9,880
Nechako						
EarlyNadina	205	821	0	1,337	806	18
Late Nadina River	58	1,024	194	3,035	659	1,516
Channel	3,021	17,892	2,156	23,843	6,413	12,291
Stellako	72,073	22,021	69,434	121,739	60,973	42,296
Miscellaneous	306	921	0	2,910	806	18
Early Stuart						
Ankwill	0	8,497	46	44	99	12,012
Driftwood	0	47,298	29	P	N.O.	93,959
Forfar	2,328	12,228	676	3,591	6,848	19,433
Gluske	1,049	10,741	452	3,781	6,943	17,381
Kynoch	10,661	13,452	1,170	7,822	16,933	20,347
Narrows	257	3,583	78	895	1,568	4,209
Rossette	2,054	8,018	1,300	3,304	8,147	15,704
Miscellaneous	677	25,681	809	4,437	4,709	51,486
Late Stuart						
Kuzkwa	25	20,520	1,237	700	198	2,624
Middle	198	125,630	7,450	639	184	114,122
Tachie	756	94,050	7,528	853	810	155,655
Miscellaneous	0	9,499	543	54	36	2,241
Northeast						
Bowron	2,894	1,170	1,647	6,451	10,461	6,395
Miscellaneous					2	24
TOTALS	848,525	1,443,496	4,024,261	977,238	932,477	2,138,618

[f] **Includes Lower McKinley Creek**

A P P E N D I X   E

Pink Salmon Escapements by Area, Stream and Year
1957 - 1985

Area and Stream	1957	1959	1961	1963	1965	1967
EARLY RUNS						
Lower Fraser						
Main Fraser R.	1,073,904	733,933	547,850	516,831	543,757	785,797
Harrison						
Chehalis R.	9,336	6,729	11,921	12,394	7,621	5,625
Fraser Canyon						
Coquihalla R.	6,601	16,088	7,316	14,971	3,845	3,045
Jones Cr. & Channel	2,034	2,604	5,088	3,500	3,000	3,162
Miscellaneous	4,025	10,170	2,886	3,361	953	1,735
Upper Fraser						
Quesnel R.	—	—	—	—	3,000	2,302
Miscellaneous	263	62	83	723	180	713
Seton-Anderson						
Seton Upper Channel	—	—	6,711	14,106	7,000	7,143
Seton Cr.	58,810	14,887	52,019	107,484	88,046	197,578
Seton Lower Channel	—	—	—	—	—	20,630
Portage Cr.	2,010	52	1,550	8,013	5,931	7,822
Miscellaneous	0	1,214	1,895	6,959	24,271	6,547
Thompson						
Thompson R.	226,329	86,342	69,179	282,240	230,364	448,032
Nicola R.	1,560	806	216	1,196	894	385
Miscellaneous	1,443	76	16	1,807	1,842	2,070
TOTAL	1,386,315	872,963	706,730	973,585	920,704	1,492,586
LATE RUNS						
Lower Fraser						
Miscellaneous	8,061	1,526	4,383	1,319	268	284
Harrison						
Harrison R.	585,798	110,311	186,137	645,476	69,213	64,576
Miscellaneous	346	87	539	693	562	630
Chilliwack-Vedder						
Chilliwack-Vedder	212,334	91,517	188,059	312,685	188,843	249,499
Sweltzer Creek	9,831	751	6,224	15,215	8,908	19,586
Miscellaneous	182	845	489	5,065	5,601	4,058
TOTAL	816,552	205,037	385,831	980,453	273,395	338,633
GRAND TOTAL	2,202,867	1,078,000	1,092,561	1,954,038	1,194,009	1,831,219

....
Pink Salmon Escapements by Area, Stream and Year

Area and Stream	1969	1971	1973	1975	1977
EARLY RUNS					
Lower Fraser					
Main Fraser R.	848,532	928,046	766,053	315,049	775,016
Harrison					
Chehalis R.	7,147	32,178	14,300	2,356	2,613
Fraser Canyon					
Coquihalla R.	2,093	16,778	11,994	5,933	2,821
Jones Cr. & Channel	1,779	1,304	2,544	2,645	3,350
Miscellaneous	1,022	4,467	3,699	938	3,074
Upper Fraser					
Quesnel R.	Few	4,000	0	0	1,500
Miscellaneous	—	1,346	0	36	3,444
Seton-Anderson					
Seton Upper Channel	3,975	6,007	6,708	7,995	11,122
Seton Cr.	180,011	262,534	179,691	208,193	331,844
Seton Lower Channel	14,868	24,882	23,602	23,874	37,163
Portage Cr.	1,092	1,456	13,983	28,454	19,904
Miscellaneous	13,034	13,362	25,074	12,344	35,212
Thompson					
Thompson R.	247,896	255,880	281,146	467,019	971,157
Nicola R.	—	772	1,300	6,853	988
Miscellaneous	1,044	1,551	1,058	6,478	4,694
TOTAL	1,322,453	1,554,563	1,331,152	1,088,167	2,203,902
LATE RUNS					
Lower Fraser					
Miscellaneous	—	—	—	—	—
Harrison					
Harrison R.	96,390	73,881	196,150	180,052	126,782
Miscellaneous	925	1,435	895	1,612	3,360
Chilliwack-Vedder					
Chilliwack-Vedder	91,587	160,511	209,052	80,611	48,254
Sweltzer Creek	18,923	13,122	15,265	16,121	5,093
Miscellaneous	635	1,440	1,747	526	307
TOTAL	208,460	250,389	423,109	278,922	183,796
GRAND TOTAL	1,530,913	1,804,952	1,754,261	1,367,089	2,387,698

....
Pink Salmon Escapements by Area, Stream and Year

Area and Stream	1979	1981	1983	1985
EARLY RUNS				
Lower Fraser				
Fraser River	1,521,856	2,252,368	3,307,834	5,248,742
Fraser Canyon				
Coquihalla River	16,468	24,029	29,190	118,921
Jones Creek	4,993*	4,485*	973*	658
Channel				2,437
Miscellaneous	5,751	18,105	19,458	47,842
Upper Fraser				
Quesnel R.	350	382	400	0
Miscellaneous	1,496	5,150	1,321	530
Seton-Anderson				
Seton Creek	498,874	519,393	371,882	163,337
Seton Upper Channel	9,956	10,402	9,691	4,485
Seton Lower Channel	34,494	33,846	31,045	33,807
Portage Creek	51,842	18,733	10,202	4,116
Miscellaneous	117,674	44,028	78,655	68,375
Thompson				
Thompson River	882,207	1,159,922	502,339	192,747
Nicola River	1,981	910	4,100	265
Miscellaneous	7,003	5,516	5,959	436
TOTAL	3,154,945	4,097,269	4,373,049	5,886,698
LATE RUNS				
Lower Fraser				
Miscellaneous Streams	0	0	100	16,273
Harrison				
Harrison River	269,858	314,519	146,014	438,022
Miscellaneous	2,921	2,479	3,775	9,355
Chilliwack-Vedder				
Chilliwack-Vedder R.	121,927	68,601	99,240	95,556
Sweltzer Creek	8,889	5,213	9,134	14,712
Upper Chilliwack River				
Miscellaneous	2,114	255	406	0
TOTAL	405,709	391,067	258,672	573,918
GRAND TOTAL	3,560,654	4,488,336	4,631,721	6,460,616

* Includes channel

A P P E N D I X   F

. . . .

Sockeye
Canada Convention Waters Catch by Gear

Year	Purse Seine	% of Catch	Gill Net	% of Catch	Trap	% of Catch	Troll	% of Catch	Total
1935									825,508
1936									1,673,049
1937									1,075,986
1938			1,858,848	97.82	41,372	2.18			1,900,220
1939	93,301	16.40	422,949	74.34	52,693	9.26			568,943
1940			1,005,398	97.33	27,602	2.67			1,033,000
1941			1,986,820	93.86	129,903	6.14			2,116,723
1942	2,051,653	40.65	2,895,481	57.36	100,465	1.99			5,047,599
1943			330,988	94.84	18,023	5.16			349,011
1944			974,529	97.08	29,297	2.92			1,003,826
1945			939,000	96.86	30,444	3.14			969,444
1946	1,646,689	38.84	2,555,955	60.28	37,554	0.89			4,240,198
1947	44,011	12.40	307,407	86.58	3,617	1.02			355,035
1948	14,511	1.93	663,635	88.17	74,545	9.90			752,691
1949	111,834	10.96	857,902	84.04	51,063	5.00			1,020,799
1950	371,140	41.49	483,603	54.07	39,726	4.44			894,469
1951	214,187	16.63	1,031,963	80.11	42,012	3.26			1,288,162
1952	122,114	10.58	966,852	83.75	65,417	5.67			1,154,383
1953	600,449	30.14	1,331,823	66.85	60,071	3.02			1,992,343
1954	2,410,564	51.04	2,265,335	47.97	32,822	0.70	13,742	0.29	4,722,463
1955	462,934	41.78	625,207	56.42	18,548	1.67	1,392	0.13	1,108,081
1956	216,388	24.18	678,074	75.78			374	0.04	894,836
1957	522,426	38.39	820,850	60.32	15,759	1.16	1,725	0.13	1,360,760
1958	2,541,592	48.49	2,680,914	51.15	14,241	0.27	4,870	0.09	5,241,617
1959	516,585	32.66	1,040,916	65.80			24,382	1.54	1,581,883
1960	353,482	28.16	898,826	71.61			2,887	0.23	1,255,195
1961	352,883	26.00	991,972	73.10			12,244	0.90	1,357,099
1962	165,062	19.73	660,577	78.98			10,760	1.29	836,399
1963	115,115	16.76	561,345	81.75			10,221	1.49	686,681
1964	7,409	1.44	503,690	97.89			3,449	0.67	514,548
1965	85,914	8.27	944,266	90.87			9,015	0.87	1,039,195
1966	405,585	30.04	922,831	68.35			21,738	1.61	1,350,154
1967	602,495	32.12	1,111,186	59.25			161,801	8.63	1,875,482
1968	13,805	1.50	869,162	94.46			37,125	4.03	920,092
1969	340,187	20.30	1,268,525	75.71			66,824	3.99	1,675,536
1970	441,120	28.61	955,178	61.95			145,473	9.44	1,541,771
1971	1,233,531	39.61	1,689,607	54.25			191,160	6.14	3,114,298
1972	281,532	26.04	784,405	72.55			15,280	1.41	1,081,217
1973	1,126,314	43.67	1,395,085	54.09			57,571	2.23	2,578,970
1974	1,044,742	41.79	1,029,678	41.19			425,599	17.02	2,500,019
1975	43,201	6.70	550,783	85.46			50,489	7.83	644,473
1976	605,101	43.77	741,049	53.60			36,334	2.63	1,382,484

Year	Purse Seine	% of Catch	Gill Net	% of Catch	Trap	% of Catch	Troll	% of Catch	Total
1977	448,214	22.74	1,487,900	75.48			35,039	1.78	1,971,153
1978	460,603	34.97	626,506	47.57			230,006	17.46	1,317,115
1979	291,859	18.01	1,111,288	68.59			217,118	13.40	1,620,265
1980	69,853	15.31	379,367	83.17			6,938	1.52	456,158
1981	180,706	15.58	948,312	81.78			30,635	2.64	1,159,653
1982	1,525,003	46.57	924,648	28.24			824,826	25.19	3,274,477
1983	4,159	0.73	514,371	89.77			54,455	9.50	572,985
1984	423,623	26.88	1,038,316	65.89			113,910	7.23	1,575,849
1985	2,284,774	48.86	1,788,516	38.25			602,855	12.89	4,676,145

402

· · · · · · · · · · · · · · · · · · ·

Sockeye
United States Convention Waters Catch by Gear

Year	Purse Seine	% of Catch	Gill Net	% of Catch	Reef Net	% of Catch	Troll	% of Catch	Total
1935	600,366	97.54	9,606	1.56	5,530	0.90			615,502
1936	415,039	91.62	19,111	4.22	18,875	4.17			453,025
1937	794,797	88.60	43,418	4.84	58,807	6.56			897,022
1938	1,307,150	92.81	76,803	5.45	24,408	1.73			1,408,361
1939	480,217	86.49	16,264	2.93	58,752	10.58			555,233
1940	515,941	78.88	57,972	8.86	80,178	12.26			654,091
1941	1,350,600	86.66	43,275	2.78	164,680	10.57			1,558,555
1942	2,761,460	94.08	36,868	1.26	136,864	4.66			2,935,192
1943	203,228	83.95	5,688	2.35	33,161	13.70			242,077
1944	335,172	76.97	40,620	9.33	59,651	13.70			435,443
1945	605,982	85.78	32,245	4.56	68,237	9.66			706,464
1946	3,366,758	94.80	50,441	1.42	134,111	3.78			3,551,310
1947	76,692	86.93	1,770	2.01	9,758	11.06			88,220
1948	940,415	86.35	70,991	6.52	77,685	7.13			1,089,091
1949	850,451	80.47	123,048	11.64	83,293	7.88			1,056,792
1950	1,061,480	86.94	82,854	6.79	76,559	6.27			1,220,893
1951	875,922	77.05	152,376	13.40	108,497	9.54			1,136,795
1952	826,384	74.22	175,014	15.72	112,077	10.07			1,113,475
1953	1,355,739	66.71	427,834	21.05	248,864	12.24			2,032,437
1954	3,764,998	78.34	861,859	17.93	179,401	3.73			4,806,258
1955	621,527	61.74	282,995	28.11	102,088	10.14			1,006,610
1956	428,562	47.26	371,729	40.99	106,581	11.75			906,872
1957	1,237,700	73.27	286,614	16.97	164,951	9.76			1,689,265
1958	4,259,324	81.02	844,602	16.07	152,158	2.89	1,232	0.02	5,257,316
1959	1,401,819	77.42	241,163	13.32	163,093	9.01	4,663	0.26	1,810,738
1960	843,850	70.38	253,211	21.12	100,915	8.42	993	0.08	1,198,969
1961	823,956	59.78	471,464	34.20	81,826	5.94	1,146	0.08	1,378,392
1962	505,028	66.57	192,078	25.32	60,694	8.00	837	0.11	758,637
1963	862,616	65.65	365,873	27.84	85,110	6.48	446	0.03	1,314,045
1964	284,209	55.94	177,767	34.99	45,827	9.02	284	0.06	508,087
1965	740,123	72.13	236,133	23.01	49,707	4.84	155	0.02	1,026,118
1966	783,466	58.59	496,295	37.11	57,086	4.27	368	0.03	1,337,215
1967	1,387,370	66.45	595,580	28.53	104,694	5.01	182	0.01	2,087,826
1968	464,544	52.44	354,760	40.05	66,404	7.50	162	0.02	885,870
1969	991,598	62.52	517,650	32.64	76,570	4.83	358	0.02	1,586,176
1970	779,271	57.71	504,873	37.39	65,644	4.86	429	0.03	1,350,217
1971	1,607,117	57.07	1,016,984	36.11	191,682	6.81	346	0.01	2,816,129
1972	533,179	47.26	506,406	44.89	88,304	7.83	303	0.03	1,128,192
1973	1,410,499	53.68	1,075,698	40.94	140,921	5.36	463	0.02	2,627,581
1974	1,515,444	61.56	873,595	35.49	72,408	2.94	228	0.01	2,461,675
1975	896,416	57.32	615,790	39.38	51,096	3.27	549	0.04	1,563,851
1976	669,322	50.63	628,411	47.53	23,869	1.81	436	0.03	1,322,038
1977	822,995	45.99	899,757	50.28	65,984	3.69	873	0.05	1,789,609
1978	694,460	50.97	635,795	46.67	31,832	2.34	359	0.03	1,362,446

Year	Purse Seine	% of Catch	Gill Net	% of Catch	Reef Net	% of Catch	Troll	% of Catch	Total
1979	942,566	53.08	779,807	43.91	52,201	2.94	1,302	0.07	1,775,876
1980	189,899	40.74	263,138	56.45	13,050	2.80	86	0.02	466,173
1981	616,846	47.65	652,674	50.42	24,856	1.92	110	0.01	1,294,486
1982	1,686,609	58.83	1,124,475	39.22	55,280	1.93	546	0.02	2,866,910
1983	165,424	44.76	189,999	51.41	14,100	3.81	76	0.02	369,599
1984	675,299	41.17	934,421	56.96	30,611	1.87	36	0.00	1,640,367
1985	1,403,512	48.01	1,423,830	48.71	95,967	3.28	34	0.00	2,923,343

United States Convention Waters
Pink Salmon Catch by Gear

Year	Troll	%	Reef Net	%	Gill Net	%	Purse Seine	%	Total
1957	165,248	5.95	149,094	5.37	246,296	8.87	2,216,728	79.81	2,777,366
1959	175,921	7.25	110,416	4.55	227,643	9.38	1,913,555	78.82	2,427,535
1961	63,893	12.56	28,513	5.61	71,924	14.14	344,214	67.69	508,544
1963	499,753	11.29	89,768	2.03	382,424	8.64	3,454,287	78.04	4,426,232
1965	77,849	13.94	21,264	3.81	48,823	8.74	410,444	73.51	558,380
1967	193,521	5.06	118,994	3.11	310,744	8.12	3,203,781	83.71	3,827,040
1969	40,324	4.26	37,331	3.95	91,609	9.69	776,533	82.10	945,797
1971	12,863	.54	118,904	5.02	334,202	14.09	1,905,182	80.35	2,371,151
1973	14,126	.64	101,729	4.57	323,370	14.53	1,785,699	80.26	2,224,924
1975	23,164	1.85	55,223	4.41	196,726	15.70	978,042	78.04	1,253,155
1977	163,416	7.54	30,069	1.39	197,178	9.09	1,777,767	81.98	2,168,430
1979	260,735	6.44	42,771	1.06	388,723	9.61	3,354,044	82.89	4,046,273
1981	176,493	4.55	80,148	2.07	308,649	7.96	3,310,281	85.41	3,875,571
1983	104,428	5.58	87,024	4.65	155,889	8.34	1,523,289	81.43	1,870,630
1985	94,546	2.45	75,888	1.96	347,958	9.00	3,346,263	86.59	3,864,655

....

Canada Convention Waters
Pink Salmon Catch by Gear

Year	Purse Seine	%	Gillnet	%	Troll	%	Traps	%	Total
1945			1,057,978	82.66			221,871	17.34	1,279,849
1947	1,819,356	52.11	1,503,768	43.07			168,292	4.82	3,491,416
1949	2,293,198	71.89	787,160	24.68			109,304	3.43	3,189,662
1951	1,970,996	68.30	776,160	26.90	123,080	4.27	15,278	0.53	2,885,514
1953	2,950,595	71.23	1,030,194	24.88	75,108	1.81	86,220	2.08	4,142,117
1955	2,931,552	71.00	1,039,406	25.17	32,069	0.78	126,036	3.05	4,129,063
1957	1,435,924	54.50	1,126,085	42.74	41,402	1.57	31,309	1.19	2,634,720
1959	1,357,088	58.68	693,977	30.00	261,841	11.32			2,312,906
1961	313,636	57.53	142,518	26.14	88,974	16.33			545,128
1963	2,936,194	70.36	797,385	19.10	439,709	10.54			4,173,288
1965	336,478	56.79	182,059	30.73	73,930	12.48			592,467
1967	2,289,207	55.07	892,447	21.47	975,268	23.46			4,156,922
1969	277,592	32.22	366,005	42.49	217,908	25.29			861,505
1971	939,737	44.97	775,663	36.29	421,937	19.74			2,137,337
1973	1,246,204	60.48	395,901	19.21	418,574	20.31			2,060,679
1975	639,026	50.88	376,511	29.98	240,353	19.14			1,255,890
1977	807,194	38.89	280,674	13.52	987,610	47.59			2,075,478
1979	2,480,864	60.05	103,738	2.51	1,546,753	37.44			4,131,355
1981	2,794,078	66.72	347,984	8.31	1,045,791	24.97			4,187,853
1983	70,520	6.59	850,590	79.51	148,750	13.90			1,069,860
1985	2,123,414	65.82	378,158	11.72	724,489	22.46			3,226,061

A P P E N D I X G$_1$

Convention Waters Sockeye Catch
1946-1985

Year	U.S. Catch	% of Total	Canada Catch	% of Total	Convention Total Catch
1946	3,551,310	45.58	4,240,198	54.42	7,791,508
1947	88,220	19.90	355,035	80.10	443,255
1948	1,089,091	59.13	752,691	40.87	1,841,782
1949	1,056,792	50.87	1,020,799	49.13	2,077,591
1950	1,220,893	57.72	894,469	42.28	2,115,362
1951	1,136,795	46.88	1,288,162	53.12	2,424,957
1952	1,113,475	49.10	1,154,383	50.90	2,267,858
1953	2,032,437	50.50	1,992,434	49.50	4,024,780
1954	4,806,258	50.44	4,722,463	49.56	9,528,721
1955	1,006,610	47.60	1,108,081	52.40	2,114,691
1956	906,872	50.33	894,836	49.67	1,801,708
1957	1,678,265	55.39	1,360,760	44.61	3,050,025
1958	5,257,316	50.07	5,241,617	49.93	10,489,933
1959	1,810,738	53.37	1,581,883	46.63	3,392,621
1960	1,198,969	48.85	1,255,195	51.15	2,454,164
1961	1,378,392	50.39	1,357,099	49.61	2,735,491
1962	758,637	47.56	836,399	52.44	1,595,036
1963	1,314,045	65.68	686,681	34.32	2,000,726
1964	508,087	49.68	514,548	50.32	1,022,635
1965	1,026,118	49.68	1,039,195	50.32	2,065,313
1966	1,337,215	49.76	1,350,154	50.24	2,687,369
1967	2,087,826	52.68	1,875,482	47.32	3,963,308
1968	885,870	49.05	920,092	50.95	1,805,962
1969	1,586,176	48.63	1,675,536	51.37	3,261,712
1970	1,350,217	46.69	1,541,771	53.37	2,891,988
1971	2,816,129	47.49	3,114,298	52.51	5,930,427
1972	1,128,192	51.06	1,081,217	48.94	2,209,409
1973	2,627,581	50.47	2,578,970	49.53	5,206,551
1974	2,461,675	49.61	2,500,019	50.39	4,961,694
1975	1,563,851	70.82	644,473	29.18	2,208,324
1976	1,322,038	48.88	1,382,484	51.12	2,704,522
1977	1,789,609	47.59	1,971,153	52.41	3,760,762
1978	1,326,446	50.85	1,317,115	49.15	2,679,561
1979	1,775,876	52.29	1,620,265	47.71	3,396,141
1980	466,173	50.54	456,158	49.46	922,331
1981	1,294,486	52.75	1,159,653	47.25	2,454,139
1982	2,866,910	46.68	3,274,477	53.32	6,141,387
1983	369,599	39.21	573,985	60.79	942,584
1984	1,640,367	51.00	1,575,849	49.00	3,216,216
1985*	2,923,343	38.47	4,676,145	61.53	7,599,488
TOTAL	66,605,899	49.64	67,585,133	50.36	134,191,032

* **Management by Pacific Salmon Treaty, not under 50:50 sharing basis**

A P P E N D I X G₂

Convention Waters Pink Catch
1957-1985

Year	Canada	% of Total	United States	% of Total	Total
1957	2,634,720	48.68	2,777,366	51.32	5,412,086
1959	2,312,906	48.79	2,427,535	51.21	4,740,441
1961	545,128	51.72	508,544	48.28	1,053,672
1963	4,173,288	48.53	4,426,232	51.47	8,599,520
1965	592,467	51.48	558,380	48.52	1,150,847
1967	4,156,922	52.07	3,827,040	47.93	7,983,962
1969	861,505	47.67	945,797	52.33	1,807,302
1971	2,137,337	47.41	2,371,151	52.59	4,508,488
1973	2,060,679	48.08	2,224,924	51.92	4,285,603
1975	1,255,890	50.05	1,253,155	49.95	2,509,045
1977	2,075,478	48.90	2,168,430	51.10	4,243,908
1979	4,131,355	50.52	4,046,273	49.48	8,177,628
1981	4,187,853	51.94	3,875,571	48.06	8,063,424
1983	1,069,860	36.38	1,870,630	63.62	2,940,490
1985*	3,226,061	45.50	3,864,655	54.50	7,090,716
TOTAL	35,421,449	48.81	37,145,683	51.19	72,567,132

* Management by Pacific Salmon Treaty, not under 50:50 sharing basis

A P P E N D I X H$_1$

. . . .
1982 Cycle
Cyclic Landings of Sockeye from Convention Waters

	United States	Canada	Total
1946-1982			
Total Landings (No. Sockeye)	61,672,590	60,760,154	122,432,744
Share in Fish	50.37%	49.63%	
1982	2,866,910	3,274,477	6,141,387
1978	1,362,446	1,317,115	2,679,561
1974	2,461,675	2,500,019	4,961,694
1970	1,350,217	1,541,771	2,891,988
1966	1,337,215	1,350,154	2,687,369
1962	758,637	836,399	1,595,036
1958	5,257,316	5,241,617	10,498,933
1954	4,806,258	4,722,463	9,528,721
1950	1,220,893	894,469	2,115,362
1946	3,551,310	4,240,198	7,791,508
1942	2,935,192	5,047,599	7,982,791
1938	1,408,361	1,900,220	3,308,581
1934	3,590,058	1,430,300	5,020,358
1930	3,544,718	1,043,318	4,588,032
1926	469,900	912,566	1,382,466
1922	513,848	580,144	1,093,992
1918	569,094	242,275	811,369
1914	3,555,890	2,137,177	5,693,067
1910	2,765,726	1,690,091	4,455,817
1906	2,030,550	2,066,604	4,097,154
1902	4,001,717	3,177,538	7,179,255

1983 Cycle
Cyclic Landings of Sockeye from Convention Waters

	United States	Canada	Total
1947-1983			
Total Landings (No. Sockeye)	62,042,189	61,333,139	123,375,328
Share in Fish	50.29%	49.71%	
1983	369,599	572,985	942,584
1979	1,775,876	1,620,265	3,396,141
1975	1,563,851	644,473	2,208,324
1971	2,816,129	3,114,298	5,930,427
1967	2,087,826	1,875,482	3,963,308
1963	1,314,045	686,681	2,000,726
1959	1,810,738	1,581,883	3,392,621
1955	1,006,610	1,108,081	2,114,691
1951	1,136,795	1,288,162	2,424,957
1947	88,220	355,035	443,255
1943	242,077	349,011	591,088
1939	555,233	568,943	1,124,176
1935	615,502	825,508	1,441,010
1931	975,591	458,048	1,433,639
1927	1,069,557	713,930	1,783,487
1923	495,490	361,463	856,953
1919	778,669	470,199	1,248,868
1915	736,939	1,088,524	1,825,463
1911	1,447,919	730,714	2,178,633
1907	1,030,359	691,210	1,721,569
1903	1,911,127	2,341,492	4,252,619

....
1984 Cycle
Cyclic Landings of Sockeye from Convention Waters

	United States	Canada	Total
1948-1984			
Total Landings (No. Sockeye)	63,682,556	62,908,988	126,591,544
Share in Fish	50.31%	49.69%	
1984	1,640,367	1,575,849	3,216,216
1980	466,173	456,158	922,331
1976	1,322,038	1,382,484	2,704,522
1972	1,128,192	1,081,217	2,209,409
1968	885,870	920,092	1,805,962
1964	508,087	514,548	1,022,635
1960	1,198,969	1,255,195	2,454,164
1956	906,872	894,836	1,801,708
1952	1,113,475	1,154,383	2,267,858
1948	1,089,091	752,691	1,841,782
1944	435,443	1,003,826	1,439,269
1940	654,091	1,033,000	1,687,091
1936	453,025	2,126,074	2,579,099
1932	853,406	733,735	1,587,141
1928	630,457	311,226	941,683
1924	772,056	442,250	1,214,306
1920	677,690	532,039	1,209,729
1916	909,425	376,891	1,286,316
1912	2,005,869	1,357,425	3,363,294
1908	1,879,268	870,612	2,749,880
1904	1,506,137	892,934	2,399,071

··················

1985 Cycle
Cyclic Landings of Sockeye from Convention Waters

	United States	Canada	Total
1949-1985			
Total Landings (No. Sockeye)	66,605,899	67,585,133	134,191,032
Share in Fish	49.64%	50.36%	
1985*	2,923,343	4,676,145	7,599,488
1981	1,294,486	1,159,653	2,454,139
1977	1,789,609	1,971,153	3,760,762
1973	2,627,581	2,578,970	5,206,551
1969	1,586,176	1,675,536	3,261,712
1965	1,026,118	1,039,195	2,065,313
1961	1,378,392	1,357,099	2,735,491
1957	1,689,265	1,360,760	3,050,025
1953	2,032,437	1,992,343	4,024,780
1949	1,056,792	1,020,799	2,077,591
1945	706,464	969,444	1,675,908
1941	1,558,354	2,116,723	3,675,077
1937	897,022	1,075,986	1,973,008
1933	1,724,127	726,309	2,450,436
1929	1,334,141	725,037	2,059,178
1925	1,375,012	453,704	1,828,716
1921	1,199,929	486,312	1,686,241
1917	5,005,609	1,877,792	6,883,401
1913	21,736,398	9,606,641	31,343,039
1909	13,664,988	7,261,486	20,926,474
1905	10,330,277	10,350,959	20,681,236
1901	13,694,032	12,065,999	25,760,031

*** Data for 1985 taken from Washington Department of Fisheries, Department of Fisheries and Oceans and the IPSFC for the Fraser River Panel Area.**

A P P E N D I X H$_2$

Total Fraser River Sockeye Commercial Catch in all Waters 1946 - 1989

Year	Sockeye
1946	8,161,508
1947	632,255
1948	1,986,782
1949	2,391,591
1950	2,471,362
1951	2,716,957
1952	2,574,311
1953	4,445,398
1954	9,655,306
1955	2,273,656
1956	1,929,028
1957	3,642,369
1958	14,828,998
1959	3,755,172
1960	2,688,808
1961	3,326,556
1962	1,750,852
1963	2,193,925
1964	1,119,816
1965	2,194,030
1966	3,386,504
1967	5,342,772
1968	2,204,765
1969	3,774,948
1970	4,069,900
1971	6,794,375
1972	2,742,982
1973	5,533,332
1974	6,636,261
1975	2,424,709
1976	3,283,900
1977	4,478,000
1978	6,680,000
1979	4,726,000
1980	2,069,000
1981	5,840,000
1982	9,504,000
1983	3,886,000
1984	4,608,000
1985	11,235,000
1986*	11,546,000
1987*	5,199,000
1988*	1,855,000
1989*	14,351,000

* Fisheries under management of Pacific Salmon Commission.

Total Fraser River Pink Salmon Commercial Catch in all Waters 1959-1987*

Year	Pink
1959	5,364,360
1961	766,630
1963	3,373,118
1965	1,079,809
1967	11,019,125
1969	2,319,313
1971	7,871,510
1973	4,955,594
1975	3,455,809
1977	5,756,657
1979	10,654,416
1981	14,196,627
1983	10,592,889
1985	12,100,354
1987**	3,666,000

*   Adjusted catch data from Pacific Salmon Commission.
** Fisheries under management of Pacific Salmon Commission.

Landings of Pink Salmon from Convention Waters

	United States	Canada	Total
1957-1985			
Total Landings (No. Pinks)	37,145,683	35,421,449	72,567,132
Share in Fish	51.19%	48.81%	
1985*	3,864,655	3,226,061	7,090,716
1983	1,870,630	1,069,860	2,940,490
1981	3,875,571	4,187,853	8,063,424
1979	4,046,273	4,131,355	8,177,628
1977	2,168,430	2,075,478	4,243,908
1975	1,253,155	1,255,890	2,509,045
1973	2,224,924	2,060,679	4,285,603
1971	2,371,151	2,137,337	4,508,488
1969	945,797	861,505	1,807,302
1967	3,827,040	4,156,922	7,983,962
1965	558,380	592,467	1,150,847
1963	4,426,232	4,173,288	8,599,520
1961	508,544	545,128	1,053,672
1959	2,427,535	2,312,906	4,740,441
1957	2,777,366	2,634,720	5,412,086
1955	4,685,984	4,129,063	8,815,047
1953	4,951,429	4,142,117	9,093,546
1951	5,086,284	2,885,514	7,971,798
1949	6,235,400	3,189,662	9,425,062
1947	8,801,595	3,491,416	12,293,011
1945	5,458,890	1,279,849	6,738,739

* **Data for 1985 taken from Washington Department of Fisheries, Department of Fisheries and Oceans and the IPSFC for the Fraser River Panel Area. Not under 50:50 sharing basis**

International Pacific Salmon Fisheries Commission Publications

BULLETINS

I   Effect of the obstruction at Hell's Gate on the sockeye salmon of the Fraser River by W.F. Thompson. 1945.

II   1. A study of the tagging method in the enumeration of sockeye salmon populations by Gerald V. Howard. 2. A mathematical study of confidence limits of salmon populations calculated from sample tag ratios by D.G. Chapman. 1948

III   1. A biological study of the effectiveness of the Hell's Gate Fishways by G. B. Talbot. 2. Variations in flow patterns at Hell's Gate and their relationships to the migration of sockeye salmon by R.I. Jackson. 1950.

IV   A study of the spawning populations of sockeye salmon in the Harrison River system, with special reference to the problem of enumeration by means of marked members by M.B. Schaefer. 1951.

V   The selective action of gillnets on Fraser River sockeye salmon by A.E. Peterson. 1954.

VI   An investigation of the effect of Baker Dam on downstream-migrant salmon by J.A.R. Hamilton and F.J. Andrew. 1954.

VII   The chronological order of Fraser River sockeye salmon during migration, spawning and death by S.R. Killick. 1955.

VIII   An investigation of the problem of guiding downstream-migrant salmon at dams by F.J. Andrew, L.R. Kersey and P.C. Johnson. 1955.

IX   Collection and interpretation of sockeye salmon scales by R.I. Clutter and L.E. Whitesel. 1956

X   Character of the migration of pink salmon to Fraser River spawning grounds in 1957 by F.J. Ward. 1959.

XI   Sockeye and pink salmon production in relation to proposed dams in the Fraser River system by F.J. Andrew and G.H. Geen. 1960.

XII   Racial identification of Fraser River sockeye salmon by means of scales and its applications to salmon management by K.A. Henry. 1961.

XIII   Marine tagging of Fraser River sockeye salmon by L.A. Verhoeven and E.B. Davidoff. 1962

XIV   The age, sex ratio and size of Fraser River sockeye salmon 1915 to 1960 by S.R. Killick and W.A. Clemens. 1963.

XV The migration and exploitation of pink salmon runs in and adjacent to the Fraser River convention area in 1959 by E.H. Vernon, A.S. Hourston and G.A. Holland. 1964.

XVI Limnology of Kamloops Lake by F.J. Ward. 1964.

XVII The migration, composition, exploitation and abundance of odd-year pink salmon runs in and adjacent to the Fraser River convention area by A.S. Hourston, E.H. Vernon and G.A. Holland. 1965.

XVIII The effect of transported stream sediments on the survival of sockeye and pink salmon eggs and alevin by A.C. Cooper. 1965.

XIX Enumeration of migrant pink salmon fry in the Fraser River estuary by E.H. Vernon. 1966.

XX Histological and hematological changes accompanying sexual maturation of sockeye salmon in the Fraser River system by G.S. Colgrove. 1966.

XXI Mechanisms controlling migration of sockeye salmon fry by E.L. Brannon. 1973.

XXII Energy sources and expenditures in Fraser River sockeye salmon during their spawning migration by Philip Gilhousen. 1980.

XXIII The 1983 early run Fraser and Thompson River pink salmon: morphology, energetics and fish health. I.V. Williams, J.R. Brett, G.R. Bell, G.S. Traxler, J. Bagshaw, J.R. McBride, U.H.M. Fagerlund, H.M. Dye, J.P. Sumpter, E.M. Donaldson, E. Bilinski, H. Tsuyuki, M.D. Peters, E.M. Choromanski, J.H.Y. Cheng and W.L. Coleridge. 1986.

XXIV Studies of the lacustrine biology of the sockeye salmon (*Oncorhynchus nerka*) in the Shuswap system. I.V. Williams, P. Gilhousen, W. Saito, T. Gjernes, K. Morton, R. Johnson and D. Brock. 1989.*

XXV Homing behavior of adult sockeye salmon in response to a hydroelectric diversion of homestream waters at Seton Creek. M.R. Fretwell. 1989.*

XXVI Prespawning mortalities of sockeye salmon in the Fraser River system and possible causal factors. P. Gilhousen. 1990.*

PROGRESS REPORTS

1. Further experiments with an electric screen for downstream-migrant salmon at Baker Dam by F.J. Andrew, P.C. Johnson and L.R. Kersey. 1956.

2. Electric screens for adult salmon by F.J. Andrew, P.C. Johnson and L.R. Kersey. 1956.

3. Seasonal and annual changes in availability of the adult crustacean plankters of Shuswap Lake by F.J. Ward. 1957.

4. Sockeye and pink salmon investigations at the Seton Creek hydroelectric installation by F.J. Andrew and G.H. Geen. 1958.

5. An examination of factors affecting the abundance of pink salmon in the Fraser River by E.H. Vernon. 1958.

---

* Completed after dissolution of IPSFC.

6.	The energy expenditures of Fraser River sockeye salmon during the spawning migration to Chilko and Stuart Lakes by D.R. Idler and W.A. Clemens. 1959.

7.	Migratory behavior of adult Fraser River sockeye by P. Gilhousen. 1960.

8.	Limnological changes in Seton Lake resulting from hydroelectric diversions by G.H. Geen and F.J. Andrew. 1961.

9.	Origin and treatment of a supersaturated river water by H.H. Harvey and A.C. Cooper. 1962.

10.	The history of the early Stuart sockeye run by K.A. Henry and A.C. Cooper. 1962.

11.	Cyclic dominance in Adams River sockeye by F.J. Ward and P.A. Larkin. 1964.

12.	The influence of physical factors on the development and weight of sockeye salmon embryos and alevins by E.L. Brannon. 1965.

13.	Toxicity and treatment of kraft pulp bleach plant waste by J.A. Servizi, E.T. Stone and R.W. Gordon. 1966.

14	Effects of log driving on the salmon and trout populations in the Stellako River prepared by the technical staffs of the Canada Department of Fisheries and the International Pacific Salmon Fisheries Commission in collaboration with the Fish and Wildlife Branch, British Columbia Department of Recreation and Conservation. 1966.

15.	Occurrence and control of *Chondrococcus columnaris* as related to Fraser River sockeye salmon by D.J. Colgrove and J.W. Wood. 1966.

16.	Genetic control of migrating behavior of newly emerged sockeye salmon fry by E.L. Brannon. 1967.

17.	Toxicity of two chlorinated catechols, possible components of kraft pulp mill bleach waste by J.A. Servizi, R.W. Gordon and D.W. Martens. 1968.

18.	Lamprey parasitism on Fraser River sockeye and pink salmon during 1967 by I.V. Williams and P. Gilhousen. 1968.

19.	Responses of young pink salmon to vertical temperature and salinity gradients by D.A. Hurley and W.L. Woodall. 1968.

20.	Comparison of sockeye salmon fry produced by hatcheries, artificial channels and natural spawning areas by R.W. Mead and W.L. Woodall. 1968.

21.	Effect of feeding before and after yolk absorption on the growth of sockeye salmon by D.A. Hurley and E.L. Brannon. 1969.

22.	Implication of water quality and salinity in the survival of Fraser River sockeye smolts by I.V. Williams. 1969.

23.	Marine disposal of sediments from Bellingham Harbor as related to sockeye and pink salmon fisheries by J.A. Servizi, R.W. Gordon and D.W. Martens. 1969.

24.	Effects of decaying bark on incubating salmon eggs by J.A. Servizi, D.W. Martens and R.W. Gordon. 1970.

25.	Toxicity and treatment of de-inking wastes containing detergents by D.W. Martens and R.W. Gordon. 1970.

26.   Detoxification of kraft pulp mill effluent by an aerated lagoon by J.A. Servizi and D.W. Martens. 1972.

27.   I. Temperature control during sockeye spawning period in McKinley Creek in 1969 by A.C. Cooper. 1973.

      II. Investigation of the prespawning mortality of sockeye in Horsefly River and McKinley Creek in 1969 by I.V. Williams. 1973.

28.   Tests with Ni-Furpirinol (P7138) to control prespawning mortalities of Fraser River sockeye by I.V. Williams. 1973.

29.   Acute toxicity of municipal sewage to fingerling sockeye salmon by D.W. Martens and J.A. Servizi. 1974.

30.   Preliminary survey of toxicity of chlorinated sewage to sockeye and pink salmon by J.A. Servizi and D.W. Martens. 1974.

31.   Acute toxicity and detoxification of kraft pulp mill effluent by R.W. Gordon and J.A. Servizi. 1974.

32.   Dechlorination of municipal sewage using sulfur dioxide by D.W. Martens and J.A. Servizi. 1975.

33.   Acute toxicity at three primary sewage treatment plants. D.W. Martens and J.A. Servizi. 1976.

34.   Resistance of adult sockeye salmon to acute thermal shock. J.A. Servizi and J.O.T. Jensen. 1977.

35.   I. Investigation of the prespawning mortality of sockeye in Chilko River in 1971. I.V. Williams.

      II. Investigation of the use af antibiotics to control the prespawning mortality of the 1971 Chilko population. I.V. Williams and D. Stelter. 1977.

36.   Evaluation of the production of sockeye and pink salmon at spawning and incubation channels in the Fraser River system. A.C. Cooper. 1977.

37.   Investigation of prespawning mortality of 1973 Horsefly River sockeye salmon. I.V. Williams, U.H.M. Fagerlund, J.R. McBride, G.A. Strasdine, H. Tsuyuki and E.J. Ordal. 1977.

38.   Acute toxicity at Annacis Island primary sewage treatment plant by J.A. Servizi, D.W. Martens and R.W. Gordon. 1978.

39.   Effects of selected heavy metals on early life of sockeye and pink salmon. J.A. Servizi and D.W. Martens. 1978.

40.   Toxicity of Butoxyethyl Ester of 2, 4-D to selected salmon and trout. D.W. Martens, R.W. Gordon and J.A. Servizi. 1980.

41.   Application of dual-beam acoustic procedures to estimate limnetic juvenile sockeye salmon. R.L. Johnson and J.J. Burczynski. 1985.

42.   Wounds, scars and marks on Fraser River sockeye salmon with some relationships to predation losses. P. Gilhousen. 1989.*

43.   Historical efficiency of the Fraser River gillnet sockeye fishery and calculation of annual escapements, 1893 to 1944. P. Gilhousen. 1990 MS*. In preparation for Progress Report.

---

* Completed after dissolution of the IPSFC.

SPECIAL PUBLICATIONS

1949   Report on the Chilko River watershed.

1951   Report on the fisheries problems created by the development of power in the Nechako-Kemano-Nanika River systems. Prepared by the Technical Staffs of the Department of Fisheries of Canada, the Fisheries Research Bd. of Canada and the IPSFC.

1952   Report on the fisheries problems created by the development of power in the Nechako-Kemano-Nanika River systems - Supplement No. 1 - Temperature changes in the Nechako River and their effects on the salmon populations. Prepared by the Technical Staffs of the Department of Fisheries of Canada and the IPSFC.

1953   A review of the sockeye salmon problems created by the Alcan Project in the Nechako River watershed.

1953   A report on the wastes from oil refineries and recommendations for their disposal - with particular reference to the proposed Kamloops refinery to be located adjacent to the Thompson River. Prepared by the Technical Staffs of the Department of Fisheries of Canada, the Fisheries Research Bd. of Canada and the IPSFC.

1955   A report on the fish facilities and fisheries problems related to the Fraser and Thompson River dam site investigations. Prepared by the Technical Staffs of the Department of Fisheries of Canada and the IPSFC. In collaboration with the B.C. Fisheries Department and the B.C. Game Commission.

1955   A study of the feasibility and costs of electric energy transmission to the Vancouver Load Centre from the Mica Creek dam site on the Columbia River and sites on the Fraser River below Lytton.

1958   A preliminary review or pertinent past tagging investigations on pink salmon and proposal for a co-ordinated research program for 1959. Prepared by the Pink Salmon Co-Ordinating Committee.

1958   The salmon spawning grounds of the Fraser River below Hope and of the Harrison River in relation to the dredging of shipping channels. Prepared by the Technical Staffs of the Department of Fisheries of Canada and the IPSFC.

1958   Fisheries problems related to the proposed Moran Dam on the Fraser River. Prepared by the Technical Staffs of the Department of Fisheries of Canada and the IPSFC in collaboration with B.C. Game Commission and B.C. Fisheries Department.

1959   A plan for an artificial spawning channel for pink salmon at Seton Creek. Prepared by Technical Staffs of the IPSFC and the Department of Fisheries of Canada.

1960   2nd progress report on the co-ordinated research program for 1959. Prepared by the Pink Salmon Co-Ordinating Committee.

1960   3rd progress report on the co-ordinated research program for 1959. Prepared by the Pink Salmon Co-Ordinating Committee. 1960.

1961   Interim report on proposed kraft pulp mills on the Fraser River near Prince George with recommendations for the treatment and disposal of wastes. Prepared by the Technical Staffs of the Department of Fisheries of Canada and the IPSFC in collaboration with Fish and Game Branch, B.C. Department of Recreation and Conservation.

1962   Report on the fisheries problems associated with the proposed Stuart Lake storage dam. Prepared by the Technical Staffs of the Department of Fisheries of Canada and the IPSFC.

1964   Proposed artificial spawning channel for Weaver Creek sockeye salmon.

1965   Report on fish disease as a possible cause of prespawning mortalities of Fraser River sockeye. J. Wood.

1965   An examination of factors affecting sockeye and pink salmon in the Fraser and Thompson Rivers at low river levels.

1966   A plan for a 2nd artificial spawning channel for pink salmon at Seton Creek.

1966   Proposed artificial spawning channel for Gates Creek sockeye salmon.

1969   Problems in rehabilitating the Quesnel sockeye run and their possible solution. (Horsefly report).

1969   Proposed artificial spawning channel for Chilliwack River pink salmon.

1969   Report on the fisheries problems associated with the proposed diversion of water from Shuswap River to Okanagan Lake. Prepared by the Technical Staffs of the Department of Fisheries and Forestry of Canada and the IPSFC.

1970   Proposed artificial spawning channel for Nadina River sockeye salmon.

1970   Selected measurements of water quality and bottom dwelling organisms of the Fraser River system 1963-1968. J.A. Servizi and R.A. Burkhalter.

1971   Fisheries problems related to Moran Dam on the Fraser River. Prepared by the Technical Staffs of the Canada Department of Environment, Fisheries Service, and the IPSFC.

1972   Proposed program for restoration and extension of the sockeye and pink salmon stocks of the Fraser River.

1976   Tailrace delay and loss of adult sockeye salmon at the Seton Creek hydroelectric plant.

1977   An examination of erosion in the Chilliwack-Vedder system.

1979   Salmon studies associated with the Potential Kemano II Hydroelectric Development. Sockeye studies on the Nechako River.

1983   Potential effects of the Kemano Completion Project on Fraser River sockeye and pink salmon.

. . . .
Sockeye Transplants

DONOR - BOWRON RIVER

Brood Year	Type of Transplant	Numbers Taken	Area Reared	Area of Transplant Release	Numbers (Type) Released	Adult Escapement Returning
1947	Eggs	678,000	Samish Marblemount Hatcheries	Horsefly River	162,549 (fingerling)	51 (1951)
1947	Eggs Eggs	48,000	Marblemount Leavenworth Hatcheries	Samish Horsefly River	39,358 (fingerling)	

DONOR - SEYMOUR RIVER

Brood Year	Type of Transplant	Numbers Taken	Area Reared	Area of Transplant Release	Numbers (Type) Released	Adult Escapement
1949	Eggs	158,000	Horsefly	Outlet Upper Adams River	84,000 (fingerling)	0 (1953)
1950	Eggs	667,000	Seymour-eyed	Upper Adams	667,000 eyed eggs	205
1951	Eggs	325,000	Horsefly Lake[1]	Shuswap Lake Mouth-Salmon R.	28,000 (fingerling)	0
1951	Eggs	from above lot	" "	Shuswap Lake Mouth-Anstey R.	23,000 (fingerling)	0
1952	Eggs	356,000	" "		131,000 fry	5
1952	Eggs	495,000	Seymour-eyed	Upper Adams	495,000 eggs	7,205
1955	Eggs	780,000	Seymour-eyed Upper Adams	Upper Adams River	780,000 "	0
1956	Eggs	253,000	" "	"	253,000 "	Present
1957	Eggs	520,000	" "	"	520,000 "	0
1958	Eggs	483,000	" "	"	483,000 "	85
1958	Eggs	273,000	Seymour-eyed Eagle River	Eagle River	273,000 "	169[1]
1958	Eggs	283,000	Seymour-eyed Salmon River	Salmon River	283,000 "	0
1959	Eggs	900,000	Seymour-eyed Upper Adams	Upper Adams	900,000 "	±100
1962	Eggs	1,023,000	Seymour-eyed Scotch Creek	Scotch Creek	1,023,000 "	459[2]
1962	Eggs	2,757,400	Seymour-eyed Eagle River	Eagle River	2,757,400 "	338[3]
1974	Eggs	1,551,000	Eyed Upper Adams	Upper Adams	1,374,000 "	0[4]
1975	Eggs	2,187,313	" " "	" "	2,140,000 "	0[5]

1 31 spawners in brood year.	4 13 spawners in 1974.
2 7 spawners in brood year.	5 23 spawners
3 169 spawners in brood year	

DONOR - STELLAKO RIVER

Brood Year	Type of Transplant	Numbers Taken	Area Reared	Area of Transplant Release	Numbers (Type) Released	Adult Escapement Returning
1950	Eggs		Horsefly Lake[1]	Horsefly Lake	72,000 fingerling	7 (1954)
1955	Eggs Adults	400,000 100	Horsefly Lake[1]	Horsefly Lake	317,000 eggs	27
1956			Horsefly Lake[1]	Horsefly Lake	387,000 fry	23
1958	Eggs	3,429,000	Horsefly Lake[1]	Horsefly Lake	3,003,000 eggs	72
1972	Eggs	1,020,000[2]	Horsefly River	Horsefly River	1,020,000 eggs	0
1976	Eggs	4,500,000[3]	Nadina	Nadina	4,500,000 eggs	?

[1] Quesnel Field Station
[2] Stellako eggs x Horsefly 3_2's males
[3] Stellako eggs x Late Nadina Stellako males

DONOR - GLUSKE CREEK

Brood Year (Year)	Type of Transplant	Numbers Taken	Area Reared	Area of Transplant Release	Numbers (Type) Released	Adult Escapement Returning
1961	Eggs	1,370,000	Eyed Gluske Creek	Hatdudatehl Creek	534,000 eggs	0 (1965)

DONOR - TASEKO LAKE

Brood Year	Type of Transplant	Numbers Taken	Area Reared	Area of Transplant Release	Numbers (Type) Released	Adult Escapement Returning
1958	Eggs	850,000	?	Upper Adams	850,000 eggs	85 (1962)
1959	Eggs	600,000	Eyed Taseko Lake	Upper Adams	600,000	0
1960	Eggs	702,000	Eyed Taseko Lake	Upper Adams	702,000	162

DONOR - RAFT RIVER

Brood Year	Type of Transplant	Numbers Taken	Area Reared	Area of Transplant Release	Numbers (Type) Released	Adult Escapement Returning
1956	Eggs	316,000	Barriere Lake	Barriere River	316,000 eyed eggs	23 (1960)
1957	Eggs	550,000	Barriere Lake	Barriere Lake	550,000 " "	335[1]
1958	Eggs	582,000	Barriere Lake	Barriere Lake	582,000 " "	14
1959	Eggs	490,000	Eyed Raft River	Fennell Creek	490,000 " "	439[2]
1960	Eggs	1,083,000	Eyed Raft River	Barriere River	1,083,000 " "	85[3]

[1] 38 sockeye spawned in brood year
[2] 27 sockeye spawned in brood year
[3] 23 sockeye spawned in brood year (plus 146 sockeye in 1964 went upstream to Barriere River, where no spawners were observed in 1960)

DONOR - LOWER ADAMS RIVER

Brood Year	Type of Transplant	Numbers Taken	Area Reared	Area of Transplant Release	Numbers (Type) Released	Adult Escapement Returning
1950	Eggs	300,000	Eyed Lower Adams Portage Creek	Portage Creek	300,000 eggs	3,505 (1954)
1950	Eggs	400,000	Green Horsefly Lake[1]	Seton Lake Outlet Portage Creek	193,000 fingerling	0
1950	Eggs	From Above Lot	Horsefly Lake[1]	Lac La Hache Lake	15,000 fingerling	0
1951	Eggs	625,000	Horsefly Lake[1]	Mable Lake	269,000 fingerling	0
1951	Eggs	From Above Lot	Horsefly Lake[1]	Little Horsefly River	131,000 fingerling	0
1954	Eggs	1,396,000	Adams Lake	Shuswap River Middle	1,396,000 eggs	499
1954	Eggs	(From Above Lot) 390,000	Horsefly Lake[1]	Little Horsefly River	390,000 "	14
1954	Eggs	(From 1,396,000 Lot)	Seymour River	Salmon River	258,000 "	0
159	Eggs	Eyed Lower Adams 622,000	Middle Shuswap River	Middle Shuswap River	622,000 "	0

[1]   Quesnel Field Station

DONOR - FORFAR CREEK

Brood Year	Type of Transplant	Numbers Taken	Area Reared	Area of Transplant Release	Numbers (Type) Released	Adult Escapement Returning
1956	Eggs	316,000	Eyed Forfar Creek	X Creek Nadina Lake	316,000 eggs	0 (1960)

DONOR - MOMICH RIVER

Brood Year	Type of Transplant	Numbers Taken	Area Reared	Area of Transplant Release	Numbers (Type) Released	Adult Escapement (Year)
1980	Eggs	500,000	Momich	Adams Lake	334,000 Fry	(1984)
1984	Eggs	1,071,000[1]	Momich	Adams Lake	1,071,000 eggs	

[1] Momich females x Momich males for 950,000 eggs
Momich females x Upper Adams males for 121,000 eggs

DONOR - HORSEFLY RIVER

Brood Year	Type of Transplant	Numbers Taken	Area Reared	Area of Transplant Release	Numbers (Type) Released	Adult Escapement Returning
1949	Eggs	302,000	Horsefly Lake[1]	Quesnel Lake Mouth - Horsefly R.	94,000 (fingerling)	7 3$_2$'s (1952) 218 (1953)
1957	Eggs	2,946,000	Horsefly Lake[1]	Horsefly Lake[1]		?
1953	Eggs	283,000	Horsefly Lake[1]	Horsefly Lake[1]	131,000	38 (1957)
1955	Eggs	522,000	"	"	280,000	27
1956	Eggs	1,445,000	"	"	311,000	23
1957	Eggs	3,824,000	"	"	3,259,000	40
1958	Eggs	3,873,000	"	"	3,003,000	72

[1] Quesnel Field Station

A P P E N D I X   K

. . . .

Chilko River and North End Lake Sockeye Production Data*

Brood Year Year i	4(2) Return Year Year i+4	Total Escapement Year i	Total Adult Escapement Year i	Total Female Escapement Year i	Females In River Only Year i
1948	1952	671,025	670,622	392,885	392,885
1949	1953	58,310	58,247	34,191	34,191
1950	1954	26,447	17,308	7,493	7,493
1951	1955	118,110	100,116	58,134	58,134
1952	1956	490,065	485,585	261,329	261,329
1953	1957	201,245	200,691	109,341	105,515
1954	1958	37,743	34,296	21,837	21,837
1955	1959	132,146	121,167	80,589	80,589
1956	1960	647,768	646,906	386,381	341,585
1957	1961	140,765	138,464	83,512	82,501
1958	1962	137,081	120,104	70,504	70,504
1959	1963	471,162	463,060	273,383	241,266
1960	1964	426,607	426,546	247,337	202,622
1961	1965	40,315	39,101	23,586	20,592
1962	1966	92,467	77,713	49,501	49,501
1963	1967	1,002,252	998,231	543,272	45,454
1964	1968	238,601	238,272	134,495	129,698
1965	1969	39,902	35,335	23,041	22,063
1966	1970	226,702	209,619	114,698	106,191
1967	1971	176,337	174,715	102,152	101,573
1968	1972	414,446	413,862	240,624	237,140
1969	1973	76,518	70,902	42,411	42,411
1970	1974	145,049	135,388	71,905	69,426
1971	1975	174,266	157,193	99,466	88,705
1972	1976	564,465	562,650	336,715	272,281
1973	1977	61,707	55,675	30,889	30,667
1974	1978	128,131	109,563	72,994	70,526
1975	1979	220,554	199,739	118,054	118,054
1976	1980	364,311	361,752	215,328	172,261
1977	1981	54,322	49,539	28,868	28,368
1978	1982	151,835	143,402	83,133	75,868
1979	1983	240,294	234,924	154,223	151,530
1980	1984	468,658	467,812	298,375	290,135
1981	1985	35,909	34,360	21,441	20,741

1959 - Includes South End lake spawners
1963 - Includes lake population of 924,884
1964 - Includes South End lake spawners
1971 - Not known if South End lake spawners were included
1972 - Includes South End lake spawners

* Appendix K does not include production from spawning at the south end of Chilko Lake. All data are tabulated by brood year, not by year of return.

Brood Year Year i	4(2) Return Year Year i+4	% Success of Spawning Year i	Total Year i	In River Only Year i	Mean Fecundity Year i	Egg Deposition (millions) In River Only Year i
1948	1952	92.80%	364,597	364,597	-	NA
1949	1953	96.60%	33,029	33,029	2,682	88.584
1950	1954	87.48%	6,555	6,555	3,079	20.183
1951	1955	99.02%	57,563	57,563	3,069	176.661
1952	1956	89.40%	233,628	233,628	3,058	714.434
1953	1957	86.40%	94,471	91,165	2,985	272.128
1954	1958	97.30%	21,247	21,247	2,922	62.084
1955	1959	94.10%	75,834	75,834	2,724	206.572
1956	1960	95.40%	368,607	325,872	3,134	1,021.283
1957	1961	99.54%	83,128	82,121	2,668	219.099
1958	1962	99.90%	70,433	70,433	2,731	192.353
1959	1963	99.82%	272,891	240,832	2,665	641.817
1960	1964	99.00%	244,864	200,595	2,815	564.675
1961	1965	63.76%	15,038	13,129	2,956	38.809
1962	1966	85.10%	42,125	42,125	2,970	125.111
1963	1967	38.32%	57,207	17,418	2,929	51.017
1964	1968	97.84%	131,590	126,897	2,956	375.108
1965	1969	90.33%	20,813	19,930	3,074	61.265
1966	1970	93.76%	107,541	99,565	3,356	334.140
1967	1971	88.11%	90,006	89,496	2,770	247.904
1968	1972	75.60%	181,912	179,278	2,917	522.954
1969	1973	60.17%	25,519	25,519	2,885	73.622
1970	1974	70.82%	50,926	49,167	3,183	156.499
1971	1975	91.13%	90,643	80,834	3,102	250.747
1972	1976	98.70%	332,338	268,741	3,069	824.766
1973	1977	97.87%	30,231	30,014	3,238	97.185
1974	1978	96.98%	70,790	68,396	3,170	216.815
1975	1979	86.42%	102,022	102,022	3,152	321.573
1976	1980	98.26%	211,581	169,264	2,981	504.576
1977	1981	68.70%	19,832	19,489	2,946	57.415
1978	1982	99.51%	82,726	75,496	2,985	225.356
1979	1983	87.44%	134,853	132,498	2,930	388.219
1980	1984	92.52%	276,086	268,433	2,935	787.851
1981	1985	93.67%	20,083	19,428	2,948	57.274

Brood Year Year i	4(2) Return Year Year i+4	Fry In River Only i+1 (millions)	Survival From Egg To Fry (%)	Estimated Fry Production Above Canoe Cross (millions)	Number Chilko Age 1 Smolts (millions)	Egg To Smolt Survival (%)	Survival* From Fry To Smolt (%)
1948	1952	-	-	NA	NA	-	-
1949	1953	6.298	7.11%	6.298	3.147	3.55%	49.97%
1950	1954	2.613	12.95%	2.613	1.170	5.80%	44.79%
1951	1955	22.933	12.98%	22.933	11.505	6.51%	50.17%
1952	1956	53.166	7.44%	53.166	24.491	3.43%	46.07%
1953	1957	16.119	5.92%	16.704	7.690	2.73%	46.04%
1954	1958	6.680	10.76%	6.680	2.853	4.60%	42.71%
1955	1959	21.662	10.49%	21.662	9.159	4.43%	42.28%
1956	1960	54.780	5.36%	54.780	28.242	2.77%	51.56%
1957	1961	19.135	8.73%	19.341	9.458	4.27%	48.90%
1958	1962	26.012	13.52%	26.012	6.895	3.58%	26.51%
1959	1963	47.356	7.38%	48.455	32.165	4.90%	66.38%
1960	1964	48.957	8.67%	59.551	33.780	4.92%	56.72%
1961	1965	4.333	11.16%	4.963	1.592	3.58%	32.08%
1962	1966	16.126	12.89%	16.126	8.813	7.04%	54.65%
1963	1967	13.561	26.58%	44.151	9.270	5.58%	21.00%
1964	1968	52.654	14.04%	52.997	23.665	6.27%	44.65%
1965	1969	5.745	9.38%	5.745	2.346	3.83%	40.84%
1966	1970	23.265	6.96%	23.930	17.355	5.05%	72.52%
1967	1971	14.204	5.73%	14.285	9.148	3.67%	64.04%
1968	1972	NA	-	NA	31.542	5.94%	-
1969	1973	5.825	7.91%	5.825	3.586	4.87%	61.57%
1970	1974	9.581	6.12%	9.923	3.833	2.36%	38.62%
1971	1975	10.624	4.24%	11.913	5.672	2.02%	47.61%
1972	1976	31.386	3.81%	38.262	20.006	1.99%	52.29%
1973	1977	5.604	5.77%	5.644	4.169	4.26%	73.87%
1974	1978	8.135	3.75%	8.388	7.057	3.16%	84.13%
1975	1979	20.058	6.24%	20.058	13.013	4.05%	64.88%
1976	1980	15.553	3.08%	19.442	25.413	4.03%	130.71%
1977	1981	3.764	6.56%	3.831	2.608	4.46%	68.08%
1978	1982	14.426	6.40%	15.807	18.140	7.35%	114.76%
1979	1983	11.837	3.05%	12.048	20.720	5.24%	171.98%
1980	1984	24.378	3.09%	25.073	31.983	3.95%	127.56%
1981	1985	3.451	6.03%	3.567	1.672	2.82%	46.86%

*By mid 1970s large numbers of sockeye spawned at the south end of Chilko Lake. Survival data of fry to smolts not accurate thereafter.

Brood Year Year i	4(2) Return Year Year i+4	Jack 3(2) Return Year i+3	Adult 4(2) Return Year i+4	Adult 5(2) Return Year i+5	Total Sub-2 Return	Survival Age I Smolt to Age 4(2) (%)	Survival Age I Smolt to Total Sub-2 Return (%)
1948	1952	32,000 e	1,643,062	11,182	1,686,244	NA	NA
1949	1953	3,732	560,635	10,368	574,735	17.82%	18.26%
1950	1954	1,489	183,278	4,705	189,472	15.66%	16.19%
1951	1955	3,925	644,911	20,763	669,599	5.61%	5.82%
1952	1956	18,732	1,763,929	35,975	1,818,636	7.20%	7.43%
1953	1957	1,172	514,554	14,004	529,730	6.69%	6.89%
1954	1958	12,234	632,132	4,415	648,781	22.15%	22.74%
1955	1959	32,418	1,407,963	31,011	1,471,392	15.37%	16.06%
1956	1960	13,905	2,379,854	16,684	2,410,443	8.43%	8.53%
1957	1961	76	117,362	3,149	120,587	1.24%	1.27%
1958	1962	4,055	278,320	13,613	295,988	4.04%	4.29%
1959	1963	23,792	2,080,497	18,659	2,122,948	6.47%	6.60%
1960	1964	5,472	958,877	5,980	970,329	2.84%	2.87%
1961	1965	256	52,713	11,583	64,552	3.31%	4.05%
1962	1966	10,657	960,609	13,582	984,848	10.90%	11.17%
1963	1967	37,579	1,112,861	4,045	1,154,485	12.01%	12.45%
1964	1968	7,252	1,818,921	55,810	1,881,983	7.69%	7.95%
1965	1969	1,787	138,555	2,360	142,702	5.91%	6.08%
1966	1970	26,456	744,469	27,636	798,561	4.29%	4.60%
1967	1971	28,734	1,933,329	23,351	1,985,414	21.13%	21.70%
1968	1972	46,952	2,335,183	21,925	2,404,060	7.40%	7.62%
1969	1973	4,126	369,954	15,839	389,919	10.32%	10.87%
1970	1974	16,775	627,337	1,084	645,196	16.37%	16.83%
1971	1975	24,353	571,260	0	595,613	10.07%	10.50%
1972	1976	37,001	1,861,092	12,635 e	1,910,728	9.30%	9.55%
1973	1977	7,640	180,432 e	4,843 e	192,915	4.33%	4.63%
1974	1978	18,667 e	547,007 e	4,748 e	570,422	7.75%	8.08%
1975	1979	8,131 e	1,406,022 e	7,375	1,421,529	10.80%	10.92%
1976	1980	7,798 e	1,572,025	25,168	1,602,174	6.19%	6.30%
1977	1981	2,708	189,574	2,387 e	195,909	7.27%	7.51%
1978	1982	8,307	1,083,536 e	77,743	1,180,308	5.97%	6.51%
1979	1983	4,119 e	1,459,307	33,679 p	1,497,105	7.04%	7.23%
1980	1984	8,842	3,772,379 p	473,961	4,255,182	11.79%	13.30%
1981	1985	1,211 p	177,561	8,073 p	186,845	10.62%	11.17%

1971 - River only, total return = 43,201
1975 - River only, total return = 17,376
1979 - River only, total return = 6,680
1983 - River only, total return = 9,228

e - Estimating
p - Preliminary

. . . .
Fraser River Sockeye Indian Subsistence Catches
1946-1985

Year	Indian Catch
1946	50,127
1947	42,275
1948	62,265
1949	79,762
1950	87,786
1951	85,043
1952	83,986
1953	113,031
1954	95,200
1955	65,504
1956	64,918
1957	96,839
1958	82,632
1959	64,999
1960	83,649
1961	137,194
1962	137,448
1963	190,002
1964	145,639
1965	120,570
1966	154,059
1967	107,173
1968	124,191
1969	159,105
1970	150,555
1971	154,461
1972	133,003
1973	163,866
1974	222,430
1975	253,151
1976	233,462
1977	244,427
1978	237,552
1979	291,709
1980	185,966
1981	440,664
1982	429,799
1983	361,556
1984	356,000
1985	442,000

A P P E N D I X M

Total Runs of Fraser Sockeye and Pink Salmon (Millions)

	Sockeye	Pink
1946	11.128	
47	1.325	
48	3.068	
49	3.559	
50	4.351	
51	3.408	
52	3.505	
53	5.854	
54	12.198	
55	2.747*	
56	2.867	
57	5.401	
58	18.779	
59	4.770	6.460
60	3.421*	
61	4.714	1.889
62	3.512	
63	3.985	5.478
64	1.825*	
65	3.167	2.323
66	5.460	
67	6.804	12.968
68	2.956	
69	4.941	3.930
70	6.164	
71	7.696	9.768
72	3.708	
73	6.878	6.789
74	8.616	
75	3.684	4.894
76	4.341	
77	5.835	8.249
78	9.432	
79	6.426	14.404
80	3.133	
81	7.743	18.685
82	13.989	
83	5.242	15.346
84	5.918	
85	13.881	18.864
86	15.904**	
87	7.694**	7.136**
88	3.762**	
89	18.403**	

*   Includes fish which failed to reach spawning grounds.
** From Pacific Salmon Commission.

Sockeye - Pink Salmon Escapements (Net)*

	Sockeye	Pink
1938	885,595	
1939	157,763	
1940	552,541	
1941	541,608	
1942	2,894,594	
1943	242,724	
1944	315,961	
1945	554,363	
1946	2,916,266	
1947	650,116	
1948	1,015,790	
1949	1,087,887	
1950	1,791,496	
1951	606,262	
1952	846,435	
1953	1,288,181	
1954	2,447,802	
1955	383,266	
1956	872,742	
1957	1,661,829	2,202,867
1958	3,867,190	
1959	949,413	1,078,000
1960	628,025	
1961	1,250,087	1,092,561
1962	1,624,004	
1963	1,606,924	1,954,038
1964	426,459	
1965	852,271	1,194,099
1966	1,919,608	
1967	1,353,640	1,831,219
1968	626,706	
1969	1,006,972	1,530,913
1970	1,943,221	
1971	747,523	1,804,952
1972	830,128	
1973	1,181,093	1,754,261
1974	1,757,474	
1975	992,520	1,367,089
1976	823,453	
1977	1,113,453	2,387,698
1978	2,514,318	
1979	1,407,828	3,560,654
1980	848,525	
1981	1,443,496	4,488,336
1982	4,024,261	
1983	977,238	4,631,721
1984	932,477	
1985	2,138,618	6,460,616

* **Escapements for earlier years may differ from those shown in Annual Reports as a result of later year examination of methods and calculations.**

Fraser River Pink Salmon Production*

BROOD YEAR

	1961	1963	1965	1967	1969	1971	1973	1975	1977	1979	1981	1983
Total Spawners (millions)	1.093	1.954	1.194	1.831	1.531	1.805	1.754	1.367	2.388	3.561	4.488	4.632
Female Spawners (millions)	0.654	1.216	0.692	0.973	0.957	1.096	1.009	0.781	1.362	2.076	2.560	2.931
Potential Egg Deposition (billions)	1.569	2.435	1.488	2.132	2.018	1.923	1.865	1.493	2.960	3.787	4.814	4.702
Fry Production (millions)	143.6	284.2	274.0	237.6	195.6	245.0	292.4	279.2	473.3	341.5	590.2	554.8
Adult Return Catch + Escapement (millions)	5.478	2.323	12.968	3.930	9.768	6.789	4.894	8.249	14.404	18.685	15.346	18.864
Freshwater Survival	9.2%	11.7%	18.4%	11.1%	9.7%	12.7%	15.7%	18.7%	16.0%	9.0%	12.3%	11.8%
Marine Survival	3.8%	0.8%	4.7%	1.7%	5.0%	2.8%	1.7%	3.0%	3.0%	5.5%	2.6%	3.4%

* Fry Production Data not available prior to 1961

A P P E N D I X   P

. . . .
International Pacific Salmon Fisheries Commission Summary of Operating and Construction Expenditures, November 1937 to December, 1985.

| | CANADA | | UNITED STATES | |
FISCAL YEAR	OPERATING	CONSTRUCTION	OPERATING	CONSTRUCTION
1937-38	7,739		14,890	
1938-39	20,856		18,167	
1939-40	34,735		35,985	
1940-41	34,623		34,576	
1941-42	39,811		39,550	
1942-43	41,644		43,624	
1943-44	41,580		39,425	
1944-45	38,635	91,727	41,382	91,727
1945-46	40,678	370,063	40,792	370,063
1946-47	38,859	169,481	38,231	169,481
1947-48	91,198	126,972	96,207	126,972
1948-49	96,905	125,620	99,546	125,619
1949-50	124,403	58,175	141,100	58,176
1950-51	131,232	54,945	132,633	54,944
1951-52	127,895	27,863	131,825	27,864
1952-53	130,953		135,928	
1953-54	134,350		127,853	
1954-55	126,924		133,220	
1955-56	132,516		134,143	
1956-57	166,084		181,965	
1957-58	241,181		233,333	
1958-59	219,540		210,007	
1959-60	219,594		238,120	
1960-61	230,262		227,673	
1961-62	253,308	105,754	264,083	
1962-63	248,500		243,936	
1963-64	282,937	33,519	298,963	106,217
1964-65	281,362	78,113	280,655	122,928
1965-66	310,241	155,898	294,976	127,223
1966-67	286,047	105,386	291,949	128,062
1967-68	296,929	126,293	303,238	98,600
1968-69	305,904	128,951	316,346	93,184
1969-70	351,274	102,443	338,512	64,720
1970-71	354,743	53,133	365,228	43,558
1971-72	399,070	118,582	394,647	125,805
1972-73	393,133	153,132	400,569	146,609
1973-74	508,478	87,267	533,648	62,136
1974-75	571,030	82,155	587,379	79,474
1975-76	633,853	14,096	633,853	14,096
1976-77	682,987		682,986	
1977-78	760,773		760,772	

1978-79	789,567		789,566	
1979-80	849,069		849,069	
1980-81	975,264		975,264	
1981-92	1,066,411		1,066,411	
1982-83	1,190,139		1,190,140	
1983-84	1,405,697		1,405,697	
1984-85	1,626,600		1,450,000	
1985-86	1,553,548		1,802,842	
TOTAL	18,889,061	2,369,568	19,090,904	2,237,458
ADD CONSTRUCTION	2,369,568		2,237,458	
ADD CANADA ONLY			21,328,362	
SWELTZER CR. LAB.	132,107			
	21,390,736			

S U B J E C T   I N D E X

N O T E S

N O T E S

NOTES